P9-DUQ-376

UNM - GALLUP

3 7996 1004 4165 4

Social Issues in Science and Technology

Social Issues in Science and Technology

An Encyclopedia

David E. Newton

ABC-CLIO

Santa Barbara, California
Denver, Colorado
Oxford, England

Library of Congress Cataloging-in-Publication Data
  Newton, David E.
    Social issues in science and technology : an encyclopedia / David
  E. Newton.
      p.   cm.
    Includes bibliographical references and index.
    ISBN 0-87436-920-7 (alk. paper)
    1. Science—Social aspects encyclopedias.   2. Technology—Social
  aspects encyclopedias.   I. Title.
  Q175.5.N49   1999
  303.48'3'03—dc21   99-30274
                                                                CIP

  04   03   02   01   00   99   9 8 7 6 5 4 3 2 1

ABC-CLIO, Inc.
130 Cremona Drive, P.O. Box 1911
Santa Barbara, California 93116-1911

This book is printed on acid-free paper ∞.

Manufactured in the United States of America

For Delores and Reg

Contents

Preface, ix

Social Issues in Science and Technology

Abortion, 1
Alar, 5
Alternative Therapies, 7
Anencephalic Babies, 9
Animal Rights, 10
Appropriate Technology, 13
Assisted Reproductive
  Technologies, 14
Barrier Islands, 19
Behavior Modification, 21
Biological Determinism, 24
Biometric Identification, 26
Bison, 27
Bovine Somatotropin, 29
Captive Breeding Programs,
  31
Cassini Spacecraft, 31
Chemical Castration, 33
Chernobyl, 35
Chlorofluorocarbons, 36
Clearcutting, 37
Cloning, 39
Cold Fusion, 42
Conservation and
  Preservation, 45
Creatine, 47
Creationism, 48
Criminality and Heredity,
  51
Cyclamates, 54
DDT, 57
Deicing Roads, 59
Delaney Clause, 60
DNA Fingerprinting, 61
Dobson Unit, 64
Drug Testing, 64

Drug Testing in the
  Workplace, 67
Early-Term Surgical
  Abortions, 71
Echinacea, 72
Electrical Stimulation of the
  Brain, 73
Electroconvulsive Shock
  Therapy, 74
Endangered Species Act, 76
Environmental Justice, 78
Ethanol as a Fuel, 81
Eugenics, 83
Euthanasia, 85
Faith Healing, 89
Feedlot Pollution, 91
Fetal Tissue Research, 93
Fluoridation, 95
Formaldehyde, 98
Genetic Testing, 101
Genetically Manipulated
  Foods, 104
Global Warming, 106
Grazing Legislation, 111
Halons, 115
Hazardous Wastes,
  International Dumping,
  115
Headwaters Forest, 118
Health Hazards of
  Electromagnetic Fields,
  121
HIV and AIDS, 124
Homosexual Behavior, 129
Human Experimentation,
  132

Human Gene Therapy, 135
Human Genome Project,
  138
Hydrochlorofluorocarbons,
  141
Indoor Air Pollution, 143
Intact Dilation Evacuation,
  145
IQ, 146
Irradiation of Food, 149
Kennewick Man, 151
Legalization of Drugs, 153
The Limits to Growth, 155
Logging Roads, 156
Masked Bobwhite Quail,
  159
Medical Uses of Marijuana,
  160
Mediterranean Fruit Fly, 163
Methyl Bromide, 164
Montreal Protocol, 166
"Morning-After" Pill, 166
MTBE, 167
Multiple Use/Sustained
  Yield, 169
Nature versus Nurture, 173
Needle Exchange Programs,
  175
Noise Pollution, 178
Northwest Forest Plan, 179
Nuclear Power Plants, 182
Nuclear Wastes, 185
Nuclear Weapons, 188
Nuclear Winter, 192
The Oceans, 195
Olestra, 197

Organ Transplantation, 199
Ozone Depletion, 202
Ozone Depletion Potential, 206
Particle Accelerators, 207
Pollutant Standard Index, 211
Polybrominated Biphenyls, 211
Polychlorinated Biphenyls, 212
Polygraph, 214
Population, 216
Prescribed Burn, 220
Privacy and the Internet, 222

Psychosurgery, 225
Radon, 229
Rain Forests, 230
Research, Basic and Applied, 234
Right to Die, 236
Ritalin, 239
RU486, 240
Saccharin, 243
Sagebrush Rebellion, 244
Salton Sea, 246
Savage Rapids Dam, 248
Science and Religion, 250
Search for Extraterrestrial Intelligence, 253
Secondhand Smoke, 255

Slash-and-Burn, 257
Sociobiology, 258
Space Station, 261
Sterilization, Human, 264
Steroids, 265
Stream Channelization, 267
Superfund, 267
Three Mile Island, 271
Tris, 272
Vitamins and Minerals, 275
Water Rights, 277
Wetlands, 279
White Abalone, 282
Wild Horses and Burros, 283
Wildlife Management, 285

Selected Bibliography, 291
Index, 297

Preface

Science and technology have revolutionized human civilization. Today, we have drugs and medicines that cure diseases that once killed humans on a widespread scale. We have synthetic materials that are stronger and longer lasting than any natural product. We have ways of generating energy that have the potential to free humans from their centuries-long dependence on wood and fossil fuels. We have medical techniques that have extended the average human lifetime by many years. Most of us are aware, consciously or not, of the enormous benefits that science and technology have provided.

But progress in science and technology is seldom a matter of unadulterated improvement. For each gain that can be attributed to science and technology there is a new problem or a new challenge that did not exist previously. For example, new drugs and medicines can be very expensive, available only to those who are able to pay for them. Synthetic materials may be *so* long lasting that they do not even break down in nature, causing problems for the environment. Nuclear power plants may be able to replace fossil-fueled power plants, but they also pose a serious threat to humans and the environment. And medical advances may extend human life, but questions may also arise as to the *quality* of that life that remains.

This book provides an overview of about 100 areas in which advances in science and technology have raised new issues for human societies. A three-part approach is taken with each of these issues. The entry provides a brief background as to the science or technology involved. Then, some history is usually provided to show how the particular scientific or technological advance has led to social, economic, political, religious, ethical, or other issues for the general society. Finally, a summary of the various positions that exist on each issue is offered.

The issues presented in this book include some typically found in books on environmental science. But this book does not deal primarily with such issues. One reason for this decision is that a number of excellent books, dictionaries, and encyclopedias on the environmental sciences already exist. Another reason is that advances in science and technology almost without exception create new problems for the environment, problems that are to many people already familiar.

Thus, the reader will find no entry on air pollution, water pollution, or other topics of environmental importance. The reason is that such problems often have a common theme: A development in science or technology makes possible some kind of improvement in human life. Some corporation will be interested in promoting that development among the general public. It will, that is, hope to make a profit on the new development.

The same development may have deleterious effects on the environment or on human health. Other groups of people will make an effort to monitor the marketing of the new development in order to minimize the harm it causes to humans and the environment. A battle ensues that pits commercial exploitation of a scientific or technological development with "environmentalists" who try to limit its harmful effects and, almost by definition, the corporation's profits.

It has not seemed worthwhile to retell that story over and over again in this book. Some elements of that pattern do appear, of course, in issues such as global warming and ozone depletion. But the greater emphasis in selecting topics for the book has been to seek out issues in which two or more positions on some issue can be defined and elaborated. In this way, the reader can have an even more extended view of the complexity of science-based issues in our society.

In many cases, the author has chosen to discuss a relatively specific topic in order to highlight more general principles. For example, the masked bobwhite quail is only one of hundreds of endangered species. Yet, the issues that have grown up around efforts to protect its survival are similar to those that arise for many other endangered species.

In some regards, this book is—and cannot be other than—"a work in progress." Many of the issues discussed here are not yet settled and may never be so. The best that one can do is to offer a "snapshot" of the current status of affairs. In some cases, issues are at a crisis point in their development, and important decisions may occur any week or month. In other cases, issues are just beginning to catch the public notice, and critical debates may be months or years into the future. In still other cases, some resolution has been achieved for an issue, although it is unlikely that controversies about that issue are settled forever.

Each entry concludes with a list of references, books, magazines, and web sites concerned with the issue discussed. They are simply references that show some promise of continuing the discussion begun in this book and leading the reader to additional information on the issue.

Social Issues in Science and Technology

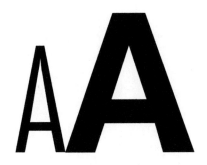

Abortion

Abortion is any procedure used to induce the expulsion of an embryo or fetus before it would be delivered naturally. Abortion has been used to prevent unwanted births among humans for untold centuries. The number of abortions carried out worldwide has been estimated to be as high as 50 million annually. There may be as many as one abortion for each four live births each year.

Historically, abortion has been illegal in many, if not most, cultures. Under somewhat unusual circumstances, the practice has been permitted or encouraged as a method of birth control. The 1960s saw a trend worldwide toward more permissive attitudes and policies about abortion. By 1983, according to some estimates, 90 percent of the world's population lived in countries where abortion was legal under at least some circumstances.

Illegal Abortions

This fact does not mean that safe abortions have been readily available to women in these countries. In many places, adequate medical assistance has not been available, or women have been unable to pay for the legal abortion they wanted to have. As a result, the rate of illegal abortions has tended to be high in countries whether the practice was legal or illegal.

About half of all illegal abortions are self-induced. Women may take harmful chemicals with the hope of inducing an abortion, or they may engage in physical activity designed to produce the same result. The other half of all illegal abortions are performed by individuals who may or may not have adequate medical background to carry out the procedure. However performed, illegal abortions tend to be high-risk procedures for women. The death rate from such procedures is estimated to be at least fifty times that for legal abortions. In the United States, the death rate from illegal abortions is about 50 per 100,000 cases, whereas in developing countries the rate may be as high as 1,000 for 100,000 cases.

Abortion in the United States

Attitudes and policies toward abortion in the United States have traditionally been relatively negative. Until 1973, the practice was regulated by state laws that were, in most cases, highly restrictive. Abortion was permitted in most states, if at all, only under conditions such as danger to a woman's life or health.

This situation changed dramatically in 1973 when the U.S. Supreme Court was asked to rule on a case involving a then-anonymous woman, "Jane Roe," who wished to have an abortion in the state of Texas but was prevented from doing so by state law. A suit was filed on her behalf against the attorney general of Texas, Henry Wade.

The Court's ruling in *Roe v. Wade* was dramatic and unexpected because it went far beyond the specific case before it. The Court took the very general stance that a woman's right to personal privacy was a constitutional right. No state law or action could deprive her of that right. In taking

More than 25 years after the Supreme Court made most abortions legal in the United States, public debates over the issue have not become any less bitter. (Corbis/Bettmann)

this position, the Court removed the issue of abortion from state control and made it a federal issue.

Specifically, the Supreme Court ruled that states could make no laws at all regarding abortion during the first trimester (first third, or about thirteen weeks) of pregnancy. States could make such laws during the second trimester only if there were compelling reasons to ensure a woman's heath. And states were allowed to make laws governing abortion once the fetus was viable (capable of living outside the mother's body), usually after the second trimester.

Right to Life versus Freedom of Choice

One might have expected the *Roe v. Wade* decision to end the debate over abortion in the United States. Nothing could be less true, however. In fact, the decision marked the beginning of perhaps the most acrimonious period in the dispute over abortion

in U.S. history, if not in all of world history. Those who opposed access to legal abortion realized that they would have to focus their efforts on a national level by trying to influence members of the U.S. Congress, presidential candidates, and other national figures and national opinion. They chose as their campaign slogan "Right to Life," referring to the life of an unborn embryo or fetus, not to the life of a pregnant woman.

Those who favored access to abortion came to a similar conclusion as did those in the Right to Life movement, only somewhat later. Having thought that the right to abortion was won for women in the United States in *Roe v. Wade*, they tended to "rest on their oars." Only somewhat later did those who favored access to abortion realize that abortion rights had not necessarily been won for all time but that the battle to protect those rights would never really be over. Eventually, individuals in this camp organized around the rubric, "Freedom of Choice."

Perhaps the most basic issue involved in the abortion debate has to do with the question as to when life begins. Many people argue that life begins when a sperm penetrates an egg and causes fertilization. The fertilized egg, they say, has all the potential needed to grow into an embryo, a fetus, and eventually a baby. Destroying the fertilized egg at the very beginning of its existence or at any point in its development is the same as killing a human. It is an act of murder. Since the fertilized egg, the embryo, or the fetus cannot speak or act for itself, humans who oppose abortion must speak and act for it.

Other individuals feel that life does not begin until some later point in the development of a fertilized egg. That point may be designated as implantation, formation of certain organs within the fetus, spontaneous movement of the embryo (quickening), the ability of the fetus to live on its own outside the mother's body (viability), or some other point. Those who favor access to abortion may do so unconditionally, recommending that women alone should be able to decide at what stage an abortion should occur. Others who favor abortion believe that, except in the case of a medical emergency, the procedure should not be allowed after some given point, such as the moment of viability. In this debate, some will argue, the "rights" of two individuals may conceivably be involved, those of the mother and those of the fetus. In such a case, they suggest, the rights of the mother should prevail.

The Evolving Debate over Abortion

In the three decades since *Roe v. Wade,* abortion has become far more than a medical question involving a woman, her doctor, and perhaps her mate. It has become a political issue at the local, state, and national levels. It has become politically divisive within both parties, but more so within the Republican Party, where a Right to Life contingent has sought to make its views on abortion a "litmus test" for all party candidates. In the 1996 presidential political campaign, for example, the Christian Coalition suggested withholding financial contributions to Republican candidates who refused to oppose certain abortion procedures. That proposal was not adopted by the Republican Party, however.

The debate over abortion continues to change and evolve over time. One reason for this is the intensity of feelings by those on both sides of the issues. Efforts by one person or one group to change minds by reasoned argument tend to be rare. By contrast, violent behavior that includes the murder of abortion providers and the bombing of abortion clinics tends to be more common.

Another factor in the evolution of the abortion debate has been progress in medical technology. Since 1973, a number of new techniques have become available that make abortion safer and easier at earlier stages of pregnancy. The distinctions between very early abortions and very late contraceptive measures have become more uncertain, with a consequent blurring of the debate over abortion itself.

Attitudes and Practices in the United States

Perhaps the two most important trends in abortion statistics are those concerning public attitudes and practices. With regard to the former, the division among those who support access to abortion and those who oppose such access has changed remarkably little since *Roe v. Wade.* In 1975, for example, 21 percent of all Americans interviewed in one survey agreed that abortion should be legal "under any circumstances." In 1997, the number who agreed with that position was 22 percent. In 1975, the number who felt that abortion should be legal under at least some circumstances was 54 percent. By 1997, that number had risen to 61 percent.

On the other hand, Americans had begun to feel that abortion, while legal, should perhaps be less readily available. A *New York Times*/CBS News poll announced in 1998, for example, suggested that many Americans were moving toward a position articulated by President Bill Clinton, that abortion should be "safe, legal, and rare." The poll indicated that among those who favored access to abortion, the mood was toward greater restrictions on the practice. For example, nearly 80 percent of those polled favored a waiting period for those requesting an abortion and favored a requirement that parental consent be obtained for underage women seeking an abortion. These two policies are among the most common to be adopted by states seeking to regulate abortion after the first trimester.

Trends in the actual number of abortions performed in the United States had begun to follow the patterns indicated above. The number of legal abortions in the United States peaked at 1,429,577 in 1990 and has been decreasing ever since. By 1995, the last year for which data are available, the number of legal abortions had fallen to just over 1,200,000. An important change in that overall trend has been toward earlier abortions. In 1995, 54 percent of all abortions were performed during the first eight weeks, compared to 34 percent during the same period in 1972.

One factor in the declining number of abortions may have been the increasing difficulty of finding abortion providers. The number of medical facilities offering abortion dropped from nearly 3,000 in 1982 to less than 2,500 in 1992. That change was even more dramatic in rural areas, where the number of institutions offering abortion services fell from just over 400 to less than 100.

This trend almost certainly reflects to some extent the physical dangers to which abortion providers are exposed from those opposed to the practice, dangers that may include loss of their own lives. The decreasing number of physicians trained in abortion procedures has also been a factor. In 1976, less than 10 percent of all obstetrical/gynecological programs offered no training in abortion procedures at all. By 1992, that number had risen to 30 percent.

Recent Court Decisions

The most recent landmark decisions on abortion by the U.S. Supreme Court were announced in 1992. Both sides in the abortion debate had looked forward to these decisions, expecting that the Court might either reaffirm or overthrow *Roe v. Wade.* The decisions came in a series of cases involving state laws placing a variety of restrictions on free and unlimited access to abortion.

The Court did reaffirm in its decisions a woman's constitutional right to abortion. But it no longer held that right to be "fundamental." Instead, it said that state laws could pose no "undue burden" on a woman's right to abortion. Within that context, it said that a state law requiring a woman to notify her husband that she was going to have an abortion was unconstitutional. On the other hand, the Court upheld a state law requiring that a pregnant woman wait twenty-four hours after requesting an abortion and being told of its possible consequences (except in case of a medical emergency).

For those in the Right to Life movement, these decisions were a disappointment. They meant that a woman's constitutional right to abortion was intact as the law of the land. A spokesperson for the National Right to Life Committee called one of the 1992 rulings "a major blow." A decision to overturn *Roe v. Wade,* she said, would have "ended the killing and millions [more] would be alive today" (Deibel, A17). By contrast, a representative of the Center for Reproductive Law and Policy said that laws like those upheld in the 1992 decisions "crippled choice." She said that the Court's decision only confirmed that the effort to promote access to abortions was "behind where we should be" (Deibel, A17).

Abortion and International Politics

For those who oppose access to abortion, this issue permeates nearly every aspect of their political views. Financial contributions by the United States to the United Nations and its agencies are often "held hostage" because of abortion issues. Antiabortion legislators often refuse to vote for funding measures that may bring money to nations where abortions are permitted.

As an example, the U.S. Senate passed a bill in April 1998 authorizing payment of more than $800 million in back dues owed by the United States to the United Nations. The bill contained an amendment that prohibited U.S. aid to any international family planning group that performed or provided information on abortions. President Bill Clinton chose not to sign the bill because of the amendment. As a result, the United States fell even farther behind in its financial organizations to the United Nations.

See also: Anencephalic Babies; Early-Term Surgical Abortions; Fetal Tissue Research; Intact Dilation Evacuation; "Morning-After" Pill; RU486

References

Andryszewski, Tricia, and Victoria Sherrow. *Abortion: Rights, Options, and Choices.* New York: Millbrook Press, 1996.

Butler, J. Douglas. "Abortion and Reproductive Rights: A Comprehensive Guide to Medicine, Ethics, and the Law" (CD-ROM). Phoenix: Oryx Press, 1997.

Day, Nancy. *Abortion: Debating the Issue.* Hillside, NJ: Enslow Publishers, 1995.

Deibel, Mary. "25 Years of Contention over Roe vs. Wade." *San Francisco Examiner,* 18 January 1998, A13+.

Gorney, Cynthia. *Articles of Faith: A Frontline History of the Abortion Wars.* New York: Simon and Schuster Books, 1998.

Koch, Wendy. "Veto of U.N. Dues Bill Likely." *USA Today,* 29 April 1998, A1.

Reagan, Leslie J. *When Abortion Was a Crime: Women, Medicine, and Law in the United States, 1867–1973.* Berkeley: University of California Press, 1997.

Rubin, Eva R., ed. *The Abortion Controversy: A Documentary History.* Westwood, CT: Greenwood Publishing Group, 1994.

Solinger, Rickie, Faye Ginsburg, and Patricia Anderson. *Abortion Wars: A Half Century of Struggle, 1950–2000.* Berkeley: University of California Press, 1998.

Steffen, Lloyd, ed. *Abortion: A Reader.* Cleveland, OH: Pilgrim Press, 1996.

A group that supports access to abortion is the National Abortion and Reproductive Rights Action League (NARAL), whose web page is at http://www.naral.org/. A web page that focuses on opposition to access to abortion is located at http://www.prolife.org/ultimate/.

AIDS

See HIV and AIDS

Alar

Alar is a trade name for the chemical compound daminozide, succinic acid–2,2-dimethylhydrazide. The compound is also marketed under other trade names, such as Kylar, B-Nine, and B-9.

Alar is a plant growth regulator. It retards growth in a plant, causing flowers and fruit to develop more slowly. Trees sprayed with Alar yield fruit that are firmer and more resistant to bruising than are fruit not sprayed with the product. Growers can use Alar on their trees to make fruiting occur at nearly the same time in all parts of an orchard. This makes harvesting a simpler and more efficient process.

Patents for the production of Alar and related compounds were granted to the U.S. Rubber Company (now Uniroyal) in 1962 and 1966. The chemical rapidly became very popular, particularly among apple growers.

Health Issues

Throughout the 1970s, a number of research studies on the health effects of Alar began to appear. In 1973, for example, a study published in the *Journal of the National Cancer Institute* reported that mice whose diet included a compound related to Alar—(unsymmetrical 1,1-dimethylhydrazine, or UDMH)—developed cancer of the lungs, kidneys, liver, lymph system, and blood vessels.

UDMH is used in the manufacture of Alar. It is also formed when Alar is heated. Because of this relationship between the two compounds, it is common to find UDMH, often in very small amounts, in association with Alar in foods processed from apples.

Over time, a number of regulatory agencies examined the accumulating mass of evidence and began to regard Alar as a possible carcinogen. The International Agency for Research on Cancer listed UDMH as a probable carcinogen in 1982, as did the Carcinogen Assessment Group of the U.S. Environmental Protection Agency (EPA) and the U.S. National Toxicology Program in 1984.

Over the next five years, the EPA continued its study of Alar and UDMH to decide whether the growth regulator should be banned from use on apples and other fruits. By 1989, the agency had taken no action on the issue. At that point, the subject of Alar was raised by three important organizations: the Natural Resources Defense Council (NRDC), Consumers Union (CU), and CBS Television's *Sixty Minutes* news program. These three organizations raised the question as to whether the U.S. public was being exposed to an unnecessary cancer risk because of the use of Alar by apple growers.

The mention of this issue by three major groups created an extraordinary response in the general public. People became concerned about the possibility of being harmed by eating and drinking some well-known and well-liked products, such as applesauce and apple juice. Questions were raised in particular about possible harmful effects on babies and young children, among whom apples are a popular food.

One factor that contributed to public concerns about Alar was that the National Food Processors Association and the Gerber Baby Food Company had both reported earlier finding traces of Alar in their products. It seemed clear that the compound used to treat apple trees in the orchard had survived the processing routine or, even worse, had broken down to produce the even more dangerous UDMH.

Reaction

The Alar story has been cited many times since 1989 by those who feel that (1) an aroused public can cause regulatory agencies to become more conscientious about their responsibility of protecting the public health and (2) "scare" tactics, like those supposedly used by NRDC, CU, and CBS, can cause enormous economic damage to an industry even when there is little or no real health threat to the general public.

The response to the Alar news from NRDC, CU, and CBS was a dramatic drop in the purchase of apples and apple products by the general public. People were apparently convinced that the threat posed by Alar and UDMH was real, and they chose to avoid that threat by discontinuing or reducing their purchase of apple products.

Apple growers and producers of foods that contain apples were thrown into disarray. Some expressed concern about their ability to survive in the face of massive public rejection of their products and/or in the face of having to give up on the use of Alar in apple orchards.

Faced with the potential loss of huge revenues, many apple growers voluntarily agreed to stop using Alar on their orchards. In 1989, Uniroyal announced that it would no longer manufacture the product for use on apples and other fruits. Many health experts and environmentalists felt that these decisions represented a triumph of sound public health policy over the need for corporate profit.

Many scientists and corporate leaders felt otherwise. They thought that the health risks posed by Alar were vanishingly small. It was absurd, they argued, to discontinue use of a very valuable agricultural product simply because of a publicity uproar caused by a small number of people who actually knew very little about the subject.

References

Montague, Peter. "How They Lie." *Rachel's Environment and Health Weekly,* 23 and 30 January 1997 and 20 and 27 February 1997. Also at http://www.monitor.net/rachel/r530.html. 20May 1999.

Negin, Elliot. "The Alar 'Scare' Was for Real." *Columbia Journalism Review,* September/October 1996, 13–15.

Rosenberg, Beth. "The Story of the Alar Ban: Politics and Unforeseen Consequences." *New Solutions,* Winter 1996, 34–50.

Smith, Kenneth. *Alar: Five Years Later.* New York: American Council on Science and Health, 1994.

Alternative Therapies

When a person in the United States has a physical illness or injury today, that person often goes to a medical doctor for treatment. The medical problem could be a broken bone, a sore throat, an upset stomach, or a more serious, even life-threatening, condition. If the doctor who is consulted is typical of most U.S. physicians, he or she is likely to prescribe drugs (most commonly, synthetic chemicals) or suggest a surgical procedure. The modern medical community has, of course, a large and impressive arsenal of tools with which to diagnose and treat physical disorders. These tools tend, however, to fall within the very general category of "chemicals and surgery" as a way of treating human medical disorders.

The mind-set that drives modern medical practice has been, to a very large extent, adopted by the general public in many industrialized nations, including the United States. Most people expect and are comfortable with the treatments recommended by their white-coated scientific physician. Indeed, when ill, they often seek out from their doctor or the local pharmacist "a pill" that will cure their problem.

Alternative Forms of Diagnosis and Treatment

The philosophy and tools of modern medical science are not, however, the only conceivable way of thinking about and dealing with human health problems. Cultures in other times and other places have had and do have different views as to how the

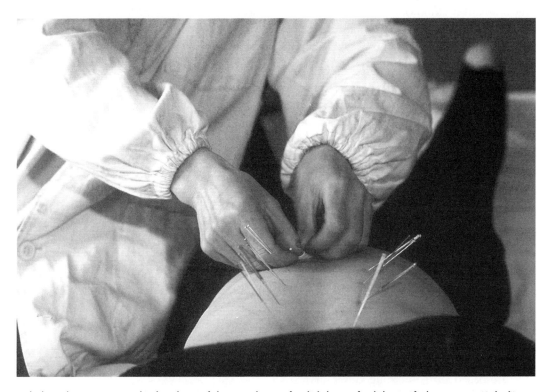

Medical providers continue to wonder about the use of alternative therapies for which there is often little scientific documentation. (Corbis/Kevin R. Morris)

human body works, what causes disease, and how illness can be cured. For example, homeopathy is a form of therapy that treats the body as a complete system. A disfunction in one part of the body, such as a sore wrist, cannot be treated in isolation, according to this view. Instead, the relationship of the wrist to the rest of the body must be taken into consideration before developing a treatment for the wrist itself.

Some of the better-known alternative therapies in use today include acupuncture, aromatherapy, biofeedback training, chelation therapy, magnetic-field therapy, naturopathic medicine, qigong, and yoga. Some of these techniques, such as acupuncture and qigong, have been around for centuries, while others, such as chelation therapy, are relatively new. Some therapies are embraced by relatively small groups of devoted believers, while others are standard medical procedures for millions of people. Indeed, among the latter group, modern Western medical science is often the alternative therapy for treating disease, not the predominant method of treatment.

Americans have been demonstrating an increased interest in alternative therapies A 1990 study showed that Americans spend about $13.7 billion on "unconventional" medical treatments. Some authorities believe that figure to be a gross underestimate because it was based on information collected from English-speaking respondents only. Yet, alternative therapies are likely to be even more popular among those with non-Western backgrounds.

Effectiveness of Alternative Therapies

One of the most fundamental questions about alternative therapies, of course, is how effective they are. That question itself poses something of a dilemma. In Western medical science, the way to answer that question is to conduct careful scientific studies to find out how many people are cured, and how many are not cured, by any given therapy. But some alternative therapies are based on something other than a Western scientific philosophy. If a therapy

works for anyone, or a few people, or many people, according to these views, it is a legitimate form of treatment. In this context, personal testimonials and traditional reports of success may be more influential than statistical data about cure rates.

It can probably be said with some confidence that most medical scientists today have doubts about the use of at least some, and perhaps most, forms of alternative therapies. One speaker at a symposium sponsored by the American Association for the Advancement of Science in 1997, for example, said that "essentially, the public is spending billions of dollars on sugar pills to cure their sniffles, hand-waving to speed recovery from operations, and good thoughts to ward off illnesses, all with assurances that it's based on science."

Still, other medical practitioners are taking a somewhat more measured view. They are becoming more open to the possibility that some aspects of some therapies may have some role in the treatment of some disorders. For example, chiropractic techniques and acupuncture tend to be more accepted by medical doctors than was once the case.

The dispute over alternative therapies has some very specific and concrete implications for practitioners in those fields, for medical doctors, and for the general public. Companies that pay for medical treatments may be expected to pay for *any* treatment that works. If politicians, health insurance companies, and the general public become convinced that one or another alternative therapy is actually effective, it will probably have to be included along with medical procedures for payment by insurance companies. That decision is likely to have an impact on everyone involved in the society, from the U.S. Social Security Administration down to the individual citizen.

References

"The alternative medicine home page." http://www.pitt.edu/~cbw/altm.html. 7 April 1998.

Hall, Carl T. "Navigating New Age Medicine." *San Francisco Chronicle,* 9 December 1996, C1+.

Hill, Richard L. "Scientists Debunk Alternative Therapies." *Oregonian,* 15 February 1997, A1+.

Office of Alternative Medicine (of the National Institutes of Health). http://altmed.od.nih.gov/. 7 April 1998.

Anencephalic Babies

Anencephaly is a medical condition in which a child is born with an incompletely developed brain. An anencephalic baby typically lacks a skull, a scalp, and all or most of a forebrain. The forebrain is the region of the brain responsible for cognitive, or "thought," processes. The hindbrain, responsible for automatic responses, is generally present in the anencephalic child. Anencephalic children seldom live longer than a few days.

Organ Transplantations with Anencephalic Babies

Questions have arisen about the transplantation of organs from anencephalic babies to other very young babies who may have disorders of the heart, kidney, or liver or have other life-threatening medical conditions. The argument is as follows: Anencephalic babies will not survive in any case. In most instances, their organs (other than their brains) are healthy enough to use for transplantations. Physicians should be allowed to remove those organs for use in other ill babies.

This proposal is complicated by two factors. First, organ transplantations normally occur only after a donor patient has died. With babies under the age of seven days, predicting the likelihood of survival is very difficult, although admittedly less difficult with anencephalic babies. Declaring that a child is "dead," therefore, can be very risky. Second, death is usually determined clinically on the basis of brain activity. In an anencephalic baby, the brain may be the last organ to die. That is, by the time the baby had been declared legally dead, its organs may no longer be usable for transplantation.

These conditions have led to proposals that an anencephalic baby's organs be removed for transplantation *before* the baby can be declared legally dead. The child will die anyway, according to this argument, and its organs may very well save the life of another child.

Other authorities reject this view. The anencephalic baby must be viewed and treated as would be any other human being, they say. No one would consider removing the organs from a terminally ill adult, they point out, and the same policy should be applied to anencephalic babies. Besides, the adoption of this practice would make it easier to apply the same policy to other "near-dead" patients, such as those in a persistent vegetative state. Finally, removing organs from a still-living anencephalic baby would negatively affect the public's view of the medical profession and the view of both medical workers and the general public of anencephalic babies themselves.

References

Diamond, Eugene F. "Anencephalic Infants as Donors for Organ Transplantation." *Linacre Quarterly,* 56:1989.11.

———. "Management of a Pregnancy with an Anencephalic Baby." http://www.asfhelp.com/asf/mc/diamond.htm. 3 April 1998.

Harrison, M. R. "The Anencephalic as an Organ Donor." *Hastings Center Report.* April 1986, 21–30.

McCullagh, Peter. *Brain Dead, Brain Absent, Brain Donors: Human Subjects or Human Objects?* New York: John Wiley and Sons, 1993.

Nolan, K. "Anencephalic Infants: A Source of Transplantable Organs." *Hastings Center Report.* October–November 1988, 28–30.

Peabody, J. "Plan to Use Anencephalic Organs Being Reevaluated" *Pediatric News* 22.1988.17.

Shewmon, D. A. "Anencephaly: Selected Medical Aspects." *Hastings Center Report,* October–November 1988, 11–19.

"Transplantation of Organs from Newborns with Anencephaly." Canadian Paediatric Society. http://www.cps.ca/english/statements/B/b90–01.htm. 3 April 1998.

Animal Rights

Biologically, human beings are animals. Beyond that taxonomic fact, however, humans are almost always regarded as different from and, generally, superior to other kinds of animals. It is common practice for humans to keep animals in cages, force them to perform for our amusement, experiment on them for noble or less noble purposes, and even raise them and then kill them for food. It would be considered immoral and/or illegal to use humans for almost any of these purposes. Is this distinction between humans and other animals ethically correct?

The Animal Rights Movement

That question forms the basis of an intense debate that has raged through U.S. culture—and that of some other nations—for centuries. The debate has become distinctly more intense in recent decades. As with many social issues, the debate over animal rights is often conducted not in a calm and reasoned atmosphere but with language designed to arose passions. Those who believe in animal rights are labeled as "humaniacs," "zealots," and "wackos" who wander in an "ethical wilderness." On the other hand, hunters, animal researchers, and even zookeepers are branded as "monsters" who inflict "needless pain" and use "senseless cruelty" in dealing with animals.

Those who believe in animal rights have sometime resorted to violent action to support their views. They have broken into research laboratories to release animals involved in scientific experiments. They have harassed men and women who wear fur coats, leather, or other animal products. And they have harassed hunters in the field. In response, those who oppose animal rights have not hesitated to defend their believes and practices with equally violent behavior.

The animal rights debate includes those who hold the most extreme positions on the use of animals by humans as well as every conceivable position between these extremes. For example, some people object to eating any animal or animal product; others refuse to keep pets;

The raising of animals specifically for the purpose of using their furs raises questions as to the relative "rights" of humans versus other animals. (Courtesy of PETA)

still others object to captive breeding programs to save endangered species; and some argue that zoos are nothing other than "animal prisons." By contrast, many individuals have no problems in killing animals "for fun."

Animal Rights and Animal Welfare

Perhaps the most difficult issues in the animal rights debate are closer to the center of this wide spectrum of positions, the dispute over animal *rights* versus animal *welfare*. Animal welfarists could be characterized as those who believe that it is legitimate for humans to use animals for certain purposes provided that they are treated as humanely as possible. Some of the "legitimate" uses recognized by welfarists include medical experimentation, a source of food and/or clothing, and a source of amusement, as in zoos and circuses. Welfarists would argue that animals used in medical experiments, for example, must be fed properly, housed in clean and spacious cages, and protected from unnecessary suffering. Animals killed for food or clothing, they say, must be destroyed as quickly, painlessly, and efficiently as possible.

The philosophical underpinning for the animal welfare position is that animals do not and cannot have "rights" in the sense used by humans. Rights arise only when an organism has the capacity to make free moral judgements about right and wrong. And animals do not have this capacity. As one writer has put it:

> The holders of rights must have the capacity to comprehend rules of duty, governing all including themselves. In applying such rules, the holders of rights must recognize possible conflicts between what is in their own interest and what is just. Only in a community of beings capable of self-restricting moral judgments can the concept of a right be correctly invoked.
>
> Humans have such moral capabilities. They are in this sense self-legislative, are members of communities governed by moral rules, and do possess rights. Animals do not have such moral capacities. They are not morally self-legislative, cannot possibly be member of a truly moral community, and therefore cannot possess rights. (Cohen 1986, 866)

This position does not suggest that animals can or should necessarily be treated as inanimate objects. Obviously, they do share some biological characteristics with humans, such as the ability to experience emotions and feel pain. Animals might, therefore, be thought to have certain "interests" about which humans should be cognizant and concerned. This view distinguishes an animal welfarist from someone who, for example, feels that no justification is needed to shoot ducks purely as a source of amusement.

The welfarist viewpoint suggests that humans should be able to justify using animals in ways that they would not use other humans. In general, such justifications are couched in terms of "the greater good." That is, one might argue that causing pain and suffering to experimental animals is justified if a cure for some human disease results. Or breeding, raising, and killing animals for food can be justified since meat is an essential component of the human diet.

The justifications used for the whole range of animal uses, however, are very diverse. At one end of the spectrum is the "human disease" argument. Many thoughtful people might argue that sacrificing the lives of rats and mice to find a cure for lung cancer is reasonable and acceptable. The same argument might hold less force if the object of the research is to develop a new kind of cosmetic.

Similarly, one might be hard-pressed to argue against hunting among people who depend on animal meat for their survival. But the decision might be more difficult if the hunter's object is to bring home an animal head to hang on the den wall.

The Animal Rights View

Animal rights advocates begin from a very different philosophical position. They argue that the capacity for moral decision making is not necessarily the primary basis for assigning rights. Very young children, some elderly people, and the mentally incompetent, for example, do not have or have lost the capacity for moral decision making. Yet, they are still accorded most of the same "rights" as other humans. It is not too much of a stretch, they say, to accord the concept of "rights" to other animals.

Issues about the treatment of animals among animal rightists, then, have quite a different basis. The question is not whether it is legitimate to use animals for human purposes but how to balance the rights of one group (humans) against the rights of another group (for example, race horses).

Such dilemmas are common in human societies. Societies have to balance the rights of individuals to own guns against the rights of the general populace to be protected from violent action by gun owners. The rights of a pregnant woman have to be balanced against the rights of the child-to-be she may be carrying.

In cases such as these, one solution may be for an individual or group to argue that no "rights" exist on one side or the other. But for those who acknowledge some legitimacy on both sides, the debate becomes much more difficult and more complex. In fact, in many situations, it can be resolved only on a case-by-case basis.

At the present time, most formal law appears to be based on an animal welfare position. For example, the U.S. Congress adopted a series of amendments in 1985 to the federal Animal Welfare Act. Those amendments require the creation of animal care committees who are charged with ensuring that animals used in research are treated humanely. The underlying philosophy of the amendments, however, appears to be that any use of animals in medical research is to be permitted as long as it can be justified as scientifically necessary.

References

Baird, Robert M., and Stuart E. Rosenbaum, eds. *Animal Experimentation: The Moral Issues.* Amherst, NY: Prometheus Books, 1991.

Cohen, Carl. "The Case for the Use of Animals in Biomedical Research." *The New England Journal of Medicine,* 2 October 1986, 866.

Day, Nancy. *Animal Experimentation: Cruelty or Science?* Hillside, NJ: Enslow Publishers, 1994.

Francione, Gary L. "Animal Rights and Animal Welfare: Five Frequently Asked Questions." http://arrs.envirolink. org/ar-voices/fire_qs.html. 20 May 1999.

Guither, Harold D. *Animal Rights: History and Scope of a Radical Social Movement.* Carbondale: Southern Illinois University Press, 1998.

"Hunting vs. animal 'rights.'" http://www. acs.ucalgary.ca/~powlesla/personal/hun ting/rights/index.html. 4 April 1998.

Leahy, Michael P. T. *Against Liberation: Putting Animals in Perspective.* New York: Routledge Press, 1994.

Loew, Franklin M. "Animals in Research." In *Birth to Death: Science and Bioethics,* edited by David C. Thomasma and Thomasine Kushner, 301–312. Cambridge: Cambridge University Press, 1996.

Moros, Daniel A. "Taking Duties Seriously: Medical Experimentation, Animal Rights, and Moral Incoherence." In *Birth to Death: Science and Bioethics,* edited by David C. Thomasma and Thomasine Kushner, 313–324. Cambridge: Cambridge University Press, 1996.

"Project Vote Smart: Issues: Animal." http://www.vote-smart.org/ issues/ANIMAL/. 4 April 1998.

"Reference Materials" (on the animal rights movement). http://arrs.envirolink.org/ Faqs+Ref/. 20 May 1999.

Regan, T. *The Case for Animal Rights.* Berkeley: University of California Press, 1983.

Rowan, A., and F. M. Loew. *The Animal Research Controversy.* Boston: Tufts Center for Animals and Public Policy, 1995.

Sherry, Clifford J. *Animal Rights.* Santa Barbara, CA: ABC-CLIO, 1994.

Singer, Peter, and Susan Reich. *Animal Liberation.* New York: New York Review of Books, 1990.

Information about the animal rights movement can be obtained from People for the Ethical Treatment of Animals, 501 Front Street, Norfolk, VA, 23501, 757–622-PETA.

A detailed description of the Animal Welfare Act of 1985 can be found on the Internet at http://www.aphis.usda.gov.80/oa/creg.html#toc. 20 May 1999.

Appropriate Technology

Appropriate technology is a form of technology that is small, simple, and inexpensive, in contrast to "hard" (or "high") technology, which is large, complex, and expensive. Appropriate technology was promoted by a number of groups and individuals in the late 1970s as a way for people in developing nations to deal with a number of problems in their daily lives. Its most persuasive spokesperson was probably E. F. Schumacher, who wrote about the practice in his book, *Small Is Beautiful: Economics as if People Mattered.*

In most industrialized nations, the tasks of everyday life are handled through large, complex, centralized systems. Small family farms, for example, have been largely replaced by megafarms covering thousands of acres. Operating these farms requires the use of large, complex machinery that is expensive to operate and maintain. By contrast, a relatively small number of humans is needed in the daily operations of such farms.

Agricultural systems of this magnitude are impossible to maintain in many developing countries. Advocates of appropriate technology point out, however, that people can still benefit from technology that is less expensive, simpler, and designed for the specific setting in which it is to be used. This kind of appropriate technology has such characteristics as the following: It is labor intensive (rather than capital intensive), providing jobs for more people; it encourages individuals and communities to be self-sufficient; it makes use of inexpensive, locally available materials and pro-duces products for local consumption; it produces little or no pollution; it requires simple machinery that is inexpensive, easy to use, and easy to repair; and it retains control of technology in the local area.

Appropriate technology has not always been received with enthusiasm by the people for whom it is intended in developing nations. Some people feel that appropriate technology is a way of preventing technological development in their nations. Many aspire to the "high tech" lifestyle of industrialized nations, even if there are severe costs to individual citizens in the process of reaching that level of development. And appropriate technology seldom, if ever, can compete with high technology systems that large multinational corporations are eager to impose on the existing cultures of developing nations.

References

A number of colleges and universities have created departments with special interests in appropriate technology. Some of those institutions are the following:

Colorado State University. http://www.colostate.edu/Orgs/ATI/. 20 June 1998.

Humboldt (CA) State University. http://www.humboldt.edu/~ccat/ccat.htm. 20 June 1998.

University of Arkansas at Fayetteville. http://www.attra.org/. 20 June 1998.

Other sources of information include:

Appropriate Technology International, 1828 L Street, N.W., Suite 1000, Washington, D.C. 20036, (202) 293–4600; (202) 293–4958 (fax); atinl@igc.apc.org.

Appropriate Technology Marketing—ATM, Division of Sequel Corporation, 119 North Jonathan Blvd., Chaska, MN 55318.

The Appropriate Technology Sourcebook, from Colorado State University (see above) and at http://www.colostate.edu/Orgs/ATI/book.htm. 20 June 1998.

"Books for Sustainable Living: Energy, Building, Transportation." http://www.jademountain.com/bookspi.html. 20 June 1998.

Development Center for Appropriate Technology, P.O. Box 41144, Tucson, AZ, 85717, (520) 624-6628, and at http://www.cyberbites.com/dcat/. 20 June 1998.

Assisted Reproductive Technologies

The process of conceiving a child is sometimes imagined to be among the simplest and most natural of all human acts. Indeed, men and women often accomplish this task without intending to do so. The world is filled with untold numbers of unwanted children who were produced in this way. The possibility of having to carry and bear unwanted children is the basis for most issues concerned with contraception and abortion.

Problems of Infertility

Yet, for many couples, childbearing may be profoundly difficult or even impossible. By some estimates, about 15 percent of all couples in the United States who want to have children are unable to do so after a year of trying. In excess of 4 million U.S. couples have infertility problems. An important field of medical research has grown up around efforts to aid couples who want, but are unable to have, children by means of normal sexual intercourse. This research can be described in general as Assisted Reproductive Technologies (ART).

Infertility may be caused by any number of factors in the male, the female, or both. For example, a man may be unable to produce sperm in the number needed to produce fertilization. A woman may be unable to produce eggs or may have infrequent periods that make fertilization difficult. Or she may have physiological or anatomical disorders, such as blocked or scarred fallopian tubes, that make ovulation or fertilization difficult or impossible. Experts estimate, however, that about a third of all infertile couples can be aided by some form of ART. The following forms of ART are some of those currently available.

In vitro fertilization (IVF) is the oldest of all forms of ART. The first child conceived by IVF was born in 1978. The first step in IVF is to administer drugs that will stimulate a woman's ovaries to produce multiple eggs. The most common drug combination used today is Lupron (leuprolide acetate) in combination with either Metrodin (urofol-

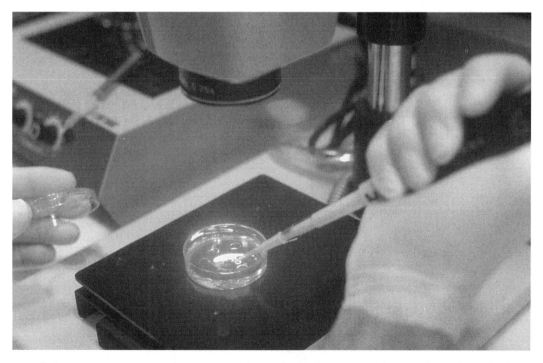

In vitro fertilization creates new issues as to who "owns" sperm and eggs and under what circumstances that "ownership" can be transferred. (Corbis/Owen Franken)

litropin) or Pergonal (menotropin). Lupron is administered first in order to block the normal process of ovulation that might interfere with the planned IVF. Pergonal and Metrodin are forms of the natural hormones luteinizing hormone (LH) and follicle-stimulating hormone (FSH). They stimulate the ovaries to produce multiple eggs.

After these drugs have been administered, blood and ultrasound tests are carried out to determine when eggs have been produced. When those eggs are mature enough, they are removed by suction with a thin needle. They are then mixed with sperm from the woman's mate and allowed to interact for about two days. The combination of eggs and sperm takes place in a petri dish or some other kind of glass container, accounting for the name of the procedure: The Latin *in vitro* means "in glass."

If fertilization occurs, three to five of the embryos formed are then returned to the woman's uterus, where, if all goes well, at least one will implant and mature. Any additional embryos that have been formed may be discarded or frozen for future use by the couple.

Gamete intrafallopian transfer (GIFT) and zygote intrafallopian transfer (ZIFT) are two procedures roughly comparable to IVF and to each other. The initial stages of both GIFT and ZIFT are similar in that drugs are administered to stimulate the production of eggs. In GIFT, those eggs are then removed, mixed with the male's sperm in vitro, and reintroduced immediately into the woman's fallopian tubes. If the procedure is successful, the eggs are fertilized and implantation and development of the embryo(s) proceed as in a natural pregnancy.

GIFT requires that a woman have at least one functioning fallopian tube and that the man have viable sperm capable of penetrating an egg. When a man's sperm does not meet this condition, ZIFT may be employed. The early stages of this procedure—stimulation of egg production and harvesting of eggs—are the same in ZIFT as in IVF and GIFT. The harvested eggs are then fertilized in vitro and returned to the woman's fallopian tubes. A successful procedure results in the maturation of an embryo, its implantation, and development along the lines that occur naturally.

Donor egg IVF (DEIVF) is a variation of IVF used when a women is unable to produce viable eggs. For example, a woman who has passed menopause no longer produces eggs, but she may still wish to have children. In such cases, some other woman is found who is willing to donate eggs to the postmenopausal woman. Those eggs are then combined by standard IVF techniques with the sperm of the infertile woman's mate. Any embryo(s) produced by this method are then introduced into the uterus of the first woman. A successful procedure results in the implantation and development of an embryo along the lines of a natural pregnancy.

Male factor procedures are methods for inducing pregnancy when a male has an insufficient number of sperm or sperm incapable of penetrating an egg. One of the most promising of these techniques is known as intracytoplasmic sperm injection (ICSI). ICSI involves the introduction of a single sperm into a single egg.

Under natural conditions, more than a million sperm may be released during ejaculation. Only one of these sperm, however, is actually able to penetrate an egg. After that one sperm has entered the egg, the egg's membrane undergoes a chemical change that makes it impermeable to other sperm. In ICSI, very delicate instruments are used to collect a single egg and a single sperm. A micropipette (a small, thin glass tube) is used to puncture the egg membrane and introduce the sperm. If and when fertilization occurs, the embryo is placed into the woman's uterus by conventional IVF or ZIFT techniques.

Surrogacy is the procedure in which some woman agrees to have a fertilized egg implanted into her uterus and then to carry that pregnancy to completion. After birth of the baby, that woman also agrees to give up the baby to some other couple. The biological mother, the woman who actually carries and bears the child, is the surrogate ("in place of") mother.

In principle, the fertilized egg introduced into the surrogate's uterus could be one from (1) the woman to whom the child is given after birth, the nonbiological mother; (2) the surrogate herself; or (3) some third woman. The sperm could come from (1) the nonbiological mother's mate or (2) any other man. Deciding who the "mother" and "father" are of a child born under these circumstances can be very complicated.

At one time, it was possible to hire a women to be a surrogate mother. That practice is no longer allowed because of the abuses practiced by some women who chose to become surrogates. As a result, most surrogates today are friends or relatives of women who wish to have children but are biologically unable to do so themselves.

Fertility drugs provide one of the most direct means of inducing pregnancy. In some cases, pregnancy does not occur because women produce too few eggs or produce eggs at irregular intervals that make conception difficult. In such instances, it may be possible to administer drugs that will increase the frequency of ovulation and the number of eggs produced. In many cases, the use of fertility drugs has resulted in a pregnancy after six months or less.

Problems Associated with ART

All forms of ART are beset with a number of social and ethical questions in addition to the biological problems that accompany any pregnancy. For example, one can imagine complex situations in which up to five individuals may be involved in a pregnancy and birth. As an example, a man incapable of producing sperm and a woman incapable of producing eggs could arrange to obtain viable sperm from a second man and viable eggs from a second woman that, after fertilization, could be introduced into the uterus of a third woman, the surrogate mother. The child born of this arrangement

could be said to have two fathers and three mothers.

Ironically, the only people in this arrangement who could *not* be called *biological* parents are those who will actually raise the child as their own. A number of legal cases have arisen in cases simpler than this when one or another "parent" has decided to exercise his or her "right of parenthood" over the child.

Questions often arise also when drugs used to stimulate ovulation are even more successful than planned. For example, the rate of multiple births due to the use of fertility drugs has begun to climb precipitously. The number of multiple births with three or more children in the United States has increased from less than 1,000 in 1975 to about 3,000 in 1990 to more than 5,000 in 1996. This increase appears to be directly related to the increasing use of fertility drugs.

Frozen Sperm and Frozen Embryos

The ultimate fate of frozen sperm and embryos (unfertilized eggs cannot be frozen) can also be the subject of intense dispute. People for whom fertility is an important issue are unlikely to permit the destruction of sperm or, especially, embryos that were obtained with such difficulty and often at great expense.

As a consequence, the number of sperm and embryos now stored in liquid nitrogen is growing rapidly. That number is not known with certainty but is believed to be in the tens of thousands. Experts say that the supply of frozen embryos has been increasing at a rate of about 10,000 per year since 1992.

Many questions can arise about these materials. For example, what happens when a married couple for whom embryos have been produced by ART is divorced? Who "owns" these embryos? Or what becomes of frozen embryos when both parents of the embryos have died? Or can a woman have frozen embryos implanted into her uterus even though she and her pervious mate are no longer married to each other but *are* married to new spouses? Or what happens when storage facilities become overcrowded and parents of frozen embryos refuse to continue paying for storage of their embryos or even decline to respond to requests for information about those embryos?

The legal status of frozen sperm and embryos in the United States is unclear. In Louisiana, frozen embryos have been given the status of living persons and cannot be destroyed. They must be kept in cold storage until implanted into a woman. In Illinois, a woman who has produced an embryo now kept in cold storage is regarded as being legally pregnant. Other states and the federal government have no specific laws dealing with frozen sperm and eggs, and issues about their disposition are still being decided on a case-by-case basis.

The complexity of these issues is illustrated by a recent case in England, where frozen sperm and embryos are regulated by the national government. In late 1996, England's highest court ruled that a women may not be impregnated with frozen sperm from her dead husband. The sperm had been collected and frozen specifically because the husband had contracted bacterial meningitis and fallen into a coma. His wife asked that sperm samples be collected so that she could bear his child if and when he died. The nation's Human Fertilization and Embryology Authority ruled, however, that the husband had not given written approval for this procedure, and it denied the woman permission for use of the sperm. The High Court agreed with this decision and ruled that the sperm could not be removed from storage to be used anywhere else in the world.

References

Alpern, K. D., ed. *The Ethics of Reproductive Technology.* New York: Oxford University Press, 1992.

"Assisted Reproductive Technologies." http://www.ihr.com/bafertil/assistre.html. 14 April 1998.

The Assisted Reproductive Technology Workbooks. Somerville, MA: Resolve, 1993.

Dusky, Lorraine. "Who Owns Embryos When a Marriage Goes Bad?" *USA Today,* 14 May 1998, 13A.

Frerking, Beth. "The Fertility Dance." *Oregonian,* 21 November 1997, A18.

Harden, Blaine. "N.Y. Trial Highlights Dilemmas Posed by Artificial Conception." *San Francisco Chronicle,* 10 April 1998, A12.

Hull, R. T. *Ethical Issues in the New Reproductive Technologies.* Belmont, CA: Wadsworth, 1989.

Koglin, Oz Hopkins. "Doctors Give in Vitro a New Twist." *Oregonian,* 17 June 1998, A17+.

McClure, Michael E. "The 'ART' of Medically Assisted Reproduction: An Embryo Is an Embryo Is an Embryo." In *Birth to Death: Science and Bioethics,* edited by David C. Thomasma and Thomasine Kushner, 35–49. Cambridge: Cambridge University Press, 1996.

Rosenbert, H. S., and Y. M. Epstein. *Getting Pregnant When You Thought You Couldn't.* New York: Warner Books, 1993.

Stolbert, Sheryl Gay. "Recipients of Donated Eggs Wrestle with Disclosure." *San Francisco Examiner,* 18 January 1998, A19.

"Surrogacy Questions & Answers." http://www.surrogacy.com/tasc/q_and_a.html. 14 April 1998.

Zadig, David. "The New Fertility." *Harvard Magazine,* November/December 1997, 54–64+.

Attention Deficit Disorder

See Ritalin

Attention Deficit Hyperactivity Disorder

See Ritalin

B

Barrier Islands

Barrier islands are long, narrow stretches of land that lie parallel to and offshore of coastlines in various parts of the world. Some of the better known barrier islands are located along the Atlantic and Gulf Coast shores in North America. Included among these structures are Cape Hatteras, in North Carolina; Padre Island, in Texas; Sanibel and Captiva Islands, in Florida; and Ocean City, in Maryland. The city of Miami Beach is also built entirely on a barrier island.

Origin and Dynamics of Barrier Islands

Barrier islands are thought to have developed about 5,000 years ago, at the end of the last glacial period. Melting glacial ice is believed to have caused a rise in sea level of about three feet per century, a rise that continued for long periods thereafter. When the rate of sea-level rise finally slowed down, barrier islands began to be deposited along sandy beaches.

Barrier islands have been described as "nomadic landforms." That term reflects the fact that such islands are not static geologic structures but are constantly undergoing change. Their shape and position are constantly being altered by wind and waves, storms, tidal patterns, and alongshore current. The materials of which barrier islands are made are continuously being washed away and redeposited, moved toward or away from the shore, and transported up or down parallel to the coastline.

Barrier islands serve important geologic and biologic functions. For example, the islands tend to bear the brunt of heavy storms. Wind and waves lash the seaward side of the islands, so that the shore is protected from the devastating effects of these elements.

The sheltered side of the island and the water trapped between island and shore are generally protected from such storms.

What kinds of insurance and protections should governments provide to people who choose to build homes on fragile environments, such as barrier islands like Cape Hatteras, shown here in a photo taken from the *Apollo 9* spacecraft in 1969? (Corbis)

Both land and water are able, therefore, to develop rich and complex ecosystems. The bay or lagoon that forms between a barrier island and the shore is often one of the richest sources of marine life found anywhere in the world.

Human Impact on Barrier Islands

Barrier islands have long been regarded as highly desirable locations for housing and commercial developments. Entrepreneurs are able to construct homes, hotels and motels, restaurants, and other buildings almost directly on the oceanfront. They can provide homeowners and tourists with some of the most spectacular ocean views to be found anywhere.

But such developments usually do not take into consideration the fundamental nature of barrier islands. Homes and other structures are easily damaged by the severe storms that tend to strike coastlines where barrier islands exist. The materials of which the islands are made continue to shift position, as they have for centuries. In so doing, they may carry away buildings, roads, and other structures constructed on them by developers.

Over the past century, billions of dollars have been spent on the construction of new buildings and infrastructure on barrier islands and then, later, on systems for keeping the islands from washing away or moving. One technique is to build seawalls, jetties, riprap, and other structures designed to deflect the force of wind and waves on barrier islands. In many cases, these structures have the opposite effect of that intended and may actually contribute to the destruction or displacement of a barrier island.

Another technique has been to replace materials washed away or displaced on an island. The City of Miami Beach used this approach when it discovered that very large portions of the natural sandy beaches for which it had become famous had been washed away. Attempts by humans to keep barrier islands in one specific location are, however, never-ending

battles against nature. They fly in the face of the natural tendency of the islands to continuously relocate.

Many environmentalists have long argued that the best approach to use with barrier islands is preservation. That is, efforts should be made to ban new construction and allow the islands to behave geologically and biologically in ways that are natural for them. With pressures from humans for new commercial developments, however, it is unclear how successful such appeals for preservation are likely to be.

References

"Barrier Islands." http://naples.com/amainsan/barisl.htm. 17 July 1998.

"Barrier Islands: A Fool's Paradise?" Excerpt from L. Erik Calonius, "Hurricane Experts Say the State of Their Art Can't Avert a Disaster." *Wall Street Journal*, 14 October 1983, 1. In Joseph M. Moran, Michael D. Morgan, and James H. Wiersma. *Introduction to Environmental Science.* 2d ed. New York: W. H. Freeman, 1986, 500–501.

"Development of Property on Coastal Beaches, Specifically Barrier Islands." http://gurukul.ucc.american.edu/nature/16/cosent/dvlpmnt.html. 17 July 1998.

Hawes, E. "Castles in the Sand." *New York Times Magazine*, July 1993, 24–32.

Martin, Elaine L. "Barrier Islands." In *Gale Encyclopedia of Science*, edited by Bridget Travers, 411–413. Detroit: Gale Research, 1996, 411–413.

"Morphodynamics." http://coastal.er.usgs.gov/wfla/HTML/morpho.html. 17 July 1998.

"Prevention of Barrier Island Erosion." http://gurukul.ucc.american.edu/Nature/16/cosent/prevent.html. 17 July 1998.

Wagner, Richard H. *Environment and Man.* 3d ed. New York: W. W. Norton, 1978, 226–229 and 481–483.

Basel Convention (on the Control of Transboundary Movements of Hazardous Waste and Their Disposal)

See Hazardous Wastes, International Dumping

Behavior Modification

Behavior modification is the process by which some individual or group of individuals attempts to change the actions of a second individual or group of individuals. Behavior modification is one of the oldest and most common of all human activities. For example, the process of "growing up" involves, except for the physical process, little other than the child's learning what it should or should not do, which behaviors are acceptable and which are not. A child's behavior is modified in positive ways by praise and rewards and in negative ways by scoldings and spankings.

Attempts to modify adult human behavior are legion. Every advertisement is designed to influence the choices people make when they buy something. Prisons are among the most highly organized systems for changing behavior that has been defined as "antisocial."

The Scientific Basis for Behavior Modification Practices

Throughout history and in the vast majority of instances, behavior modification has been and is an unscientific, haphazard process. Teachers may tend to believe that letter grades (A, B, C, and so on) and honor rolls are powerful factors in affecting a student's learning behavior. Certainly, this form of behavior modification has an important part in most formal programs of education today. However, that assumption is not well founded in psychological or educational research.

Similarly, many features of the law enforcement process are predicated on beliefs about behavior modification techniques. We assume, for example, that incarceration in prison is an effective way of changing an individual's antisocial behavior. But rates of recidivism (return to prison) have long raised profound questions about that assumption. Also, a strong argument for the use of the death penalty has been that it serves as a deterrent to violent crime. Yet, scientific evidence to support this view is still quite controversial.

The absence of concrete data to support society's views of the behavior modification practices it uses in education, the prison system, or elsewhere is generally not regarded as a reason for reassessing or changing those practices.

Conditioning

In fact, a very large body of scientific information exists about the ways in which human behavior (and that of other organisms) can be changed. Some of these behavior modification techniques involve physical or chemical processes, such as brain surgery, drug therapy, and electrical stimulation of the brain. Those techniques are discussed in other sections of this book. Another group of behavior modification techniques involves psychological procedures; these techniques are collectively known as psychological conditioning or, more simply, conditioning.

The earliest research on psychological conditioning was conducted by Russian psychologist Ivan Pavlov in the late nineteenth century. Pavlov discovered that he could train dogs to associate the sound of a bell with the sight and smell of food. At feeding time, the dogs began to salivate naturally at the sight and smell of food. Pavlov arranged his experiments so that a bell was rung at the same time that dogs were presented with their daily meal. Eventually, he found that the dogs would salivate at the sound of the bell, even if no food was present at the time.

Pavlov called this form of learning conditioning. The term means arranging signals (the stimuli) an organism receives in such a way as to produce the behavior (the response) an experimenter wants to achieve. Pavlovian conditioning is also known as classical conditioning to distinguish it from other forms of conditioning developed later in the twentieth century.

The changes that occur during classical conditioning can be represented quite simply as follows:

UCS————→UCR

This diagram says that an unconditional (natural) stimulus, the meat, provokes from dogs an unconditional (natural) response, salivation. As conditioning occurs, this pattern changes so that now a conditional (nonnatural) stimulus (CS), the bell, provokes the same response:

$$CS----\rightarrow UCR$$

In fact, scientists later learned that the response provoked (salivation) is not precisely the same with both the unconditional and conditional stimulus. It is more correctly represented as a new, conditional (nonnatural) response (CR):

$$CS----\rightarrow CR$$

This distinction is a somewhat technical point because the unconditional and conditional response often appear to be identical.

Operant Conditioning

A second form of conditioning was developed also in the late nineteenth century by U.S. psychologist Edward Thorndike. Thorndike worked with cats who were confined to boxes that they were able to open by performing some action, such as pulling on a string. Food was placed outside the box next to the escape hatch.

Thorndike made the simple discovery that cats that were rewarded for performing some behavior were more likely to repeat that behavior. That is, a cat might discover that pulling on a string would allow it to escape from the box and collect the food placed outside. Very quickly, the cat would learn to perform the correct behavior (pulling on the string) in order to escape from the box and obtain food.

Thorndike formulated his approach to conditioning somewhat differently than did Pavlov. He focused on the importance of rewarding or reinforcing behaviors in order to encourage an organism to repeat those behaviors. That is, any behavior exhibited by a cat that resulted in its earning food would be more likely to be repeated;

any behavior that did not earn it food was less likely to be repeated.

This form of conditioning was eventually given the name operant conditioning. The term *operant* is chosen because the response being conditioned (such as pulling on a string) is itself called the operant because it is "operating" on the environment. In operant conditioning, the two crucial elements are the operant and the reinforcement. The pattern that occurs in operant conditioning can be described as:

$$Response--Reinforced---\rightarrow Strengthened$$
$$Response--Not\ Reinforced---\rightarrow Weakened$$

If a particular response is not reinforced over a sufficiently long period of time, it no longer appears. That is, it is extinguished.

Skinnerian Conditioning

The principles of operant conditioning were developed to their highest point in the twentieth century by U.S. psychologist B. F. Skinner. Skinner's research focused on finding out in great detail how various patterns of reinforcement affect the likelihood that a response will or will not be repeated. Two of the basic questions he explored were (1) the relative effectiveness of positive reinforcement ("rewards") versus negative reinforcement ("punishment") and (2) the relative effectiveness of various schedules of reward and punishment. As an example of the latter question, he attempted to discover the differential effects of rewarding a pigeon (for example) with one grain of corn after each correct peck versus one grain after every third peck versus one grain after some random number of pecks, and so on.

Skinner summarized the results of his earliest research in one of the most impressive and influential books in modern psychology, *Schedules of Reinforcement*. He reported some basic general principles discovered in his work. For example, he found that positive reinforcement is almost always more effective at producing learning than is negative reinforcement. That is, if you want to teach a dog a new trick, you will be more successful by providing food for each correct response than by scolding the dog for each incorrect response.

Skinner and his colleagues applied the knowledge he learned to develop an incredible array of behaviors in experimental animals. For example, he taught pigeons to play ping-pong, dance in figure eights, or feed themselves by pushing some correct sequence of buttons.

Behavior Modification in Humans

As fascinating as the results of conditioning have been with animals, it is their implications for behavior modification in humans that are of greatest interest. Skinner argued, for example, that his research had direct and critical significance for all aspects of human behavior. To be "human," he suggested, was nothing other than to display a certain set of very specific behaviors. Therefore, it was at least theoretically possible to design a world in which human behaviors that society thinks are desirable are reinforced while those of which society disapproves are phased out.

Skinner was extraordinarily active in translating his results into tools for the manipulation of human behavior. For example, he designed an enclosed space—often called a Skinner box—for the raising of his own children. The space was designed to reinforce the behaviors that he and his wife wanted to encourage in his children and to extinguish those they wanted to discourage. He also developed textbooks and mechanical systems for use in education institutions that he said would be more effective than the unscientific, random methods used by human teachers.

In some ways, Skinner's most influential—and certainly most controversial—application of his research appeared in his classic book, *Walden Two*. In this book, Skinner described an ideal human society in which every person was "engineered" by means of behavior modification techniques to fit a particular slot in society. Every slot was filled perfectly by individuals who had

been educated to understand their roles, have the ability to perform them, and be happy in filling those roles.

Some people were so convinced of Skinner's arguments that they actually tried to create such communities. Eventually they discovered that, for whatever reasons, such communities were not able (at least, not yet) to survive, and they were disbanded.

Practical Applications of Conditioning Today

There can be little question about the enormous impact of Pavlov, Thorndike, Skinner, and their colleagues in the field of psychology. Research on conditioning has produced some of the clearest, most specific knowledge about the behavior of organisms available today. That research, however, has had surprisingly little effect on programs for behavior modification in humans.

For example, most educational systems today incorporate only a small fraction of the information developed by Skinnerian psychology in the design of learning systems. It seems likely that many teachers operate as if they had never heard of behavior modification techniques or had chosen not to use them in their own classrooms.

In fact, there have been only two arenas in which the principles of conditioning have been attempted on a large scale: mental hospitals and prisons. Both of these institutions contain, of course, individuals whose behaviors are socially unacceptable. They are designed to change those behaviors, insofar as possible, to allow those individuals to return to and function appropriately in the general society. Some mental hospitals and prisons have used conditioning to control patient and inmate behavior without a conscious plan to do so. For example, electric shocks and beatings have historically not been an unusual way of controlling the behavior of the mentally disordered and prisoners. These forms of negative reinforcement, also called aversive conditioning, are designed, of course, to extinguish undesirable social behaviors.

On rare occasions, carefully designed programs of aversive conditioning have been put into practice in mental hospitals or prisons to correct antisocial behaviors. The maximum security prison at Somers, Connecticut, for example, introduced a program of shocking convicted child molesters while they looked at pictures of young children. The treatment was designed to eliminate the prisoners' desires to have contact with young children. The program appeared to work but was eventually discontinued, at least in part by objections that it was "cruel and inhumane" to prisoners.

Other prisons have attempted to employ programs of positive reinforcement to change human behaviors. The Special Treatment and Rehabilitative Training (START) program at the Medical Center for Federal Prisoners in Springfield, Missouri, for example, initiated such a program in October 1972. This program consisted of a number of "levels" through which prisoners could move as their behavior improved. Each time they demonstrated more acceptable social behaviors, they were allowed greater privileges and freedoms within the prison. The program lasted for sixteen months and was regarded as successful by those who ran it. It has not been expanded or repeated in other prisons, however, to any large extent.

Examples like these are, however, rare. The limited number of institutions in which scientific behavior modification is systematically used suggests that most institutions in the business of changing human behavior on a large scale (such as schools, mental hospitals, and prisons) have chosen not to or do not know how to incorporate the principles of conditioning into their operation in any organized, systematic, and effective way.

This does not mean that the scientific principles of conditioning have no place at all in the everyday world. They are often applied in more limited ways to very specific issues, such as dealing with weight loss problems or selling specific products through advertising in the mass media.

See also: Electrical Stimulation of the Brain; Electroconvulsive Shock Therapy; Psychosurgery

References

Bennett, C. M. "A Skinnerian View of Human Freedom." *The Humanist,* July/August 1990, 18–20+.

"Brainwashing." *Society,* March/April 1980, 19–50.

Chomsky, Noam., "The Case Against B. F. Skinner." *New York Review of Books,* 30 December 1971, 18–24.

Conway, Flo, and Jim Singleman. "Modified Behavior Modification.," *The Clearing House,* December 1982, 183–188.

Council on Scientific Affairs. "Aversion Therapy." *Journal of the American Medical Association,* 13 November 1987, 2562–2566.

Geiser, Robert L. *Behavior Mod and the Managed Society.* Boston: Beach Press, 1976.

Holden, Constance. "Patuxent: Controversial Prison Clings to Belief in Rehabilitation." *Science,* 10 February 1978, 665–668.

McConnell, James V. "A Case for Brainwashing Criminals." *Psychology Today,* April 1970, 74.

Skinner, B. F. *Walden Two.* New York: Macmillan, 1948.

Zimbardo, Philip. "Mind Control in 1984." *Psychology Today,* January 1984, 68–72.

Biological Determinism

Biological determinism is the notion that individuals and groups of individuals are inherently different from each other and that these differences are manifested in the social position that those individuals and groups have in a society.

Biological determinism is closely related to the historic argument over "nature versus nurture." That issue is concerned with the question as to which aspect of human personality—one's hereditary background or one's learning experiences—are more influential in determining how one turns out in life.

Inherently Equal or Inherently Unequal?

To some thinkers, all humans are created inherently equal. That concept is embodied, of course, in the Declaration of Independence of the United States. Proponents of this view say that people are what they are because they study, learn, struggle, and make the best of a genetic background that is not much different from that of everyone else in society.

In contrast to this view is the belief that humans are inherently unequal. That is, some people are born with genes that give them greater intelligence, superior moral thought, higher ethical yearnings, and other traits that make them inherently "better" than others in the same society. By extension, certain groups of individuals are genetically superior to others. Whites are superior to Blacks; men to women; and heterosexuals to homosexuals.

The latter view is one that has long been used to justify existing social structures. In earlier centuries, for example, people were classified as nobility or serfs because of their family background. No serf could ever "learn" enough or work hard enough to become a noble. Nobles *were* nobles because they shared "blue blood" or some genetic quality that automatically placed them in an upper class. Divisions in society existed because of inherent and immutable differences in people's heredity.

Similar arguments have been used throughout history to justify all manner of political actions. Any number of scholarly studies have been carried out to justify slavery, for example, because it could be shown that Blacks (or other subservient groups) were genetically inferior: not as smart, of lesser moral value, or deficient in some other way.

The debate over biological determinism is by no means a purely academic squabble among researchers. It has profound effects in many aspects of human life and human societies. For example, many scientists at the beginning of this century felt that women are inherently inferior to men. That is, women lack the genes needed for the mental capacity, the physical stamina, and other qualities that would allow them to compete in the world of business or politics

with men. These arguments were used to prevent women from enrolling in institutions of higher education. Their more fragile nature—determined by their genes, of course—made them incapable of learning as much as men could learn and predisposed them to physical weakness and illness that would prevent them from completing a course of higher eduction.

Biological determinism has also been used to explain the social differences that one finds between races. Whites tend to be in a position of authority over Blacks in the United States, according to this argument, not because one race has better homes, schools, and cultural opportunities but because Blacks are genetically inferior to Whites.

The response by critics of biological determinism has traditionally been vigorous. On the one hand, proponents of biological determinism are often criticized for designing or carrying out their research improperly, for inserting their own personal and political biases into their research, for citing data selectively, and for many other procedural errors. On the other hand, critics emphasize the importance of learning experiences in a person's development. In all but the most extreme cases, they suggest, people are enough alike genetically so that the place one achieves in society *can* be determined by educational opportunities, provided that those opportunities are afforded to them.

See also: Criminality and Heredity; Eugenics; Homosexual Behavior; IQ; Nature versus Nurture; Sociobiology

References

Birke, Lynda. "Frenetic Determinism." *Women's Review of Books,* January 1995, 24–25.

Ehrenreich, Barbara, and Janet McIntosh. "The New Creationism: Biology under Attack." *The Nation,* 9 June 1997; also at http://www.thenation.com/issue/970609/0609ehre.htm. 13 July 1998.

"Journal: Report Says Blacks More Prone to Nicotine Addiction." *Ashland Daily Tidings,* 8 July 1998, 5.

Lewontin, Richard C. "Biological Determinism as a Social Weapon." In The Ann Arbor Science for the People Editorial Collective, *Biology as a Social Weapon.* Minneapolis: Burgess Publishing, 1977, 6–18.

Plomin, Robert. "The Role of Inheritance in Behavior." *Science,* 13 April 1990, 183–188.

Rawat, Rajiv. "The Return of Determinism?" http://ice.englib.cornell.edu/scitech/u95/bell.html. 13 July 1998.

Biometric Identification

Biometric identification is a procedure by which a person's identity is verified by means of some physical characteristic, such as a fingerprint, voiceprint, retinal pattern, facial image, or hand scan.

Fingerprinting is perhaps the oldest and most widely used system of biometric identification. The unique character of fingerprints was first described in 1823 by German anatomist J. E. Purkinje. Purkinje's discovery found no practical application for three decades, however. Then, in 1858, English criminologist Sir William J. Herschel developed a system for using fingerprints to identify individuals. Today, fingerprints are used for a number of purposes, such as identifying people suspected of committing crimes and identifying bank customers.

Type of Biometric Identification

Two forms of eye scanning have been developed. In one, a beam of light is shone into an individual's eye, and the pattern of blood vessels on his or her retina is recorded. This process is known as *retinography.* In the most sophisticated forms of retinography available, an accuracy of one out of a million (one error per million trials) is not unusual.

In the second form of eye scanning, the beam of light is reflected off an individual's irises. The beam can be used to measure the color, shape, and other features of the iris. When both irises are scanned, the accuracy of the technique approaches one error in 10^{51} trials, one of the most precise measure-

Biometric devices are now used routinely to permit access to secure areas for certain authorized personnel only. (Courtesy of EYEDENTIFY, Inc.)

ments of human traits available. Eye scans are now being used or considered by some banks to verify users of automatic teller machines (ATMs).

Facial recognition is another method of biometric identification. To use facial recognition, a camera first takes a picture of an individual's face. The photographic image is digitally stored so that it records detailed features of the face, such as distance between eyes and width of the mouth. When an individual needs to be identified, he or she stands in front of a second camera. The second camera records the individual's facial features and compares them with the stored record. Facial recognition has been proposed as a method for logging on to computers on which sensitive work must be done.

Voice authentication is still another form of biometric identification. In voice authentication, various sound patterns that make up a person's speech are recorded, analyzed, and stored. A person's identity can be authenticated, then, by simply having him or her speak into a microphone. The voice pattern can be compared with the one stored in a computer.

Future Uses for Biometric Identification

Many experts believe that biometric identification will play an important role in many aspects of human society in the future. As our systems of communication become more complex and more widely distributed, it will be necessary for those systems to ensure that only legal and legitimate transactions take place. As an example, the invention of ATMs created the problem of making sure that the person asking for money from the machine was, in fact, the person whose account was being accessed. Personal Identification Numbers (PINs) were developed to provide this authentication. Today PINs have become as important to most people as their driver's license number or their Social Security number.

Dozens of other everyday applications of biometric systems have also been identified. For example, airlines would like to have better systems for ensuring that they know who boards their airplanes and that those passengers are who they say they are. One possibility is for ticket counters to have cameras that record the faces of people buying tickets. Those photographs can then be matched by face recognition systems with the passengers who actually show up at the boarding gate. One problem with this application, however, is that many passengers do not purchase their ticket in person.

Another application of biometric identification that is already in place is used at the Lancaster County Prison in Pennsylvania. Prisoners who enter the prison are required to have an iris scan. When they are ready for release from the prison, the iris scan is repeated. The two iris scans must match before the individual is allowed to leave the prison.

Issues Related to the Use of Biometric Identification

People concerned about the increasing use of biometric identification raise two fundamental questions about the procedure. First, they suggest that individuals responsible for operating biometric devices be very carefully trained. A mistaken identification can cause very serious problems for the person whose identity is missed.

The second, and far more important issue, however, is one of personal privacy. Should Americans have to submit to "body searches," albeit highly sophisticated body searches, in order to prove who they are? Some biometric devices operate over relatively long distances, of a few meters or so. That means that a person could be scanned or examined without necessarily realizing that the "search" was taking place. Constitutional experts are debating the question as to whether the benefits derived from biometric systems justify their "intrusion" into a person's body as a "reasonable" search, as required by the Fourth Amendment of the Constitution.

Some authorities argue that existing systems of personal identification, such as PINs, provide adequate security for the great majority of transactions in which most people are involved. Biometric systems are interesting and exciting new procedures, they agree, but they may be more intrusive than is necessary to obtain the level of security needed. As interactions among individuals and corporations become more extensive and more complex, questions about the use of biometric identification systems are likely to become even more important.

References

"Biometric Research." http://www-engr.sjsu.edu/~graduate/biometrics/index.html. 6 August 1998.

"Biometrics: Can I See Your Biometric Identification?" *Computer Life*, March 1998; also at http://www.zdnet.com/complife/buz/9803/biometricid–1.html. 6 August 1998.

Slater, Eric. "Not All See Eye to Eye on Identification by Anatomy." *San Francisco Chronicle*, 4 June 1998, A9.

Biotechnology

See Bovine Somatotropin; Cloning; Genetically Manipulated Foods; Human Gene Therapy

Bison

The American bison (*Bison bison*) is a herbivorous mammal native to the grasslands of North America. The animal has long repre-

Bison have made one of the most remarkable comebacks of all species once thought to be endangered or threatened. (Corbis/ W. Wayne Lockwood, M.D.)

sented one of the great struggles in this country over endangered species. At one point, more than 60 million bison are thought to have lived in the area between Texas and Montana. However, that number was drastically reduced as Europeans arrived in North America and began to explore and move into the western part of the continent. Hunters hired by railroad companies killed the animals as food for construction crews. And bison hunting expeditions were organized for rich hunters from the East who often shot the animals from trains "for sport." By 1905, the massive native herd of bison had been reduced to about 500.

Managing Bison Herds

At that point, voices were raised demanding that one of the continent's most magnificent native animals be spared from extinction. The natural solution appeared to be to nurture a small herd of bison in one or more of the nation's large national parks. As a result, 21 bison were brought from private herds in Texas and Wyoming to Yellowstone National Park. They were added to the 25 to 50 wild bison still living in the park. The National Park Service began a policy and practice of "wildlife management" in which humans began to determine the size of the bison herd.

These management practices were largely successful. The initial group of 50 to 75 bison grew to nearly 1,500 animals in the mid-1950s. At the time, the bison were treated essentially like domestic cattle. They were allowed to graze on their own during the summer but brought into pens and hand-fed during the winter. By 1966, however, this practice had come to a halt. Strong objections were entered by hunters, among others, who wanted the opportunity to hunt bison. As the animals were left to their more natural grazing habits, the Yellowstone herd grew to more than 3,000 by the 1980s.

By the 1980s, however, problems in the National Park Service's laissez faire attitude toward bison began to develop. One of the most important of those problems was that bison often wandered out of the park into nearby National Forest lands. These lands were often occupied by herds of domestic cattle, whose owners were very much concerned about the arrival of bison among their animals, because bison carry an infectious disease known as brucellosis. Scientists believe that brucellosis can be transmitted from bison to domestic cattle, although no documented case of such a transmission has been found in nature. Cattle who develop brucellosis tend to have miscarriages. The bacterium that causes brucellosis can also cause undulant fever in humans. Because of these concerns, ranchers sought and received approval for the right to kill bisons that left park boundaries. By some estimates, at least 500 bison a year were killed in the early 1990s as a result of this policy.

Modern technology made its own contribution to the problem of the wandering bison. Yellowstone has become an extremely popular area for winter snowmobiling. More than 50,000 snowmobiles are driven on about 200 miles of groomed trails throughout the park. These trails have become favored paths for bison also. The animals find it much easier to travel on them than through deep snow, and those paths often take them out of park boundaries.

Recommendations by park personnel to close some snowmobile trails have been met by resistance from snowmobilers themselves as well as by businesspeople in towns outside the park. The sport brings millions of dollars in revenues to such towns each winter, and the lives of a few bison seem relatively unimportant when compared to this financial reality.

See also: Endangered Species; Wildlife Management

References

Angell, Jim. "Season Sharpens Park Conflicts." *Sunday Oregonian,* 21 December 1997, A24.

Blashfield, Jean F. "Bison." In *The Gale Encyclopedia of Science,* edited by Bridget Travers, 517–519. Detroit: Gale Research, 1996.

The Editors of Time-Life Books. *Lords of the Plains.* Alexandria, VA: Time-Life Books, 1993.

"Glickman Visit Catapults Bison Debate to the Fore." *Sunday Oregonian,* 23 March 1997, A27.

Hull, Dana. "Bitter Conflict over Bison Management in Yellowstone." *San Francisco Chronicle,* 29 July 1997, A4.

McMillion, Scott. "Slaughter of Park's Bison Is Old Story." *Sunday Oregonian,* 26 January 1997, A20.

Bovine Somatotropin

Bovine somatotropin is a hormone that occurs in cows that stimulates the production of milk. The hormone is a member of a larger family of hormones, the somatotropins, that occur in all animals. The somatotropins are produced in the pituitary gland and contribute to animal growth.

Bovine somatotropin is also known by other names and abbreviations, including bovine somatotropin hormone (BSH), bovine growth hormone (BGH), and just bovine somatotropin (BST). This hormone can now be produced by recombinant deoxyribonucleic acid (DNA) technology. When produced in this way, a small "r" is added to the abbreviation for the hormone to indicate its origin. Thus rBST is bovine somatotropin produced by recombinant DNA technology.

Commercial Uses for rBST

The methodology for producing BST by recombinant DNA technology was developed by the Monsanto Company in the 1980s. Monsanto proposed to sell the new product to dairy farmers to increase the amount of milk produced by their cows. Studies had shown that rBST could increase the output of milk by as much as 25 percent. After the usual series of tests to determine the product's safety and efficacy, the U.S. Food and Drug Administration (FDA) gave its approval in November 1993 for public sale of the drug. Monsanto began marketing the rBST under the trade name of Posilac®.

Issues Surrounding the Use of rBST

Posilac® quickly became a popular product among some dairy farmers in the United States because of the increased milk production it promised. But resistance to the use of the product also developed quickly among environmentalists, some dairy farmers, public health officials, political activists, and other interested individuals and groups. This resistance was based primarily on three concerns.

First, questions were raised about the effect of Posilac® on the health of cows. Monsanto's own studies had shown an elevated risk for mastitis among cows treated with Posilac®. Mastitis is an inflammation of the mammary gland that can also be associated with more serious health problems, such as disorders of the ovaries, uterus, and digestive system and enlargements of and damage to body structures, such as the knee. The rates of mastitis among cows treated with Posilac® ranged as high as 75 percent higher in some studies than among those not treated with the hormone.

Second, some authorities were concerned about possible health effects on humans who drank milk or ate milk products from cows treated with Posilac®. The reason for this concern is that rBST may also stimulate the growth of another hormone, insulinlike growth factor-1 (IGF-1). Some scientists believe that IGF-1 is associated with breast and colon cancer. Higher levels of IGF-1 in cow milk might, according to these scientists, lead to higher rates of cancer in humans.

Finally, some activists point out that increased milk production is hardly a major problem for the U.S. dairy industry. For more than a decade, the industry has been producing more milk than can be sold to the public. In fact, the U.S. government maintains a price support system for milk that results in the destruction of hundreds of thousands of tons of milk each year. Under those circumstances, critics ask, what is the point of encouraging farmers to produce even more milk by injecting their cows with Posilac®? The only economic effect of this practice, they say, is to put out of business smaller farms that cannot afford the new technology provided by Monsanto.

The FDA has confirmed on more than one occasion that milk produced by the use of Posilac® is safe for human consumption. Other nations have been somewhat slower to reach the same conclusion. The Agricultural Council of the European Union, for example, withheld its approval for the use of Posilac® until extensive scientific tests had been completed in 1999.

References

Adler, T. "Debating BST 'til the Cows Come Home." *Science News*, 27 January 1996, 52–53.

Centner, T. J., and K. W. Lathrop. "Legislative and Legal Restrictions on Labeling Information Regarding the Use of Recombinant Bovine Somatotropin." *Journal of Dairy Science*, 80:1997, 215–219.

Corey, Beverly. "Bovine Growth Hormone: Harmless for Humans." *FDA Consumer*, 1 April 1990, 16–18.

Crooker, B. A., et al. "Dairy Research and Bovine Somatotropin." http://www.mes.umn.edu/Documents/D/I/DI6337.html. 8 July 1998.

Greger, Michael. "Bovine Growth Hormone." http://arrs.envirolink.org/AnimaLife/spring95/BGH.html. 8 July 1998.

Captive Breeding Programs

Captive breeding is a "last ditch" method of trying to save an endangered species from extinction. This method is used with a species whose numbers have been reduced to a point at which it cannot survive on its own in nature. It involves the removal from the wild of the last remaining members or eggs of that species. Those individuals are then brought to a zoo or other area where they are encouraged to breed under carefully controlled conditions. Under the most favorable conditions, an endangered species may actually be saved from extinction and returned to the wild.

As an example, the state bird of Hawaii, the nene (Hawaiian goose) had become nearly extinct in the early 1940s. Only thirty-five members of the species were known to exist. Captive breeding saved the species from extinction, and several thousands of the birds now exist in their natural habitat. Similar programs have been able to save from extinction the whooping crane, the elephant seal, the Arabian oryx, and the California condor.

Captive breeding programs are not without some serious problems, however. They can be used only in a limited number of situations and are accompanied by difficult ecological and economic problems. They tend to be very expensive and require more individuals of a species than most zoos are able or willing to maintain. Also, as the number of individuals in a species diminishes, so does the genetic diversity in the population. With reduced genetic diversity, a species is less likely to be able to survive in the wild, even in its own specific ecological niche.

References

Batra, Puja. "Captive Breeding and Reintroduction." In *Gale Encyclopedia of Science*, edited by Bridget Travers, 661–664. Detroit: Gale Research, 1996.

Carcinogens

See Cyclamates; DDT; Formaldehyde; Polybrominated Biphenyls; Polychlorinated Biphenyls; Saccharin; Tris

Cassini Spacecraft

The *Cassini* spacecraft is a National Aeronautics and Space Administration (NASA) space probe designed to visit Saturn and study that planet and its largest moon, Titan. The launch of *Cassini* in late 1997 was accompanied by outspoken and insistent demands by a number of scientists and lay persons that the mission be modified or canceled because of the presence of radioactive plutonium on board the spacecraft.

The *Cassini* Mission

The *Cassini* space probe was designed to approach Saturn via a roundabout route. After leaving Earth, its orbit was to take it around Venus and back past Earth again before being carried to Jupiter and, finally, Saturn. *Cassini* is scheduled to reach Saturn on 1 July 2004. The circuitous route was designed to take advantage of planetary grav-

The *Cassini* space probe lifted off from Cape Kennedy without incident, although some critics still worry about the risk of future flybys that will carry it past the Earth. (Courtesy of NASA)

itational fields to "sling-shot" *Cassini* toward the outer planets.

Cassini itself has a mass of 4,750 pounds with a payload of twelve scientific experiments. The spacecraft also carries a smaller Huygens probe, designed to be dropped from *Cassini* about five months after it reaches Saturn. The Huygens is designed to study Titan, Saturn's largest moon.

Power for *Cassini*'s operation is provided by three canisters containing about 72 pounds of plutonium. Plutonium is a radioactive element that gives off heat as it undergoes radioactive decay. This heat is to be used to provide power in *Cassini*, since the solar panels often used on spacecraft are ineffective at the great distances from the Sun to Saturn and Titan.

Issues Surrounding the *Cassini* Flight

NASA's decision to use plutonium generators instead of solar panels for *Cassini*'s power needs was the source of considerable concern for many observers. Plutonium is one of the most toxic materials known to humans. The inhalation of minute amounts of plutonium dust can cause serious health problems and even death. What would happen, critics asked, if the rocket carrying the *Cassini* probe were to explode during launch? The plutonium carried within the spacecraft would be vaporized, only to settle back to Earth eventually. There, it would present a serious threat to the health and lives of untold numbers of people.

NASA scientists were less concerned about this eventuality. The Titan rocket to be used for the *Cassini* launch was the "workhorse" of the U.S. space fleet. It had carried hundreds of space vehicles into orbit and could be depended upon to function safely and efficiently. If there should happen to be an accident, NASA calculations suggested that the level of plutonium produced would be so small as to be of no threat to humans on the Earth's surface.

Critics were not satisfied with this explanation. They continued to speak out, to write to Congress, to hold public rallies, and, finally, to protest at the *Cassini* launch itself, which occurred on schedule in October 1997. Even after the spacecraft had reached its planned flight path on its way to Venus, critics did not become silent. They pointed out that the probe was due to pass close to Earth again in August 1999, on its way to Jupiter and Saturn. It was still not too late, they said, for NASA to acknowledge and then deal with the safety issues posed by the plutonium-carrying spacecraft. As of August 1999, NASA officials were still expressing confidence in *Cassini*'s ability to carry out its mission with little or no danger to the general public.

References

"An Answer to Jeff Cuzzi." http://www.animatedsoftware.com/cassini/jc9709pu.htm. 12 August 1998.

"Flight Operations." http://www.jpl.nasa.gov/cassini/FlightOps. 12 August 1998.

"Much Ado about Cassini's Plutonium." http://www.cnn.com/TECH/9710/10/cassini.advancer/. 12 August 1998.

"Statement to President Clinton Signed by 15 Congresspersons Calling for Postponement of Cassini." at http://www.animatedsoftware.com/cassini/hr9709sg.htm. 12 August 1998.

Stop Cassini Web Site. http://www.animatedsoftware.com/cassini/cassini.htm. 12 August 1998.

Chemical Castration

Chemical castration is a procedure used with convicted sex offenders, usually child molesters, in which a person is injected with a drug that has the effect of reducing that person's sex drive. Since the very large majority of convicted sex offenders are men, the treatment is used exclusively with males.

Depo Provera

The drug used in chemical castration is usually Depo Provera, the trade name for medroxyprogesterone acetate. As its name suggests, Depo Provera is a form of progesterone, a naturally occurring hormone responsible for the development of female sexual characteristics. The drug was first developed in the 1960s for the treatment of endometriosis and spontaneous miscarriages. Endometriosis is a painful condition resulting from the growth of endometrial tissue in sites where it is not normally located. The most common use of Depo Provera today is to prevent pregnancy. The drug prevents ovulation from occurring and, therefore, can be used as a birth control technique.

Depo Provera has been used with sex offenders because it also tends to reduce blood testosterone levels. Testosterone is a so-called male sex hormone because it controls the development of male secondary sexual characteristics, such as increase in body muscle mass, growth of facial and body hair, and an increase in aggressive behavior. Although it is generally called a male sex hormone, it occurs naturally in both males and females, although to a lesser degree in the latter.

The theory behind using Depo Provera to treat sex offenders is that injections of the drug will reduce men's aggressive sexual behavior, thus making it less likely that they will repeat offenses like those for which they were convicted. The procedure (chemical castration) gets its name from its similarity to surgical castration, in which a man's testicles—the site of testosterone production—are removed. In fact, convicted sex offenders are sometimes offered the option of chemical or surgical castration.

Chemical Castration as a Legal Issue

The popularity of chemical castration as a way of treating sex offenders mushroomed in the late 1990s. The first state to pass a law requiring the procedure was California. In 1996, the state legislature passed and Governor Pete Wilson signed a law requiring sex offenders to have weekly injections of Depo Provera as a condition of their parole. State representative Bill Hoge (R-Pasadena) called his bill "the toughest anticrime measure in the country."

That description was probably apt since the legislation could be interpreted to require sex offenders to remain on Depo Provera therapy not just for the length of their parole but for life. An option to a lifelong regimen of Depo Provera was surgical castration. In addition, the law applied to all cases of sexual offense, even if the act was consensual.

Four more states—Florida, Georgia, Louisiana, and Montana—followed California's example by passing chemical castration laws in 1997. By mid-1999, about six dozen more states were in some stage of considering similar chemical castration statutes. This movement appears to have been driven by a combination of factors, including a small number of highly publicized child molestation cases that ended in abduction and/or murder, a more conservative and aggressive anticrime drift in national policies, and, perhaps, personal factors among legislators.

Objections to Chemical Castration

While support for the use of chemical castration among state legislators was very strong, it was not unanimous. In addition, serious doubts about the procedure were expressed by civil libertarians and by professional counselors who treat sex offenders. One concern was that the procedure by itself simply was unlikely to work as expected. Physicians and psychologists pointed out that sexual crimes are usually complex phenomena, caused by a host of different factors. While Depo Provera may be one element useful in the treatment of sex offenders, it is hardly a cure-all.

Depo Provera treatments are likely to be more effective, professionals pointed out, if they are accompanied by counseling. But the California bill did not provide for such counseling, and the Georgia bill required offenders to pay for their own counseling, a condition thought by some to be too restrictive for most parolees to meet. For these reasons, the California Psychiatric Association took a stand in opposition to that state's bill.

In addition, questions were raised as to whether Depo Provera was a form of cruel and unusual punishment, a situation prohibited by the U.S. Constitution. One reason this question was raised was that there appeared to be almost no way an offender could ever have his treatment come to an end. An offender would have to prove that he would *not* perform some act in the future (an act of sexual molestation), a challenge that is always difficult to meet.

Yet another objection to the use of Depo Provera for chemical castration was the possible side effects of the drug. Long-term use of the drug leads to the development of secondary female characteristics, such as enlarged breasts and hips, loss of facial and body hair, and a higher voice pitch. In addition, extended use of the drug has been implicated in modest health problems, such as nausea, insomnia, and hair loss and depression, as well as more serious problems, such as hepatitis, stroke, abnormal blood clotting, liver diseases, breast cancer, and gonadal tumors.

References

"Barbarism, California-style." http://www.guidemag.com/newsslant/barbarism.html. 4 January 1998.

"Castration Isn't Enough." http://www.oracle.usf.edu/archive/199706/19970602/19970602-comment1.html. 4 January 1998.

"Chemical Castration Declared Dead." http://www.mdn.org/1997/STORIES/CASTRA.HTM. 4 January 1998.

Miller, Adam. "Chemical Castration: Political Reaction or Reasoned Response?" http://www.foxnews.com/front/050697/castration.sml. 4 January 1998.

"Montana Law to Allow Injections for Rapists." *New York Times,* 27 April 1997, A32.

Seligman, Katherine. "Chemical Castration: Will It Work?" *San Francisco Chronicle,* 15 September 1996, C1+.

The full text of the Georgia and Missouri bills can be found on the Internet at, respectively, http://www.state.ga.us/legis/1997_98/fulltext/hb211.htm and http://www.housestate.mo.us/bills97/bills97/HB753.htm#text.

Chernobyl

Chernobyl is a region in Ukraine where the worst nuclear power plant accident in history occurred. On 26 April 1986, plant operators reduced power for a series of routine tests on Unit 4 of the V. I. Lenin Atomic Power Station at Chernobyl. Over a period of twenty-four hours, a series of human and mechanical errors occurred. Heat built up in the reactor core faster than it could be removed. Steam collected within the reactor and eventually blew a hole in the plant roof. Pulverized fuel, by-products, and other materials were carried out of the reactor and high into the atmosphere.

At the same time, the graphite core of the reactor caught fire, further contributing to the release of radioactive materials from the plant. Winds carried these materials across the region surrounding Chernobyl, into southern Belarus (north of Ukraine), and then westward across Europe. Elevated levels of radiation were detected as far away as Great Britain.

Health and Environmental Effects

The precise health and environmental effects of the Chernobyl disaster will probably never be known. About 30 people were killed immediately, and perhaps 500 more died shortly after the explosion. Some Ukrainians claim that 150,000 died in the decade after the accident and another 55,000 became invalids. Other health authorities question the accuracy of these estimates. They say that those numbers may have been exaggerated in order to qualify for more international assistance. One statistic that is suggestive is the rate of thyroid

The disaster at the Chernobyl nuclear reactor remains a symbol for those opposed to the development of nuclear power of the greatest harm that can come as a consequence of depending on this source of energy. (Corbis/Caroline Penn)

cancer in children. In the decade prior to the Chernobyl accident, there were seven such cases in nearby Belarus. Between 1990 and 1996, more than 300 cases of thyroid cancer were reported.

The Chernobyl explosion also devastated the environment around the plant. Thousands of villages were abandoned and their inhabitants relocated to new homes. Whole herds of cattle were slaughtered, and farmland was declared unusable because of high levels of radioactivity. More than ten years later, twenty of twenty-one farming districts in southern Belarus are still abandoned. The area within an eighteen-mile radius of the Chernobyl plant has been declared uninhabitable.

As soon as it was safe, a huge concrete and steel sarcophagus was built over the damaged power plant. The purpose of the sarcophagus was to prevent the further release of radioactive gases into the atmosphere. Topsoil around the plant was scooped off and buried in underground "graves," some of which were covered, and some not. Radioactive materials continue to leach out of the soil, drain into groundwater, and eventually work their way into rivers, streams, and lakes in the region.

Some experts have called for more aggressive methods for dealing with the Chernobyl legacy. Officials in Ukraine and Belarus say that they are already doing as much as they can. About 15 percent of the gross natural product of the Belarus economy, for example, is already spent on dealing with the Chernobyl problem. The nation has spent nearly $250 billion since the 1986 accident and can do no more, it says.

References

Charlton, Angela. "Flames at Chernobyl." *San Francisco Chronicle,* 24 April 1996, A1.

Cheney, Glen Alan. *Chernobyl: The Ongoing Story of the World's Deadliest Nuclear Disaster.* New York: New Discovery Books, 1993.

Chernousenko, V. M. *Chernobyl: Insight from the Inside.* New York: Springer Verlag, 1991.

Hoversten, Paul. "Robot to Map No Man's Land in Chernobyl Plant." *USA Today,* 23 April 1998, 10A.

Medvedev, Zhores A. *The Legacy of Chernobyl.* New York: W. W. Norton, 1992.

Specter, Michael. "A Wasted Land—A Special Report. 10 Years Later, Through Fear, Chernobyl Still Kills in Belarus." *New York Times,* 31 March 1996, A1.

Yaroshinska, Alla, et al. *Chernobyl: The Forbidden Truth.* Lincoln: University of Nebraska Press, 1995.

Childhood Hyperactivity

See Ritalin

Chlorofluorocarbons

Chlorofluorocarbons (CFCs) are chemical compounds having molecules that contain carbon, chlorine, and/or fluorine atoms only. A typical CFC is monofluorotrichloromethane, $CFCl_3$, also known as CFC-11. The "-11" suffix is a designation developed by industry as a shorthand for different CFCs. CFC-113, for example, is the industry designation for the compound 1,2,2-trichloro–1,1,2-trifluoroethane ($CClF_2CCl_2F$).

CFCs were discovered in 1928 by U.S. chemical engineer, Thomas Midgley, Jr. Midgley was working at the time for the Fridgidaire Corporation, a company that manufactured refrigerators, air conditioners, and other cooling systems.

The CFCs have a number of physical and chemical properties that make them very useful in a number of commercial applications. They are very stable (that is, they do not break down very easily), nonflammable, nontoxic, and easily liquefied. Before long, they were put to use in a large variety of applications, including heat pumps, refrigerators, freezers, and air conditioners. They were also used as cleaning and degreasing agents in the manufacture of metals and electronic components; as dry-cleaning fluids; as "blowing agents" in the manufacture of plastic foam products (such as styrofoam) and insulation; and as propellants in aerosol products. They became so popular that the production of the two most common CFCs (CFC-11 and CFC-12) increased from about 1 million kilograms in 1935 to 20.5 million kilograms in 1945; 83.9 million kilograms in 1955; 312.9 million kilograms in 1965; and 695.1 million kilograms in 1975.

Environmental concerns about CFCs began to develop in the 1970s. Evidence began to suggest that these compounds may break down in the atmosphere, releasing free chlorine atoms. These chlorine atoms, in turn, are able to attack and destroy ozone molecules:

$$Cl + O_3 \rightarrow ClO + O_2$$

This discovery was a considerable surprise to chemists. One reason for the popularity of CFCs had been their stability. It appeared that they no longer possessed this property in the stratosphere, however. In that region of the atmosphere, the energy of sunlight was sufficient to cause CFC molecules to break down:

$$CFC - \text{solar energy} \longrightarrow CFC^* + Cl$$

By the late 1980s, little doubt remained as to the role of CFCs in the destruction of the Earth's ozone layer that was then being observed. In a remarkably swift and effective action, the world's nations agreed to very stringent controls on both the production and use of CFCs. By the end of the century, they were no longer being produced in significant amounts and were being used only to the extent that they remained in existing systems (such as car air conditioners). To a large extent, CFCs had been replaced by a similar family of chemicals, the hydrochlorofluorocarbons (HCFCs).

See also: Halons; Hydrochlorofluorocarbons; Ozone Depletion; Ozone Depletion Potential

References

Joesten, Melvin D., et al. *World of Chemistry.* Philadelphia: Saunders College Publishing, 1991, 577–581.

Newton, David E. *The Ozone Dilemma: A Reference Handbook.* Santa Barbara, CA: ABC-CLIO, 1995.

Travers, Bridget, ed. *The Gale Encyclopedia of Science,* vol. 2. Detroit: Gale Research, 791–794.

Clearcutting

Clearcutting is the practice of cutting down all of the trees in a given area. Clearcutting is a popular technique among timber companies for harvesting trees because they can focus their work on the removal of all the trees in a given area at the same time. Fewer logging roads need to be built, and trees can be removed efficiently all at the same time. The area can then be replanted with trees of a common species that can be managed and harvested efficiently in the future.

Clearcutting can be done on a single region in which all trees are removed at the same time. It can also be done in patches or strips that leave groups of trees standing between areas that have been cut.

Environmental Issues

Clearcutting raises a number of environmental and ecological issues. Some scientists say that the practice leads to the erosion of cut areas, to the pollution of water resources by runoff, and to the loss of wildlife in areas that have been cut. Timber companies argue that clearcut areas can be restored to provide an even more favorable ecological habitat than existed before the cut. Besides, they say, economic considerations are not insignificant in areas that are, have been, or can be converted to "tree farms" rather than left as natural forests.

Whatever else may be true, clearcutting has long been a public relations problem for timber companies. In many parts of the United States, huge patches of land, sometimes hundreds or thousands of acres in size, have been totally cleared of trees, leaving barren stretches that many ordinary citizens find aesthetically offensive and disturbing. In 1992, the U.S. Congress reacted to these concerns by adopting a clearcutting reform act. That act called for a 25 percent reduction in the amount of clearcut-

The damage that can result from clearcutting is shown in this section of the Headwater region of redwood forests in Humboldt County, California. (Corbis/Galen Rowell)

ting done on federal lands, based on 1989 levels of harvesting.

By 1998, the effects of that policy were becoming evident... perhaps. According to the U.S. Forest Service and timber companies, clearcutting had been reduced by levels required by the 1992 act. Clearcut logging on national forests in Montana, as an example, were reported to have dropped from 40 percent of all tree harvesting to 17 percent. The size of clearcut regions had also been reduced, according to these reports, to a size of 40 acres or less, as required by the 1992 law.

Shelterwood and Seed-Tree Cutting

Environmentalists were not convinced that much progress had actually occurred. They pointed out that, while clearcutting might have been reduced, shelterwood and seed-tree cutting had increased substantially. And these techniques, they argued, were ecologically no better than clearcutting.

In shelterwood cutting, an areas is harvested over a period of many years. In the first stage of shelterwood cutting, dead, dying, and diseased trees are removed. Later, healthy mature trees are removed in order to allow young trees and seedlings to develop. Finally, all mature trees are cut, leaving only healthy, well-developed seedlings in the area.

Seed-tree cutting is similar to shelterwood cutting except that the early stages of cutting are completed earlier and at the same time. No more than ten to fifteen trees per acre are left standing after a seed-tree cut. This practice is regarded by lumber companies as a quick and efficient way of promoting growth of a single, economically desirable species of tree in a region.

Environmental Concerns

Environmentalists argue that both shelterwood and seed-tree cutting are forms of clearcutting "in disguise." They say that the few trees left by either practice are soon blown down, and a harvested region looks no different from a clearcut region.

With regard to meeting the 1992 legal requirements, some environmentalists claim that timber companies have simply continued to clearcut, using a different terminology for their practices. Although clearcutting in the Montana example cited above may have dropped by the amounts claimed, they point out, shelterwood and seed-tree cutting have increased by 26 percent in the same period. The lumbering efforts have been the same overall, they say, by whatever name one chooses to call them.

References

Devall, Bill. *ClearCut: The Tragedy of Industrial Forestry.* San Francisco: Earth Island Press, 1993.

Freedman, Bill. "Forestry." In *Gale Encyclopedia of Science,* edited by Bridget Travers, 1527–2531. Detroit: Gale Research, 1996.

Gallagher, Susan. "Slowdown in Clearcutting Doesn't Assuage Critics." *San Francisco Chronicle,* 26 February 1998, A4.

Horowitz, E. C. J. *Clearcutting.* Washington, DC: Acropolis Books, 1974.

Cloning

Cloning is the process by which a new organism is grown from a single cell or group of cells taken from a second organism. The new organism produced by this process is genetically identical to the organism from which the cell or cells was obtained.

Cloning is a very old process. Humans have been growing new plants from stem cuttings, for example, for more than 2,000 years. The more complex the organisms with which scientists attempt cloning, however, the more difficult the moral and ethical issues of such research become. The debate over cloning reached a fever pitch in February 1997 with the announcement of the first cloning of a mammal, a sheep named Dolly. Questions immediately began to arise as to the issues involved in cloning other kinds of mammals, especially humans.

Early Speculations about and Research on Cloning

The possibility of cloning animals was first discussed in a scientific context by German biologist Hans Spemann, often described as the Father of Modern Embryology. In 1938, Spemann described what he called "a fantastical experiment" in which the nucleus of one egg cell was transplanted into a second cell, whose own nucleus had been removed. Should the second cell then divide and mature, it would eventually exhibit the genetic properties of the first cell.

Spemann's suggestion appeared three decades before the discovery that deoxyribonucleic acid (DNA) is the genetic material in cells. Thus, scientists had only the most general insight into the technique by which Spemann's "fantastical experiment" could be carried out. Over the years, however, scientists tried out Spemann's proposal with a variety of organisms and cells. In 1952, for example, two American re-

The remarkable success associated with the cloning of Dolly has been tainted to some extent by recent questions as to the reduced life expectancy she may have. (Archive Photos)

searchers, Robert Briggs and T. J. King, successfully transplanted the nucleus from the cell of an embryo frog to a denucleated frog egg. But the egg did not develop. Twenty years later, John Gurdon of England repeated the Briggs-King experiment. This time, eggs developed and tadpoles were born, but they soon died.

Similar efforts to clone mammals were eventually developed, but they were successful only when DNA from immature embryonic cells was used. Similar experiments with DNA from more mature cells—those that had already begun to differentiate—consistently failed. Scientists became convinced that the DNA from such cells had "shut down" genes controlling more general forms of cell development and could not, therefore, be used in cloning.

The Dolly Experiment

In the late 1980s, Ian Wilmut, an embryologist at the Roslin Institute, an animal research facility near Edinburgh, Scotland, came to the conclusion that current "received wisdom" about the cloning of animals was wrong. He began a series of experiments—virtually the only person in the world to be doing so—to clone a mammal. His research was conceptually very simple: He removed an embryonic cell from one adult sheep and removed its nucleus. He then took a mammary gland cell from a second sheep and fused it with the denucleated cell from the first sheep. A very small shock of electricity made possible the fusion of the two cells. Eventually, the embryonic cell with its new DNA began to divide and mature. After a normal gestation period, the product of that process, Dolly, was born.

Dolly was the daughter of her own birth mother, of course, but also an identical twin of the donor sheep from whom her DNA was originally taken. At the time of Dolly's birth, in fact, her "donor twin" had already died. Scientists point out that Dolly is, in every way except her conception, a normal sheep. She has not been genetically engi-

neered in any way, nor is she a transgenic animal. A transgenic animal is the product of a cross between two different species.

The important feature of Wilmut's experiment was that it involved the use of a *mature* donor cell, a cell that had already differentiated as a mammary gland cell. It was precisely this kind of cell that scientists had come to believe could *not* be used in cloning. The secret to Wilmut's success in the use of these cells was a step that turned off their natural cycle of replication. The cells were stored in a nutrient-poor broth that kept them in an inactive state until the actual moment of implantation in the host cell. In this way, Wilmut's team apparently assured that donor DNA replication and host cell mechanisms were synchronized with each other, permitting a successful fusion of the two components.

Potential Consequences

If successfully replicated by other scientists, Wilmut's work would have two far-reaching consequences. In the first place, it could be used to produce identical copies (clones) of living or nonliving animals. As long as a sample of DNA can be obtained, it can be implanted into a denucleated egg cell.

This application by itself probably has relatively little significance for most mammals since existing methods of reproducing similar (if not identical) offspring are already available. The situation with humans is different. One can imagine situations (some benign, others not so benign) when a person might want to produce a familial copy of himself or herself. For infertile couples, for example, cloning might provide a method for having children generally similar to in vitro fertilization or other alternative birth technologies.

For animals other than humans, Wilmut's technology might provide a mechanism for producing "animal farms," herds of animals selectively designed to produce specific molecular products. To accomplish this goal, one modification of the Wilmut technology is needed. A gene that controls for the production of the desired product must be inserted at some point in the cloning process. For example, one might imagine that the gene that codes for the manufacture of insulin might be inserted into the DNA removed from a donor animal. If that engineered DNA were then inserted into a host egg cell, the animal produced would be a clone with genetically engineered properties, a clone whose milk or blood would contain insulin.

Wilmut and his colleagues have pointed out consistently that the second of these goals is the only one of interest to them in the development of cloning techniques. He would, he has said, "find it repugnant" to imagine the use of his technique for the cloning of humans. Indeed, Wilmut's research team reported in July 1997 that it had already achieved the goal of cloning an engineered organism. The team reported on the birth of five sheep that carried the gene for alpha-1-antitrypsin, a protein that has shown promise in the treatment of cystic fibrosis. The breakthrough promises to open doors to the inexpensive, widespread manufacture of chemical compounds needed for the treatment of a wide variety of human disorders.

Issues

The implications of cloning for the animal sciences have been largely ignored in the uproar over Wilmut's announcement of the birth of Dolly. Instead, most of the debate that has evolved centers on the possibility of cloning humans. A very basic fear and distrust of this concept date back at least to the stories of Frankenstein's monster and appear most graphically, perhaps, in Aldous Huxley's famous novel, *Brave New World*. One can scarcely read the comments of politicians, many scientists, religious leaders, and the general public without realizing that a very basic question about the limits of scientific research may have been reached. With relatively few exceptions, these comments call for an immediate ban on cloning research, some calling for a comprehensive restriction on all such research and others demanding a limit on human cloning only.

For example, President Bill Clinton spoke out strongly against research on human cloning and submitted the Cloning Prohibition Act of 1997, asking Congress to "give this legislation prompt and favorable consideration." One of the most comprehensive limitations on cloning research was that issued by the Council of Europe in November 1997. If ratified by its 40 members states, this action would prohibit "any intervention seeking to create a human being genetically identical to another human being, whether living or dead" ("Plan Enshrines Ethics of Genetics," 893).

A relatively modest number of experts called for a cooler, perhaps more reasoned approach to the issue of human cloning. They pointed to some very real benefits that might result from such research. Primary among these would be the availability of an alternative method of artificial birth technology for infertile couples. For example, John Robertson, professor at the University of Texas School of Law in Austin and cochair of the Ethics Committee of the American Society for Reproductive Medicine, suggested to a congressional hearing in June 1997 that

> Rearing healthy, biologically-related children is the source of great meaning in our lives. The freedom to do so undergirds our constitutional rights of procreative liberty and drives current medical practices in assisted reproduction and prenatal screening. Because having and rearing children is such a central part of our personal freedom, government should not ban practices necessary to achieve that goal unless it can be shown that tangible harm to others would result from the reproductive or selection technique in question. (U.S. Congress 1997)

In fact, some individuals are convinced that the potential benefits of human cloning outweigh the possible dangers they pose. For example, Richard Seed, a physicist in the Chicago area, has announced plans to set up a clinic that would clone babies for would-be parents. "It is my objective," he said, "to set up a Human Clone Clinic in greater Chicago, here, make it a profitable fertility clinic and when it is profitable to duplicate it in 10 or 20 other locations around the country and maybe five or six international" ("Chicago Scientist Plans Human Cloning" 1998).

References

"Big Uproar over Plan to Clone People." *San Francisco Chronicle*, 8 January 1998, A8.

"Chicago Scientist Plans Human Cloning." http://www.d-b.net/dti/980106seed.txt. 16 February 1998.

Kolata, Gina. "Lab Yields Lamb with Human Gene." *New York Times*, 25 July 1997, A18.

———. "With Cloning of a Sheep, a Shake-up for Ethics." *New York Times*, 24 February 1997, A1+.

"Plan Enshrines Ethics of Genetics." *Chemistry and Industry*, 17 November 1997, 893.

"Remarks by the President at Announcement of Cloning Legislation." http://www.whitehouse.gov/New/Remarks/Mon/19970609–15472.html. 15 January 1998.

Rosin, Hanna. "Cloning Gives Birth to Religious Divide." *Sunday Oregonian*, 26 July 1998, G1+.

Shapiro, Harold T. "Ethical and Policy Issues of Human Cloning." *Science*, 11 July 1997, 195–196.

"Should Cloning Be Banned?" [Debate between Barbara Koening and Sara Tobin, and Roger Pedersen] *San Francisco Chronicle*, 21 March 1997, A29+.

Specter, Michael, with Gina Kolata. "After Decades and Many Missteps, Cloning Success." *New York Times*, 3 March 1997, A1+.

U.S. Senate. "Ethics and Theology: A Continuation of the National Discussion on Human Cloning." Hearing before the Subcommittee on Public Health and Safety of the Committee on Labor and Human Resources. 105th Congress, 1st Session, June 17, 1997, 47.

Text of the proposed Human Cloning Prohibition Act can be found at http://thomas.loc.gov/cgi-bin/query.

Club of Rome

See *The Limits to Growth*

Cold Fusion

Cold fusion is a phenomenon originally reported by Stanley Pons at the University of Utah and Martin Fleischmann at the University of Southampton, in England, in 1989. The Pons-Fleischmann announcement was greeted with a flurry of excitement because it involved a reaction in which more energy was generated than was originally put into the reaction.

In a relatively short period of time, most physicists in the United States and many other countries had decided that the Pons-Fleischmann experiment involved errors that accounted for their unusual results. More than a decade later, however, a number of researchers continue to believe that the original experiment may have been correct, and similar reports of cold fusion results have been announced at many laboratories around the world.

Fusion and Cold Fusion

Nuclear fusion is the process by which two light atomic nuclei combine with each other to form one larger atomic nucleus. For example, the combination of two protons with each other can result in the formation of a larger particle called a deuteron:

$$p + p \rightarrow d$$

(This reaction is actually more complicated than shown here. It is represented in its most simple form here, and in later examples, to emphasize the principles involved.)

Nuclear fusion reactions have been of considerable interest to scientists for more than four decades for a number of reasons. First, nuclear fusion is now regarded as the process by which stars produce energy. The primary energy source in our own Sun, for example, is thought to be a series of fusion reactions, the overall result of which is the consolidation of four hydrogen atoms into a single helium atom:

4 H → He

A similar reaction has also been used by scientists to produce fusion weapons. Fusion weapons are weapons in which hydrogen atoms are combined to produce helium, with the release of very large amounts of energy. In fact, the amount of energy produced by a fusion ("hydrogen") bomb exceeds any other source of energy known to humans on Earth.

Finally, fusion reactions are of interest to scientists because they have some hope that they can be made to occur under controlled conditions, somewhat similar to the nuclear fission reactions used in nuclear power plants. Were that goal to be accomplished, the world would be assured of a cheap, inexhaustible supply of energy.

The fundamental difficulty with fusion reactions is that they occur only at very high temperatures, of the order of a few millions of degrees. The reason that high temperatures are needed is that the particles being forced together, such as two protons, carry the same electrical charge (positive, in the case of protons). Large amounts of energy must be provided to overcome the force of electromagnetic repulsion and join the two particles to each other.

The Cold Fusion Experiment

The idea that fusion could occur at room temperature is an oxymoron to most scientists. Without the energy provided by high temperatures, how could two like-charged particles be combined? Yet, in 1989, Pons and Fleischmann reported on just such a reaction.

In their experiment, Pons and Fleischmann used a variation of an experiment with which most high school chemistry students are familiar: the electrolysis of water. An electrolysis apparatus is quite simple in design. It consists of (1) a container holding the substance to be electrolyzed, water in this case; (2) two metal bars (electrodes) through which electrical current passes into and out of the container; and (3) a source of electrical current, such as a battery. When the electrolysis apparatus is operating, the electrical current flowing through the water causes it to break down into its component elements:

water — electricity — → hydrogen + oxygen

Pons and Fleischmann used palladium metal for the electrodes in their experiment. One especially interesting property of palladium is that it has a tendency to adsorb (collect) hydrogen on its surface. The primary variation employed by Pons and Fleischmann is that they used "heavy water" instead of regular water. A molecule of "heavy water" contains two atoms of "heavy hydrogen," also known as deuterium, and one atom of oxygen. Its chemical formula is D_2O, rather than the H_2O for ordinary water.

Pons and Fleischmann reported that they found evidence that fusion was occurring at the electrode where deuterium was set free as the result of the electrolysis reaction. They suggested that two deuterium atoms were combining to make a larger atom, such as:

d + d → He

Promises and Doubts

If confirmed by other scientists, the Pons-Fleischmann experiment would rank as one of the greatest discoveries of all time. The researchers reported that they were producing anywhere from 111 to more than 800 percent as much energy in the palladium-deuterium reaction as was put into the reaction. That is, they were getting more energy *out of* the reaction than they were *putting in* to the reaction. A reaction of this kind with relatively common materials might be able to provide humans with an endless supply of energy for all time.

The problem that Pons and Fleischmann faced was twofold. First, other scientists had difficulty obtaining the same results using the same equipment and the same techniques. Report after report indicated that the amount of energy produced in the palladium-deuterium reaction was just what one might expect from a chemical reaction. And it was no greater than the amount of energy actually supplied to the reaction.

In addition, scientists had difficulty imagining how a cold fusion reaction with these materials could take place. No one, including Pons and Fleischmann, had yet developed a theory that would explain the results announced by the two physicists.

Before long, most scientists decided that cold fusion was a fantasy. They suggested that Pons and Fleischmann had made errors in their work or were careless in their observations. It was not long before many scientists stopped thinking about the Pons-Fleischmann results and went on to other subjects. Almost all reputable scientific journals refused to accept papers dealing with the subject.

Hope for the Future

In the early 1990s, many scientists thought that cold fusion had become a dead issue. Little discussion about the subject could be found in traditional scientific journals. Yet, a small group of researchers had not lost hope. They were convinced that the Pons-Fleischmann results were real and that cold fusion had occurred in their original experiment. They decided to continue working on the phenomenon. By the end of the 1990s, these scientists had created at least two new journals to deal with their topic, *Cold Fusion Times* and *Infinite Energy*. They were also meeting with each other regularly in international, national, and regional conferences to exchange research results and ideas. And, most important, they continued to carry out research in their own laboratories on the subject of cold fusion.

One of the issues raised by scientists interested in cold fusion is the tendency of scientists to resist the possibility that revolutionary new ideas may arise in their fields of study. That is, scientists are fundamentally conservative, preferring to stick with traditional and established ideas until evidence in opposition to those ideas becomes overwhelming. It is certainly true that many revolutionary ideas in science have been dismissed out of hand at first both by individual scientists of note and by scientific organizations. Those who believe in cold fusion say that the response of the "scientific establishment" will go down in history as another example of this resistance to new ideas by the scientific community.

Supporters of cold fusion also believe that most of the technical issues raised by scientists about the original Pons-Fleischmann experiment have now been answered. After a decade of research, every possible source of error has been sought out and removed. And the energy magnification effects continue to appear, they say.

As the twenty-first century begins, the status of cold fusion is still unclear. Many scientists believe that it will go down in history as one of the greatest errors or hoaxes ever observed in research. Other scientists think that time will show that the phenomenon will eventually be confirmed and will earn its discoverers Nobel Prizes for their work.

References

Amato, I. "Big Chill for Cold Fusion as Energy Source." *Science News*, 3 June 1989, 341.

———. "Fusion Claim Electrifies Scientists." *Science News*, 1 April 1989, 196.

"Frequently Asked Questions [about cold fusion]." http://www.mv.com/ipusers/zeropoint/IEHTML/faq.html. 27 July 1998.

Kaplan, Stephen. "Cold Fusion Confusion Reigns in Washington." *Oregonian*, 10 June 1998, B11.

Storms, Edmund. "Review of the 'Cold Fusion' Effect." http://www.jse.com/storms/1.html. 27 July 1998.

Cold Fusion Times can be reached at P.O. Box 81135, Wellesley Hills, MA, 02181, or at

their URL at http://world.std.com/
~mica/cft.html.

Computers and Privacy

See Privacy and the Internet

Conditioning

See Behavior Modification; Electrical
Stimulation of the Brain;
Electroconvulsive Shock Therapy;
Psychosurgery

Conservation and Preservation

Preservation and conservation are two
views as to the way in which humans
should treat the Earth's natural resources.
The preservationist view argues that certain resources should be left in their natural
state, completely unhindered by human activities. The conservationist view suggests
that humans should be able to make economic use of our natural resources, but
they should be treated in such a way as to
ensure that, to some extent, they will be
conserved for future generations.

Early American Environmental Ethics

The first Europeans who migrated to the
North American continent were overawed
with the wealth of natural resources they
encountered. They had come from a continent that had already been denuded of
most of its natural resources. Forests had
been cut down, minerals had been mined,
and a number of animal species had been
hunted to extinction. By contrast, the North
American continent appeared to hold a
limitless storehouse of minerals, timber, animals, and other natural resources. The suggestion that early pioneers had to be cautious about their use of the environment in
any sense would probably have seemed absurd. The question really was whether humans could have any serious impact on an
environment so rich in resources.

It took only two centuries for that question to be answered. As more immigrants

arrived in North America and moved westward, they destroyed all kinds of natural
resources at a rate that now seems improb-
able. Lumber companies, for example, followed a common practice: They moved
into a forested area, cut down all the trees,
and then moved on to the next area.

Early in U.S. history, much of the wood
produced by this practice was used to
make coke for blast furnaces. In the late
1770s, a typical blast furnace used 5,000 to
6,000 cords of wood a year. To collect that
much wood, a lumber company would
need to harvest completely about 250 acres
of forest. A century later, railroads had become the primary consumer of wood. By
some estimate, the railroads were using
more than 3 million cords of wood annually in the mid-1800s.

The Rise of a Conservationist Ethic

By that time, a few thoughtful individuals
had become concerned about the ultimate
consequence of this "rape-the-land" ethic.
They asked whether Americans ought not
to think about setting aside certain parcels
of land in order to save trees, animals, minerals, and the land itself from total exploitation for economic gain.

Probably the single most important
statement about conservation during that
period was *Man and Nature*, a book by
George Perkins Marsh, a congressman from
Vermont, published in 1864. Marsh argued
that the continent's natural resources were
not limitless. He described in detail how
human activities were dramatically disrupting the "web of life" on the continent.
He called for governmental and private action to protect some of the nation's natural
resources.

A decade later, President Ulysses S.
Grant signed a historic act creating the nation's first national park, Yellowstone National Park. The act provided protected status for plant and animal life contained
within the park boundaries.

It was not long before a debate developed among the new "conservationists" as
to the goal of the environmental protection

movement. On the one side were those who believed that there was nothing wrong with utilizing natural resources, as long as they were not totally used up and some part of those resources were protected for use and enjoyment by future generations. Foremost among this group was Gifford Pinchot, who was appointed chief of the Division of Forestry in the Department of Agriculture in 1898. The division was later renamed the Bureau of Forestry and then the National Forest Service, and finally, the U.S. Forest Service.

Pinchot was an early proponent of the "multiple use, sustained yield" philosophy. He suggested that forest lands be managed scientifically so that they could be used for lumbering, grazing, agricultural production, and exploitation of water resources. His goal was to manage resources so carefully that they would be available for human use forever.

An Alternative to Conservation: Preservation

By the early 1900s, a somewhat different view of natural resources was being proposed. Probably the best known spokesperson for this view was John Muir, a woodsman, nature writer, and founder of the Sierra Club. When Muir died in 1914, the campaign he began was taken up by Aldo Leopold, a forester and sometime employee of the Forest Service.

The position put forward by Muir and Leopold was that certain parts of the continent should be set aside, untouched, essentially forever. They proposed the creation of certain areas where any form of development was prohibited. The only human activities allowed in these areas would be hiking, bird-watching, camping, and other forms of outdoor recreation as well as scientific research.

Balancing Conservation and Preservation

There continues today a dispute among three possible philosophies as to how humans should treat the environment: economic use; conservation; and preservation. This dispute takes place within a larger context of national social, economic, religious, and ethical philosophies. For example, the United States is a self-admitted capitalist society. Capitalist societies place high value on economic gain. The pressure to allow corporations to have access to national natural resources can, therefore, be very great.

The United States is also, to a large extent, a nation that professes to be a Judeo-Christian society. Many people believe that Americans have a biblical injunction to take whatever natural resources are available and use them for their own benefit. The first chapter of Genesis, for example, tells humans to "be fruitful and multiply, and fill the earth and subdue it; and have dominion over the fish of the sea and over the birds of the air and over every living thing that moves upon the earth." At least one important public figure—Secretary of the Interior James Watt, in the administration of President Ronald Reagan—invoked this biblical command in defending the government's right to extract and remove natural resources on all of its public lands.

At the beginning of the twenty-first century, the United States has apparently adopted a compromise viewpoint among these three philosophies. Some areas of national forests, national parks, national wilderness areas, and other preserves are, indeed, set aside and protected from any form of exploitation. Other areas, however, have been made available to hunters, mining corporations, lumbering companies, and other commercial interests. In still other areas, the land has essentially been turned over to developers for their unlimited exploitation. The dispute as to exactly how the nation's natural resources are to be apportioned among these three categories is likely to continue well into the foreseeable future.

References

Leopold, Aldo. *A Sand County Almanac.* New York: Oxford University Press, 1949.

Miller, G. Tyler. *Living in the Environment.* 4th ed. Belmont, CA: Wadsworth Publishing, 1985, 169–174.

Nash, Roderick. *The American Environment: Readings in the History of Conservation.* Reading, MA: Addison-Wesley, 1968.

Newton, David E. *Taking a Stand against Environmental Pollution.* New York: Franklin Watts, 1990.

Petulla, Joseph M. *American Environmental History.* San Francisco: Boyd and Fraser Publishing, 1977.

Controlled Burn

See Prescribed Burn

Creatine

Creatine is an organic compound that occurs naturally in animal cells. Its chemical formula is $HN\text{-}C(NH_2)N(CH_3)CH_2COOH$. Creatine was first discovered in 1927 in mammalian muscle cells. It is now known to be involved in the chemical process by which adenosine triphosphate (ATP) is generated in cells.

ATP is a molecule that provides energy for the many chemical reactions that take place in cells. For example, when a person uses his or her muscles, the energy needed to activate those muscles cells comes from ATP. The energy is produced when ATP breaks down to form adenosine diphosphate (ADP):

$$ATP \rightarrow ADP + energy$$

A form of creatine known as creatine phosphate is responsible for converting ADP back to ATP:

$$creatine\ phosphate + ADP \longrightarrow creatine + ATP$$

The presence of creatine in cells thus makes possible the continual regeneration of ATP in cells, allowing them to function more efficiently.

Creatine as a Nutritional Supplement

In the early 1990s, athletes began to hear about creatine and its potential value as a nutritional additive. Reports indicated that doses of about five grams of creatine a day had a tendency to increase muscle mass, strength, and endurance. At least part of this gain in mass can be attributed to increased retention of water in muscle cells, a change for which creatine itself is responsible. Some of the increase in muscle mass, however, appears to occur as the result of tissue growth.

Very quickly, creatine became popular among professional sports athletes. Some competitors found it to be a safer supplement than steroids, which had previously been widely popular. Creatine appeared to be as effective in increasing strength as steroids, but with fewer obvious side effects. By 1998, some authorities were estimating that as many as 25 percent of all professional baseball, basketball, and hockey players were using creatine regularly. Up to half of all National Football League players were also thought to be taking creatine to improve their performance

Over time, interest in the use of creatine began to spread downward, through the ranks of college, high school, and even elementary athletes. A number of creatine producers targeted the teenage market in particular for its sales. One company, for example, developed a creatine product called Crea-TEEN designed specifically for eight- to nineteen-year-olds.

Health Effects

Some physicians and athletic trainers have expressed concern about the growing use of creatine among athletes. They provide anecdotal reports of players developing cramps and becoming dehydrated during competition. Some worry about potential effects on the heart, liver, kidneys, and other organs.

At this point, there is little evidence to suggest that the use of creatine causes any serious short-term effects. Creatine products have not been on the market long enough, however, to know the status of any possible long-term effects. It is these effects that trouble a number of sports physicians. "My biggest nightmare," according to Larry Lilja, director of strength and conditioning at Northwestern University, "is that 10 or 15 years down the road we find there's some long-term effects" (Hellmich, C3).

The use of creatine by athletes also raises many of the same questions surrounding the use of steroids. That is, do athletes who are using creatine products have an unfair advantage over competitors who are not? Does the five-gram daily dose of creatine compare to a five-yard head start for a sprinter or a one-step advantage for a professional halfback?

At the moment, the bottom line appears to be that competitors like the body changes that accompany their use of creatine. They tend not to worry about any side effects the drug may have, now or in the future. It appears that its use will only spread and increase until or unless more negative information about its effects on the body are reported.

See also: Steroids

References

Bamberger, Michael. "Performance Enhancers: The Magic Potion, or Is It?" *Sports Illustrated*, 20 April 1998, 58–60+.

"Creatine Draws Raves, Criticism." ESPN SportsZone. http://espnet.sportszone.com/gen/features/weight/creatine.html. 28 June 1998.

Gregorian, Vahe, and Elizabethe Holland. "The Muscle Tussle." *St. Louis Post-Dispatch*, 31 May 1998, F1.

Hellmich, Nancie. "Safety Still a Creatine Issue." *USA Today*, 4 June 1998, C3.

Strauss, Gary, and Gary Mihoces. "Jury Still out on Creatine Use." *USA Today*, 4 June 1998, B1–2.

"The Use or Abuse of Creatine Phosphate." http://nimbus.ocis.temple.edu/~sklein/. 28 June 1998.

Creationism

Creationism is an idea that the world and all forms of life within it were created by God out of nothing. The method of creation is usually taken to be that described in Genesis, although two slightly different forms of that process are described in different parts of Genesis.

Until the late nineteenth century, most people in Western civilization believed in some form of creationism as the only possible explanation as to how inanimate matter and life were created. Other cultures in other parts of the world generally had very different explanations of creation. Most of them, however, also involved the intervention of some supernatural being.

Most early scientists were as likely to believe in creationism as were nonscientists. German astronomer Johannes Kepler, for example, organized much of his research around an effort to show the hand of God in the physical world around us. Kepler was typical of early scientists who used their research to prove the existence of God or, at least, to show how the physical world reflected his master plan for the world.

Evolution and Creationism

As science developed as an intellectual discipline, however, it became more and more difficult to reconcile fundamental tenets of creationism with discoveries in geology, biology, and other fields of science. The first scientific geologists, for example, uncovered more and more evidence that raised questions about the biblical description of how the Earth was formed in only seven days. Religious leaders found that they had to modify or adjust their understanding of the biblical description in Genesis to fit with new scientific discoveries. But both religious leaders and scientists were generally able to find an accommodation between the Bible and new discoveries about the Earth.

One of the most difficult problems in making such accommodation was the time span suggested by the Bible for the creation of the Earth and the rise of human beings and the time span suggested by scientific research. It seemed more and more clear to scientists that very long periods of time had been necessary for the Earth and life on it to develop. In fact, they became convinced that these processes took millions of years and occurred in very slow steps. Gradually, the concept of *evolution,* the development of the Earth's physical features and of life on Earth, became more and more popular among scientists.

The Rise of Darwinism

Much of the evidence about evolution was of relatively little consequence to those who believed in creationism. Debates as to how mountains are formed were of great consequence, in theory. But outside of scholars specifically interested in such questions, they had little or no impact on the lives of ordinary people.

That fact changed as the result of Charles Darwin's research on biological evolution in the mid-nineteenth century. Of the many accomplishments for which Darwin is remembered, one of the most important was his commitment to careful, detailed, and extensive collection of information. He spent more than thirty years collecting data and trying to make sense out of it before publishing his ideas about biological evolution. The first book to contain these ideas, *On the Origin of Species,* was to be one of the most influential books ever written. His later work, *The Descent of Man,* published in 1871, laid out a new and entirely different view of the origin of human life than had been part of creationist doctrine.

In *The Descent of Man* and other of his works, Darwin argued that the human species had evolved over time, just as had other organisms. He believed that humans had not been created at a single point in history by an act of God, as described in the Bible, but had developed from other, simpler animals over very long periods of time.

The debates that followed the publication of *The Descent of Man* were some of the most interesting and most acrimonious to have ever been conducted on any scientific topic. Those who stood on either side of the evolution versus creationism debate believed that they were, indeed, arguing about the very

souls of humans. That debate has never been resolved. It is, in fact, not clear how either side *could* "win" this debate since it involves the clash of two very different ways of looking at the world.

In that regard, the evolution versus creationism debate is nothing other than the continuation of a controversy that has been going on for more than four hundred years, since modern science first appeared. In that debate, two groups of people argue over a topic coming from two entirely different value systems and "truth" systems.

That is, those who take a creationist view of this debate tend to accept the principle that we know a statement to be "true" if we can find evidence for it in some holy book (such as the Bible) or the writings of a holy leader. Those who take the evolution view of the debate are more likely to accept a scientific interpretation of "truth." That is, they tend to accept as "true" only those statements that can be tested experimentally and that have been confirmed repeatedly by different investigators.

Evolution and Creationism in Schools

One might imagine that scholars have long understood the differing backgrounds from which people discuss evolution and creationism. Since the two sides start out using very different "ground rules" in the debate, it seems difficult to know how a common ground can be reached between them. Indeed, many scientists today have found their own accommodations between a "scientific" way of thinking and a "religious" way of thinking. They use one set of ground rules for some parts of their lives and a different set of ground rules for other parts of their lives. For example, as Christians, they may believe that God can create miracles at any time he wishes. But they do not put that belief to practice in any part of their work life, such as the research they carry out or the theories they develop.

Proponents of both evolution and creationism tend to have very strong feelings, however, as to how these two subjects are treated in educational systems. In some ways, they both believe that young children need to be exposed early to their own "correct" way of viewing the Earth and life and the way both have developed (or not developed) over time.

Thus, as early as the 1920s, those with strong religious beliefs were objecting to the teaching of evolution in high school biology classes. One reason this issue arose at the time was that professional biologists were more and more adopting evolution as one of the major guiding principles of their science. They saw evolution as one of a handful of overarching generalizations that could be used to explain very large amounts of the knowledge gained in biological research. It seemed only reasonable that students should be introduced to this principle in their science education.

Evolution, Creationism, and the Law

Evolution and creationism first clashed in the courts in the Scopes trial of 1924. In that trial, a high school biology teacher in Tennessee, John Scopes, was accused of teaching evolution, an act prohibited by state law. The Scopes trial brought together two of the most famous attorneys in U.S. history, Clarence Darrow and William Jennings Bryan. Since he never denied teaching evolution, Scopes was eventually found guilty and fined $100. His conviction was later reversed on a technicality.

For more than three decades, the issue of evolution in the schools was relatively quiet. Most high school textbook publishers were gradually incorporating the subject of evolution into their biology books. It seems likely that many high school biology teachers were including a discussion of evolution in their classes, although many also were probably including some discussion of creationist ideas.

Then, in the 1950s, the debate over evolution and creationism reappeared in public. Religious leaders in many parts of the country began to promote laws that would prohibit the teaching of evolution in schools. The state legislature of Arkansas passed a law of that kind in 1958. The law

was challenged in the courts and eventually worked its way to the U.S. Supreme Court. In 1962, the Court ruled in *Epperson v. Arkansas* that a ban against the teaching of evolution was unconstitutional.

This decision had little effect on the national effort by those who opposed the teaching of evolution. It did, however, suggest that a new approach was necessary. Thus, pro-creationists changed their positions slightly and began to develop a new viewpoint on creationism, calling it *scientific creationism*. The point of this new approach was to find scientific evidence that could be used to support the creationist theories of how the Earth was formed and how life on Earth came to be as it is.

The new approach taken by pro-creationists was to argue that scientific creationism was a legitimate and alternative view of the origins of the Earth and life on Earth. As such, they said, it deserved equal treatment in schools. The argument now was that schools should either teach *both* evolution and creation science or *neither* subject. In the early 1980s, the state legislature of Louisiana passed a law establishing just such a requirement in that state. Again, the new law was challenged in the courts and made its way to the U.S. Supreme Court. Again, the Supreme Court ruled against creationists in *Edwards v. Aguillard*. The Court's decision was based on its belief that changing the words used to describe creationist teachings did not alter the fact that they still constituted religious dogma, not scientific principles. Based on the First Amendment principle of separation of church and state, the Court ruled, laws like that passed in Louisiana were unconstitutional.

The debate over evolution and creationism continues to rage. To some extent, the creationists have made more headway than might generally be realized. For more than a decade, for example, most publishers of biology textbooks refused to include the word *evolution* or any discussion of the topic in their new texts. They were too concerned that state textbook adoption committees would not consider books containing these materials. In addition, surveys of biology teachers suggest that as many as a third of this population themselves believe in creationist arguments and make an effort to present creationism along with evolution in their own classrooms.

See also: Science and Religion

References

Boxer, S. "Will Creationism Rise Again?" *Discover,* October 1987, 80–85.

Callaghan, Catherine. "Evolution and Creationist Arguments." *American Biology Teacher,* October 1980, 422–427.

"The Creationists." *Science 81,* December 1981, 53–60.

Grobman, Arnold, and Hilda Grobman. "A Battle for People's Minds: Creationism and Evolution." *American Biology Teacher,* September 1989, 337–340.

Hughes, Liz Rank. *Reviews of Creationist Books.* 2d ed. Berkeley, CA: The National Center for Science Education, 1992.

"McLean v. Arkansas Board of Education." http://cns-web.bu.edu/pub/dorman/mva.html. 6 July 1998.

Nelkin, Dorothy." The Science Textbook Controversies." *Scientific American,* April 1976, 33–39.

Vuletic, Mark I. "Frequently Encountered Criticisms in Evolution vs. Creationism: Revised and Expanded." http://icarus.uic.edu/~vuletic/cefec.html. 6 July 1998.

Wallechinsky, David, and Irving Wallace. "The Scopes Trial" *The People's Almanac #2.* New York: Morrow, 1978, 92–94.

Information about the two opposing views on creationism and evolution can be obtained from The National Center for Science Education, Inc., P.O. Box 9477, Berkeley, CA 94709 and the Institute for Creation Research, P.O. Box 2667, El Cajon, CA 92021.

Criminality and Heredity

Criminal behavior is a problem for any society. By its very definition, a criminal act is one that violates a social more and that, therefore, contributes to instability in the society. People who carry out such acts must be identified and punished by some method of another.

Throughout history, and even today, the identification of criminals has occurred *after* the criminal act has occurred. Once a murder has been committed, for example, the police begin a search for the person who committed that murder. If convicted, that person is then punished by being sent to prison, by in turn being killed by the state, or by disciplined in some other way.

Identifying the Origins of Criminal Behavior

Beginning in the nineteenth century, a few individuals began to suggest a new approach to the treatment of criminals. That approach involved the identification of men and women who were likely to carry out criminal acts *before* those acts ever occurred. Those individuals relied heavily on the methods of science to develop this approach to crime prevention.

One of the earliest scholars to suggest preventive action against criminals was Italian criminologist Cesare Lombroso. Lombroso spent many years studying people who had been convicted of various crimes and, based on this research, proposed a typology for criminals. A typology is a set of categories that contain things or individuals of the same kind.

For example, Lombroso wrote that he had found a way to identify a criminal by certain physical characteristics. "Thieves," he said, "have shifty eyes, bushy eyebrows, receding foreheads, and projecting ears" (Pyeritz 1977, 86). He was also able to identify murderers as individuals with "cold, glassy bloodshot eyes, curly abundant hair, strong jaws, long ears, and thin lips" (Pyeritz 1977, 86). One of Lombroso's colleagues reported that his studies indicated that criminals in general tended to have "a primitive brain, an unusual cephalic index, long arms, prehensile feet, a scanty beard but hairy body, large incisors, flattened nose, furtive eyes and an angular skull" (Pyeritz 1977, 86).

The purpose of developing such typologies was, of course, to know in advance those individuals who were likely to carry out a crime. Those individuals could then be put under surveillance or, if necessary, incarcerated in a local jail or prison.

Later Developments

Most people would chuckle at the typologies developed by Lombroso and early scientific criminologists. If nothing else, television has taught us that it is impossible to know who will or will not carry out a criminal act simply by looking at the person. It would appear that such an approach to predicting criminal behavior was useless and, probably, evil. Who knows how many perfectly innocent people might have been convicted of crimes, accused of crimes, or simply harassed by the police because they looked as if they might be criminals?

But Lombroso's original goals have not been abandoned. Instead, researchers have adopted more sophisticated techniques for trying to recognize individuals who are likely to carry out acts against society. As recently as 1967, for example, three physicians suggested that people involved in the Watts riots at the time might have genetic factors that predisposed them to violence. The Watts riots were carried out by poor Black residents of Los Angeles against a local community structure with which they claimed to feel frustration and anger.

The three physicians wrote that "brain dysfunction related to a focal lesion plays a significant role in the violent and assaultive behavior of thoroughly studied patients" (Pyeritz 1977, 87). These physicians were then involved in psychosurgery experimentation and may have been thinking that surgery of that kind might help relieve the violent tendencies of those who participated in the Watts riots. In other words, they were suggesting that Blacks in Los Angeles were causing trouble not because of their environment but because of their heredity.

The XYY Male

Another example of the effort to find biological or genetic bases for criminal behavior appeared in the early 1960s. An English psychologist, Patricia Jacobs, and her colleagues reported that an unusually large proportion of males with an XYY chromosome exhibited "dangerous, violent, or criminal propensities."

The XYY chromosome is a genetic abnormality in humans. In an overwhelming number of cases, women carry two sex chromosomes, both designated as X. The sex chromosome pattern in females, then, is usually XX. By contrast, males carry only one X chromosome. That chromosome is paired with a different chromosome, designated as the Y chromosome. The sex chromosome pattern in males, therefore, is usually XY.

During the processes of mitosis and meiosis, exchanges may occur between chromosomes that result in abnormal patterns. It is possible, for example, for an individual to have only one chromosome (X or Y) or to have unusual patterns, such as two Xs and two Ys (XXYY). The condition known as trisomic XXY occurs when an individual has two X chromosomes and one Y chromosome. Individuals with this genetic makeup have male characteristics, tend to be mentally retarded, and have some female characteristics, such as breast enlargement.

The physical and biological effects produced by abnormal chromosomal patterns (such as trisomic XXY) are well known and easy to observe. Any potential psychological or emotional effects resulting from such patterns, however, are more difficult to identify and to associate with genetic patterns. Jacobs and her colleagues had become convinced that the antisocial behavior they observed in their subjects was due to genetic factors rather than to environmental factors, such as poor parenting, economic deprivation, or lack of schooling.

The consequences of such a view, if confirmed, might be enormous. It is easy enough to determine an individual's chromosomal patterns before birth. In such a case, a person identified as carrying an XYY combination might be classified *in advance* as someone who would be likely to perform antisocial acts. He might be monitored very carefully or even put in "preventive detention" so that he could not act out his genetic tendencies.

This possibility poses a very significant problem in a nation like the United States where one is not prosecuted for his or her attitudes, beliefs, or thoughts but only for illegal actions. In the case of an XYY individual, that policy might be violated and a person convicted or detained for what he or she *might* do at some time in the future.

Many professionals still disagree about the status of XYY research. Some authorities remain convinced of a correlation between XYY trisomy and criminal behavior. Professor Jeffrey Grest, of the Department of Zoology at North Dakota State University, for example, states on his web site, "In conclusion it is fairly clear that there is some link between XYY men and criminal behavor, but what that connection is and how it works is not understood" ("XYY Chromosomes," 1999).

Strong disagreements have been expressed about this position, however. For example, a group of parents with XYY children maintain an Internet bulletin board dealing with this subject. Their position is that "Since the original research into XYY incorrectly raised the possibility of a criminal link, the later studies have looked at this question, *ALL* have found there is *NO LINK!* [emphasis in original]" ("An Extra Chromosome Does Not Equal Criminal/ Violent Behavior," 1999).

Recent Research

The search for biological bases for antisocial behavior has by no means come to an end. Enormous amounts of time, energy, and money are currently being spent on efforts to identify human genes that will account for a whole range of human characteristics, from hair and eye color to intelligence to various kinds of behavior.

Some of the most interesting research focuses on possible biological causes for criminal behavior. In fact, a web site now exists for the purposes of collecting and reviewing such research. That web site, "Crime Times," is supported by the Wacker foundation.

An example of the research found on the "Crime Times" site is a 1998 study by researchers at the University of California. These researchers reported that they had found differences in brain functions among a group of thirty-eight murderers compared to similar individuals who had never been convicted of violent crime. The murderers averaged 5.7 percent less activity in the medial prefrontal cortex of their brains compared to a control group and 14.2 percent less activity in one very specific part of that region, the orbitofrontal cortex. The researchers concluded their report by saying that "there are a lot of parents out there who, despite all of their best efforts, their children go off the rails and they commit violent offenses. . . . The fact that there is an identifiable biological disposition suggests it's not how the child was raised. It's that they had a biologic dysfunction, combined with a situation, that led to the violence" (Raine et al., 1998).

One can imagine at least one possible result of this research, if it is confirmed by other studies. It should be easy enough to do brain scans on young children who appear to have violent tendencies. If those scans find patterns similar to those reported by the University of Southern California researchers, those children could be isolated and treated in some way to prevent them from acting out the violence that is apparently preprogrammed into their brains. On the other hand, one can also imagine individuals and groups who would have questions about shortcutting the civil rights of such children.

See also: Biological Determinism; Genetic Testing; IQ; Nature versus Nurture; Sociobiology

References

"An Extra Chromosome Does Not Equal Criminal/Violent Behavior" at http://

ajoupath.aju.ac.kr/Bulletin/criminal. html. 18 May 1999.

Ardrey, Robert. *The Territorial Imperative.* New York: Atheneum Press, 1966.

Chorover, S. L. "Big Brother and Psychotechnology." *Psychology Today*, May 1973, 43–54.

———. "Big Brother and Psychotechnology II; The Pacification of the Brain." *Psychology Today*, December 1974, 59–69.

Crime Times. http://www.crime-times. org.

Culliton, B. J. "Patients' Rights: Harvard Is Site of Battle over X and Y Chromosomes." *Science.* 186, 715–717.

Ellis, H. *The Criminal*, 2d ed. London: Later Scott, 1897, as cited in Pyeritz, Reed, et al., "The XYY Male: The Making of a Myth," in The Ann Arbor Science of the People Collective. *Biology as a Social Weapon.* Minneapolis: Burgess, 1977, 86–100.

Fox, R. G. "The XYY Offender: A Modern Myth?" *Journal of Criminal Law, Criminology, and Police Science*, 62, 59–73.

Haller, M. H. *Eugenics: Hereditarian Attitudes in American Thought.* New Brunswick, NJ: Rutgers University Press, 1963, as cited in Pyeritz, Reed, et al., "The XYY Male: The Making of a Myth," in The Ann Arbor Science for the People Collective. *Biology as a Social Weapon.* Minneapolis: Burgess, 1977, 86–100.

Jacobs, P. A. "Aggressive Behavior, Mental Subnormality, and the XYY Male." *Nature,* 208:1351–52.

Kevles, Betty Ann, and Daniel J. Kevles. "Scapegoat Biology." *Discover*, October 1997, 58–62.

Lombroso, Cesare. *Criminal Man.* New York: G. P. Putnam's, 1911.

Lorenz, Konrad. *On Aggression.* New York: Harcourt Brace Jovanovich, 1966.

Mark, V., W. Sweet and F. Ervin. Letter to the Editor. *Journal of the American Medical Association*, (201), 895, as cited in Pyeritz, Reed, et al., "The XYY Male: The Making of a Myth," in The Ann Arbor Science for the People Collective. *Biology as a Social Weapon.* Minneapolis: Burgess, 1977, 86–100.

Pyeritz, Reed, et al. "The XYY Male: The Making of a Myth." In The Ann Arbor Science for the People Editorial Collective, *Biology as a Social Weapon.*

Minneapolis: Burgess Publishing Company, 1977, 86–100.

Raine, Adrian, et al. "Reduced Prefrontal and Increased Subcortical Brain Functioning Assessed Using Positron Emission Tomography in Predatory and Affective Murderers." *Behavioral Science and the Law.* 16. 1998, 319–332. Also at http://www.crime-times.org/99b/ w99bp2.htm. 16 May 1999.

Stein, Rob. "Brain Study Suggests Some Are Predisposed to Violence." *San Francisco Chronicle*, 15 April 1998, A7.

"XYY Chromosomes," at http://www. ndsu.edu./instruct/gerst/z460/behave/ chrom.htm. 18 May 1999.

Cyclamates

Cyclamates are a group of organic compounds derived from cyclamic acid, whose chemical formula is $C_6H_{11}NHSO_3H$. The most common cyclamates are the sodium, potassium, and calcium salts of this acid.

The cyclamates were first synthesized in 1937 by University of Illinois graduate student Michael Sveda. Sveda is said to have discovered the sweet taste of cyclamates when he smoked a cigarette on which cyclamate powder had fallen. Later tests showed that cyclamates are about thirty times as sweet as comparable amounts of sucrose (table sugar). Sucrose is the most common sweetener used and is the standard against which other sweeteners are compared.

In 1950, the patent rights to the cyclamates were purchased by Abbott Laboratories, of Chicago. Abbott developed a product for commercial sale called Sucaryl®. Sucaryl® consisted of a mixture of cyclamate and saccharin. Saccharin is another artificial sweetener that has been widely used commercially for many years. It is about ten times as sweet as sucrose.

The cyclamates, like other artificial sweeteners, have become very popular and widely used in commercial food products and as substitutes for sugar. They have the advantage of the sweet taste that people desire, without providing any calories. Diabetics and people who are interested in weight control are able, therefore, to substitute an artificial sweetener such as Sucaryl® in their diets without fear of disrupting their metabolic system or gaining weight.

Regulatory Issues

Questions about the safety of the cyclamates began to arise in 1969. A New York laboratory reported that mice in whom cyclamate pellets had been implanted developed bladder cancer at a higher rate than control mice (those that had not received cyclamate pellets). When notified of these results, the U.S. Food and Drug Administration (FDA) announced limitations on the sale of cyclamates and products that contained cyclamates. In 1970, the FDA banned the use of cyclamates completely. A number of countries around the world followed the FDA's lead and also banned the sale or use of cyclamates.

Almost immediately, however, a number of questions were raised about the New York study. The most important issue was whether implanting the cyclamates could be compared to feeding the compounds directly to the mice. Critics pointed out that the method of introducing cyclamates might be responsible for the cancers detected rather than the compounds themselves.

In response to this concern, a number of other laboratories attempted to replicate the New York study. These studies used mice, hamster, beagles, and monkeys. None was able to replicate the New York study. This failure to replicate convinced more than fifty nations to retract their bans on cyclamates. In those nations, the use of cyclamates has now been legal for nearly three decades.

Current Status of Cyclamates

The FDA did not, however, reverse its position on the use of cyclamates. It continued to prohibit the use of these compounds as a sugar substitute or as a sweetener in foods. In 1973, Abbott Laboratories asked the FDA to reconsider its position on cyclamates.

When the agency took no action, Abbott filed another appeal in 1980. By the late 1980s, it appeared that the FDA might be willing to change its mind about its ban on the cyclamates. In 1989, the acting director of the FDA's Center for Food Safety and Applied Nutrition said, "I have no reluctance in saying that with cyclamate we made a mistake"(Gladwell 1989, A1).

A decade after that remark, however, cyclamate has still not been "cleared" by the FDA. In fact, any further decisions about the sweetener are somewhat moot and irrelevant. Abbott's patent on the compound has expired, and there would be little incentive for it to make production of the cyclamates a high priority.

The cyclamate story has now become a classic case in the debate over governmental regulation of foods, drugs, and other products used by humans. Most people would agree that the FDA has a critical role in assuring the safety of our food supply. But they may also wonder as to how conservative the agency may be in some instances, such as the cyclamate case. As one journalist has written, "What the agency's actions reflected, according to many within the scientific and regulatory community, is the extreme conservatism that public and congressional pressure has demanded from the FDA on issues involving cancer risk" (Gladwell 1989, A1).

Part of the irony surrounding the cyclamate story is the new information obtained on saccharin, its partner in Sucaryl®. Saccharin is now widely regarded as a probable carcinogen. Yet, its use has never been limited or banned by the FDA, and it is still used readily as a sugar substitute and sweetener in foods.

References

Efron, Edith. *The Apocalyptics: How Environmental Politics Controls What We Know about Cancer.* New York: Simon and Schuster, 1984.

Gladwell, Malcolm. "U.S. Expected to Lift Ban on Cyclamate Sweetener; Harmless, Most Experts Say." *Washington Post,* 16 May 1989, A1.

Inhorn, Stanley L., and Lorraine F. Meisner. "Cyclamate Ban." *Science*, 7 November 1969, 685; also see responses in the Letters to the Editor in 26 December 1969, 166, and 13 March 1970, 1436.

"The Lessons Cyclamates Teach." *Consumer Reports*, January 1970, 59–60.

"Low-calorie Sweeteners: Cyclamate." http://www.caloriecontrol.org/cyclamat. html. 5 July 1998.

Thomasson, W. A. "Cyclamate." In *Gale Encyclopedia of Science,* edited by Bridget Travers, 1052–1053. Detroit: Gale Research, 1996.

DDT

DDT is the abbreviation commonly used for an organic chemical with the full name dichlorodiphenyltrichloroethane. DDT is a member of a large and important class of organic compounds known as alkyl halides or halogenated hydrocarbons. These compounds all have a backbone consisting of carbon and hydrogen atoms to which have been attached one or more halogen atoms. The halogens include fluorine, chlorine, bromine, and iodine. The alkyl halides include many commercially important products, including carbon tetrachloride, chloroform, methylene chloride, ethylene dibromide, the chlorofluorocarbons (CFCs) and hydrochlorofluorocarbons (HCFCs), vinyl chloride (from which polyvinylchloride—PVC—plastic is made), and many insecticides and herbicides.

Discovery of DDT and Its Uses

DDT was first prepared in the chemical laboratory by a German chemist named Zeidler in 1874. No commercial applications were known for the compound for more than sixty years, and it remained a laboratory curiosity. Then, in 1939, Swiss scientist Paul Müller discovered that the compound was highly effective in killing insects. Immediately, interest developed in the use of DDT to protect agricultural crops from pests and to kill insects that transmit diseases such as typhus and malaria.

The first widespread use of DDT occurred during World War II. Allied soldiers were often called upon to move into an area that had been devastated by battles. In most cases, all normal public health facilities had been destroyed. Dustings of DDT throughout the area proved highly successful, however, in preventing the spread of typhus, typhoid, and other diseases that would otherwise have spread rapidly through the area.

Within a decade, enthusiasm for DDT as an insecticide had spread like wildfire. Put to work at first as a medical agent against typhus, DDT was soon found to be effective against an impressive array of pests, including army worms, blister beetles, caterpillars, corn borers, and more than 100 other kinds of insects that feed on crops and livestock. Primarily as a result of the use of DDT, the incidence of murine typhus (to mention only a single example) dropped from almost 6,000 cases in 1944 to less than 1,000 cases only six years later.

Transported to agricultural fields, DDT was equally impressive in increasing crop yields. It was almost totally effective at first against insects that attack corn, wheat, rice, and other essential crops. DDT soon became as crucial to the farmer as was fertilizer or the plow. By 1970, production of DDT in the United States had reached a total of about 175 million kilograms.

Environmental Issues

At the same time, however, concerns had begun to develop about the environmental effects of DDT. Those concerns were based on the persistence of DDT in the environment and its potential for biomagnification. Persistence refers to the fact that DDT does not break down very quickly in the environment. Its half life is five to ten years, meaning that only half of any given sample of the compound will break down in five to ten years. With a half life of this magnitude, a sample of ten grams of DDT sprayed on a piece of land would not break down to less than one gram in less than twenty years, and possibly not for more than forty years.

There is plenty of time, therefore, for DDT to collect in and pass through the food web before it breaks down. DDT sprayed on a farm, for example, will be washed off into lakes and rivers, where it will be ingested by plankton and other single-celled organisms. The compound is then passed on to plant-eating fish, who are eaten, in turn, by birds. The concentration of DDT may increase tens of thousands of times as it passes from the lower parts of a food web to the higher parts.

By 1970, reports were beginning to appear about the effects of DDT in the environment on a variety of animals. One of the most influential publications to appear was Rachel Carson's *Silent Spring* (1962). The eggs of certain birds were thinner in areas where DDT had been used, for example, than in areas where it had not been used. Thinner eggs meant that fewer chicks survived to hatching. A number of anatomical effects, such as deformed beaks and misshapen wings, were also attributed to the presence of DDT in the environment.

Not everyone was convinced that DDT was responsible for the effects being reported. One critic in particular reported that he saw no danger in DDT at all and announced that he ate a small amount of DDT himself every day, with no apparent effects on his health. In spite of this claim, the pesticide was soon listed as being toxic to humans, and the U.S. Food and Drug Administration set a tolerance level of five parts per million in foods.

At the same time, the U.S. Environmental Protection Agency (EPA) had become convinced that the environmental risks posed by DDT were more serious than any potential benefits it might have. It banned the use of DDT in the United States in 1973, therefore, with only a few exceptions. The chemical continued to be manufactured, however, with all of the product marked for exportation to other countries in the world.

The International Situation

Not every nation followed the EPA's lead in banning DDT. Some countries felt that the health benefits provided by the pesticide were more important than the risks it might pose. In some parts of sub-Saharan Africa, for example, malaria is still the most important single health problem. By some estimates, up to 5,000 people a day die worldwide from the disease. DDT is still the most effective and least expensive method for protecting an area against malaria. Some governments argue that the compound may be dangerous to wildlife and, conceivably, pose a very mild threat to the general human population. But these disadvantages pale in comparison to the enormous and immediate health benefits posed by the compound's use.

In the late 1990s, governments from around the world were still debating the wisdom of using DDT and related alkyl halides as pesticides. The benefits to some nations were clear, but the increasing threat to the world ecosystem was also becoming obvious. Animals in every part of the world were being exposed to DDT residues from Africa, Asia, South America, and other regions where the chemical was still in use. Whales, for example, had become traveling "storage tanks" for DDT. They have a very high concentration of body fat, and DDT is very soluble in fats. Given the chemical's persistence, whales and other animals had had plenty of time to accumulate very significant levels of DDT. They were now not only at risk themselves for damage as a result of DDT poisoning but also represented a risk to humans who depended on them for their own food.

In June 1998, representatives from 120 nations met in Montreal to begin working on a plan for dealing with DDT and a dozen related compounds. The plan would have to satisfy the concerns of nations in which DDT was still an important public health control method and nations where questions about the health effects of DDT and its relatives had reached a new level of concern.

References

"Agricultural Sources of DDT Residues in California's Environment." http://www.cdpr.ca.gov/docs/ipminov/ddt/ddt_cont.htm. 19 July 1998.

Carson, Rachel. *Silent Spring.* Boston: Houghton Mifflin, 1962.

"Chemical Backgrounder: DDT/DDE/DDD." http://www.nsc.org/ehc/ew/chems/ddt.HTM. 19 July 1998.

Dunlap, Thomas R. *DDT Scientists, Citizens, and Public Policy.* Princeton, NJ: Princeton University Press, 1981.

Freedman, Bill. "DDT (Dichlorodiphenyl-trichloroethane)." In *Gale Encyclopedia of Science,* edited by Bridget Travers, 1067–1071. Detroit: Gale Research, 1996.

"Resolving the DDT Dilemma: Protecting Human Health and Biodiversity." http://www.wwf.org/new/news/pr_ddt.htm. 19 July 1998.

"Three Decades after *Silent Spring,* DDT Still Menacing the Environment: WWF Report Details Hazards, Uses and Alternatives." http://www.wwf.org/new/news/pr151.htm. 19 July 1998.

Turk, Jonathan, and Amos Turk. *Environmental Science.* 4th ed. Philadelphia: Saunders College Publishing, 1988, Chapter 19.

Warrick, Joby. "Nations Meet to Prepare Treaty to Ban DDT, Toxic Chemicals." *Oregonian,* 29 June 1998, A4.

Deicing Roads

Deicing roads refers to the treatment of roads and highways with some kind of chemical to cause the ice to break apart and melt. The chemicals most commonly used are sodium chloride (NaCl, common table salt) and calcium chloride ($CaCl_2$).

Removal of snow and ice is an absolutely essential practice in the northern United States, Canada, and any other country that experiences severe winters. In developed nations, road transportation is the primary means by which people and goods are moved from place to place within a city and between cities. Roads that are not cleared of snow and ice after a major storm result in huge economic losses and may be the cause of serious accidents that result in further loss to life and property.

Historically, two materials have been used for deicing roads: salt (sodium chloride) and/or sand. In many instances, the two materials are used together. Sand does not actually remove snow and ice, but it does improve traction on an icy road. Salt, by contrast, actually causes snow and ice to melt.

Ecological and Economic Issues

The amount of salt used for the deicing of highways in the United States reached a peak of about 12 million metric tons in 1979. Since that time, annual use of salt for this purpose has remained relatively constant at about 8 million to 11 million metric tons per year. Severity of weather conditions is, of course, an important factor as to exactly how much salt is used in any one year.

For at least two decades, however, environmentalists have expressed concerns about the use of salt as a deicer. When snow and ice melt, the salt used for deicing is carried away with runoff water into lakes, rivers, and water systems. The presence of salt in surface or groundwater can cause serious harm to plants, although these effects may not be fatal to the plants. Higher levels of salt in public water supplies can also pose health problems to people with hypertension or other medical conditions. Generally speaking, most people would prefer to find a substitute for deicing salt that posed fewer threats to the environment and to human health.

The problem is that no such substitute has yet been found. Some areas have tried to use sand alone to make roads and highways safe to travel. But the runoff of sand into water supplies can cause even more

In spite of concerns about the use of salts on icy roads, no completely satisfactory substitute has as yet been found for the practice. (Corbis/James Marshall)

problems than the presence of salt. Chemical substitutes for salt have also been developed. One of the most popular is a compound known as calcium magnesium acetate (CMA). CMA is about as effective as salt in deicing highways, and it has many fewer environmental and health effects. The main problem is that it is about twenty times as expensive as salt.

Some communities have tried using sand, CMA, or other chemicals or methods for keeping their roads free of snow and ice in the winter. During the winter of 1985–1986, for example, the city of Tulsa, Oklahoma, abandoned the use of salt and used only sand for highway treatment. The city got through the winter but decided to go back to salting roads the next year.

References

"Highway De-icing for Safety and Mobility." http://www.saltinstitute.org/30.html. 15 July 1998.

Paddock, Todd, and Cynthia Lister. "De-icing Salt Is Here to Stay, but Can Be Used More Wisely." http://www.acnatsci.org/erd/ea/de-icing.html. 15 July 1998.

Delaney Clause

The Delaney Clause is an amendment attached to the Federal Food, Drug, and Cosmetic Act of 1958 by then Congressman James J. Delaney of New York. The amendment prohibited the presence in processed foods of any substance that had been found to produce cancer at any concentration in laboratory animals or humans. For example, if studies had shown that a rat fed food containing 0.0001 percent of compound A developed skin cancer, then compound A was not allowed in *any* concentration in *any* processed food.

The Delaney Clause applied to two situations that might occur during food processing. The first situation was one in which any chemical was added to a food product during processing. The second situation was one in which a pesticide or other chemical used on raw food increased in concentration during processing. The clause did not apply to raw foods, such as fresh fruits and vegetables.

Over time, questions began to arise about the efficacy of the Delaney Clause as a mechanism for protecting the food supply. One reason for this changing view was that analytical techniques were being dramatically improved. Chemists had developed the ability to detect additives at increasingly low concentrations, such as a few parts per billion. Thus, possible carcinogens were being detected in foods in concentrations that many experts felt were far too low ever to cause health effects.

As a consequence, government officials gradually began to reduce their enforcement of the Delaney Clause in certain circumstances. Some consumer advocates objected to this practice, however, and sued the U.S. Environmental Protection Agency (EPA), insisting that the agency continue to enforce the Delaney Clause strictly. The U.S. Supreme Court eventually ruled in favor of those advocates, and the EPA once more began applying the clause at its "zero tolerance" level.

Controversy about the Delaney Clause continued, however. Finally, in 1996, the U.S. Congress passed and President Bill Clinton signed a large scale overhaul of the nation's food safety laws, the Food Quality Protection Act. That law changed the standard for possible carcinogens in all foods, both raw and processed. The act contained two major provisions. First, it required the EPA to apply the same health standards to all foods, whether raw or processed. A relatively small number of exceptions were provided for farmers producing certain foods. But foods produced under these special conditions were required to be so labeled in the marketplace.

Second, the Delaney Clause was replaced by a more general "safe level" rule for chemicals. The "safe level" for any given additive or pesticide was to be determined by a method devised by the EPA. But that method had to take into account the greater sensitivity of infants and children to exposures and had to guarantee their safety under the standards adopted.

References

Hattis, Dale. "Drawing the Line: Quantitative Criteria for Risk Management." *Environment*, July/August 1996, 10–15+.

Meyerhoff, A., and M. W. Pariza. "Should Pesticides That Cause Cancer in Animals Be Banned from Our Foods?" *Health* (San Francisco), March/April 1996, 36.

"The Nation's New Pesticide Law." http://www.ewg.org/pub/home/reports/newpestlaw/newpestlaw.html. 5 July 1998.

"New Laws Rewrite Rules on Pesticides." *Science News*, 7 September 1996, 159.

U.S. Senate Committee on Agriculture, Nutrition, and Forestry. *S. 1166—Food Quality Protection Act, Hearing.* 104th Cong., 2d sess., 12 June 1996.

DNA Fingerprinting

Deoxyribonucleic acid (DNA) fingerprinting is a method for identifying the organism from which a particular sample of DNA has been taken. The method is also called DNA profiling. DNA fingerprinting has become a powerful tool in forensic sciences, capable of equaling or surpassing in analytical potential the technique of fingerprinting after which it was named.

Unique DNA Sequences

DNA fingerprinting is most often used in the identification of humans. The discussion that follows focuses on that aspect of DNA fingerprinting. Other applications of the technique exist, however.

Every human cell contains molecules of DNA that are unique to the given individual from whom that cell is taken. The only exception to that rule is identical twins, who have identical DNA molecules. DNA is the substance in cells that contains the chemically encoded instructions as to the functions those cells are supposed to perform.

DNA is a very long molecule consisting of two major parts. One part is a long, spaghettilike string consisting of alternate molecules of substances known as sugar (S) and phosphate (P) groups. A segment of that string looks like this:

-S-P-S-P-S-P-S-P-S-P-S-P-S-P-S-P-S-P-

Attached to each sugar group is one of four nitrogen bases. These bases are adenine (A), cytosine (C), guanine (G), and thymine (T). A complete strand might have a structure similar to the following:

Finally, a complete DNA molecule actually consists of two strands of this kind, wrapped around each other in a shape similar to that of a spiral staircase.

The arrangement of nitrogen bases along a strand of DNA is known as the base sequence. For example, the base sequence shown in the above sample is:

ATTCGACTAGCGGAAT

All of the DNA in any given human contains a total of about 30 billion nitrogen bases arranged in some unique order. That is, the sequence of nitrogen bases found in Mr. A is different from that found in Ms. B, which is different from that found in Mr. C, and so on.

There is an exception to that statement, however. A certain portion of each DNA molecule contains a base sequence that has very specific meaning to the cell in which it occurs. For example, the sequence shown above, ATTCGACTAGCGGAAT, might be part of a longer sequence that tells a cell "make insulin." Sections of a DNA molecule that provide important information of this kind are called *genes.* The base sequences in genes are identical in all human cells. That is, the instructions for making insulin in one human cell are identical to the instructions for making insulin in all human cells. The base sequences in genes are, therefore, identical in all humans.

Genes, however, make up only about 5 percent of all human DNA. The nitrogen bases not found in genes serve no purpose of which scientists are yet aware. These bases are sometimes referred to as "junk sequences" or "junk DNA." Since they have no apparent function, they can vary from individual to individual without having any effect on how a person's cell operates. It is this "junk DNA" that is used in DNA fingerprinting.

How DNA Fingerprinting Is Done

The first step in making a DNA fingerprint is to collect a sample of material that contains human cells from the individual being studied. That sample could consist of a drop of blood; a piece of hair; a tissue or bone sample; semen, saliva, or vaginal secretions; or any other body part or material produced by the body. The sample does not even have to come from a living cell. It can also be obtained from an organism that was once alive. Generally speaking, however, the fresher the sample, the more accurate the testing results will be.

The DNA molecules extracted from the sample are then treated to break them down into simpler components. Each molecule is separated into its two distinct strands, and one of the strands is straightened out. A section of the strand containing "junk DNA" is then removed from the rest of the molecule. That section is cut into even smaller pieces by means of enzymes known as *restriction enzymes.* A restriction enzyme is a chemical that cuts a DNA molecule apart at very specific locations, such as the point at which an ATA segment is joined to a CGC segment. A number of restriction enzymes are known that cut at different locations. If restriction enzyme A is used, for example, scientists will know that they have two new segments, one ending (for example) in ATA and another beginning in CGC.

These segments are then sorted by size by means of a process known as gel electrophoresis. In this process, the DNA strands are laid out on a gel-like material to which an electrical field is attached. Under the influence of the electric field, the DNA strands migrate across the gel. They move at different speeds depending on their size and composition. After a certain period of time, the gel contains a set of DNA strands sorted according to size.

At this point, radioactive probes are attached to the strands, and they are laid out on a piece of X-ray film. The radiation produced by the probes exposes the X-ray film so that the DNA strands essentially "take a picture of themselves." That picture will be very different depending on the person from whom they were taken.

If this process could be carried out with a complete DNA molecule from the person being studied, it would be possible to say with 100 percent certainty who that person is. Recall that each person has a totally unique DNA "map." But working with a complete DNA molecule is technically not feasible. Only segments smaller in size than a complete molecule can be analyzed in any reasonable period of time.

The results obtained from such an analysis depend to some extent, therefore, on the size of DNA used. It is possible to produce a range of certainties in DNA fingerprinting ranging from one in 100 to one in a few hundred million or even one in a few billion.

DNA Fingerprinting in Court

Over the past decade, the methods used in DNA fingerprinting have improved enormously. As late as the 1980s, some experts were still uncertain as to the dependability of DNA fingerprinting in identifying a person with a high degree of reliability. Those technical concerns have largely disappeared. Almost everyone familiar with the technique has a high degree of confidence that it can distinguish one person from another with a very substantial level of certainty.

When DNA fingerprinting evidence is introduced in courts today, it is more likely to be challenged on grounds other than strictly scientific criteria. Attorneys may argue that samples were improperly collected, that errors were made in analysis, that samples were contaminated during handling, or that some other human error was made to invalidate the test results.

Other Applications

In most cases of DNA fingerprinting, a sample taken from a suspect is compared with a sample taken from the crime scene. The test results are generally expressed in terms of a random probability, such as one in a million or one in a billion. These numbers mean that the chances of a match occurring between the two samples purely on a random basis are one out of a million or one out of a billion. A jury then has to decide whether those results are more likely to support or contradict the arguments of the prosecutor and defense attorney.

DNA fingerprinting can also be used on organisms other than humans. For example, seed pods from a paloverde tree were found in the truck of a man accused of murder in Arizona. The prosecutor had claimed that the man killed his victim and then transported her body in his truck. The man admitted giving a ride to the woman but denied having been near the scene of the crime. The prosecutor was able to show that DNA from the seed pods found in the truck were identical with DNA taken from a paloverde tree at the scene of the crime.

References

"After 10 Years, DNA Evidence Still Causes Confusion in Court." *San Francisco Chronicle*, 27 January 1997, A5.

"DNA and Crime." In Harold Hart, *Organic Chemistry: A Short Course*, 8th ed. Boston: Houghton Mifflin, 1991, 500–501.

Hawaleshka, Danylo. "Justice: A High-tech Tool for Police." *Maclean's*, 24 March 1997, 56.

Micklos, David. "Go with the Odds: Believe DNA." *Newsday*, 27 September 1994, A31.

Thornton, John I. "DNA Profiling.," *Chemical and Engineering News*, 20 November 1989, 18–30.

"What Is DNA Fingerprinting?" http://www.biology.washington.edu/fingerprint/whatis.html. 8 July 1998.

Dobson Unit

The dobson unit (du) is a unit of measurement of the concentration of ozone in a vertical column of air above the Earth's surface. The unit was named after English physicist George M. Dobson (1889–1976), who devoted much of his professional career to a study of the physics and chemistry of the atmosphere. A dobson unit is defined as a concentration of 2.7×10^{16} molecules of ozone in a column with a cross-sectional area of one square centimeter.

Drug Abuse

See Creatine; Legalization of Drugs; Medical Uses of Marijuana; Ritalin; Steroids

Drug Testing

Drug testing is the process by which the effects of some chemical compound on the health of laboratory animals and/or humans are tested to see if the chemical can be used as a drug. Drug testing is an important part of the modern medical system because most physicians depend on a variety of drugs to treat many of their patients' ills. Those ills can range from the relatively trivial, such as the common cold, to the much more serious, such as human immunodeficiency virus (HIV) infection and cancer.

Steps in Drug Testing

Any chemical being considered for use as a drug in humans goes through at least five steps. The first step involves testing on experimental animals, such as mice and rats, and/or on animal tissue and/or cells. The purpose of these tests is to determine whether the chemical has the anticipated biological effects on cells, tissues, and animals and whether it is safe to use with living organisms. This stage of testing takes an average of about six to seven years.

Out of every 5,000 chemicals tested at this level, about five meet the criteria of safety and efficacy. In order to proceed with testing on these drugs, a chemical manufacturer must file a petition with the U.S. Food and Drug Administration (FDA). The petition asks permission for the company to proceed with testing on human volunteers. A company must meet very stringent criteria in order to continue testing with humans.

The next stage of testing is conducted to make sure that the chemical being studied is safe to use with humans. Normally, a fairly small sample of about 100 people is used in this stage of testing. Researchers watch for allergies, illnesses, and other adverse effects caused by the chemical. If no such effects are observed, the chemical moves on to the next stage of testing. Stage two is generally referred to as Phase I of the human testing process on the chemical. It normally takes an average of one to two years.

In all stages of testing, two groups of volunteers are used. Members of one group are given the experimental treatment, while members of the second group (the "controls") are given a harmless substances, such as sugar water. The harmless substance is known as a *placebo*. Subjects are assigned randomly to one or the other group. The studies are called "blind" studies because subjects do not know to which group they belong. If researchers themselves also do not know which person belongs to which group, the studies are called "double-blind."

In the next stage of testing, Phase II, the chemical is tested to see if it produces in humans the desired biochemical effects. That is, does the chemical cause a reduction in allergic reactions? Does it relieve the symptoms of cancer? Does it result in a decrease in the number of viruses that cause acquired immunodeficiency syndrome (AIDS)? Typically, about 100 to 300 human volunteers take part in this stage of testing. The process takes about two years.

In Phase III of the testing process, the safety and efficacy of the chemical are tested again on human volunteers. This time, however, a much larger population of subjects—usually 1,000 to 3,000 people—is used. Tests are performed to see that the results obtained on a small population in Phase II hold true also for a larger population. Phase III tests typically take three to four years.

At the end of Phase III, a company must file another petition with the FDA. In this petition, the company presents all of its research data on tests that have been conducted with humans. The FDA decides whether or not to allow the company to do final testing of the drug. About one chemical out of five that enter the human testing phases receives FDA approval at this point.

The final stage of testing (often called Late Phase III or Phase IV testing) focuses on potential marketing of the chemical. It compares the chemical to other drugs already on the market, studies long-term effects of the drug on human health, examines financial aspects of marketing, and so on. At the completion of this phase, which usually takes one to two years, the FDA makes a final determination as to whether the chemical can be marketed commercially. By this time, an average of fifteen years has gone past since the earliest stages of testing on the chemical began. Most companies have spent tens or hundreds of millions of dollars in carrying out the testing. Some of that money comes from the company developing the drug, some from the federal government, and some from private organizations, such as the National Cancer Society or the American Lung Association.

Protecting Test Subjects

Questions arise at every phase of drug testing about the use of subjects. At the earliest phase, many animal rights advocates object to the use of animals under any circumstances for the testing of new drugs. Others acknowledge that experimental animals may be needed in testing programs but argue for very strict controls as to how those animals are used.

There are many concerns also about the use of humans in drug testing trials. After all, during drug testing programs, no one is yet certain that a new chemical will be entirely safe for human use or that it will have the desired effects on some medical condition.

Drug testing on humans is monitored primarily in two ways. First, committees of experts in a given area must pass on the protocol (design of the experiment) submitted by researchers. These committees have an obligation to make sure that human volunteers receive every possible protection and advantage during the testing phases. Such committees are often appointed by governmental agencies, by academic institutions at which testing is taking place, or by companies who are carrying out the tests.

The second monitoring technique, and probably the most important of all safeguards in human testing, is the rule of *informed consent*. Informed consent means that a human subject understands the experiment in which he or she has been asked to participate and that that person freely agrees to participate in the experiment of his or her own free will. That is, the person understands that there may be both benefits and dangers as a result of taking part in the experiment, and he or she is willing to accept both by participating in the research.

Issues in Drug Testing with Humans

In spite of these safeguards, problems often arise when humans are used as subjects in drug testing experiments. Reports issued by the U.S. Department of Health and Human Services in June 1998, for example, estimated that "hundreds of thousands of people who participate in medical tests each year are not being adequately protected."

The failure of traditional safeguards to protect volunteers can be attributed to at least three factors. First, many human volunteers are really not volunteers at all and/or may not have given their informed consent to participate in an experiment. Traditionally, members of the military, educational institutions, and mental facilities have been used as subjects in experiments although their willingness or ability to agree to such participation may not be possible.

For example, questions have been raised about the use of U.S. military personnel during the Persian Gulf War of 1990–1991 in the testing of a drug, pyridostigmine bromide (PB). The military had hopes that PB might be an effective counteragent against the nerve gas Soman. The U.S. Department of Defense (DOD) had asked for a waiver to the informed consent rule for troops sent to the Gulf region. Critics pointed out that DOD was aware that PB was *not* effective against other nerve agents that the Iraqis were known to have and that they might well use. They also argued that soldiers should have been informed as to which of them had received the PB vaccination and which had received a placebo.

A second factor in the failure of safeguards is that the question sometimes arises in drug testing as to whether a drug should be withheld from members of a control group when it becomes obvious that the drug really is effective against a particular condition. This situation arose during the testing of zidovudine, also known as AZT. Partway into the drug testing process, it became clear that AZT was very effective in fighting the AIDS virus. Nineteen of the people in the control group had died during the early stages of the experiment, while only one person in the experimental group had died. At this point, the experiment was terminated, and AZT was given to everyone who had been participating in the experiment.

The decision to terminate this experiment was based on the general principle that researchers must always give the best possible care to people taking part in an experiment. When it was determined that "the best possible care" meant that they all be given AZT, the experiment was brought to an end.

But not all researchers adhere carefully to that rule. In the early 1990s, for example, a number of experiments were conducted with pregnant women in Africa and Asia to determine the least expensive way to administer AZT to those infected with HIV. Many of the women used in the experiments were given placebos even though they would have been helped by having received AZT. The decision to carry out the experiment along these lines resulted, according to one observer, in "the loss of hundreds of infant lives."

The third factor in the failure of safeguards is that some drug tests on humans simply go awry because of carelessness, poor planning, or some other kind of human error. For example, one set of experiments was planned in order to test the effectiveness of a new high blood pressure medication. People with high blood pressure were invited to take part in the experiments. They were told that they might receive the new medication that might be effective in reducing their blood pressures. However, they also had a chance of receiving a placebo, which would have no effect at all on their health.

When the experiment got under way, a problem developed immediately. A man who had to give up his usual medication in order to take part in the experiment suffered a heart attack and stroke. He had been placed, without his knowing it, in the control group and was receiving the placebo. Without his usual medication, his risk of heart problems increased, and he suffered the two attacks.

See also: Animal Rights; Human Experimentation

References
"Background Information on Clinical Research." http://www.centerwatch.com/BACKGRND.HTM. 7 July 1998.
Coleman, Brenda C. "Drug Study Shelved to Protect Near-Monopoly, Journal Says." *Oregonian,* 16 April 1997, F.
Davis, Robert. "U.S.: Human Medical Tests Lack Oversight." *USA Today,* 8 June 1998, A1 and A19+.
Koglin, Oz Hopkins. "The Long Journey of a New Drug." *Oregonian,* 13 May 1998, B14+.

Lurie, Peter, and Sidney M. Wolfe. "Inappropriate Use of Placebos in Human Experiments" http://www.citizen.org/hrg/PUBLICATIONS/1438.htm. 20 May 1999.

Wolfe, Sidney, W. "Waiver for Informed Consent for Experimental Drugs—Statement." http://www.citizen.org/hrg/PUBLICATIONS/1392.htm. 7 July 1998.

Drug Testing in the Workplace

Drug abuse is currently thought to be one of the most serious problems facing Americans, according to most surveys of public opinion. Indeed, the nation has been struggling for over three decades with the spread of illegal drug use in the nation. At least one presidential administration, that of Ronald Reagan, made the "war on drugs" one of its most important programs. First Lady Nancy Reagan may well be remembered best by her suggestion that young people could work toward solving the nation's drug problems if they would "Just Say No."

Reducing the Supply of and Demand for Illegal Drugs

The campaign to reduce drug abuse in the United States has generally been aimed at one of two possible goals: reducing the supply of illegal drugs to the United States or reducing the demand for those drugs by U.S. citizens. Most antidrug programs contain, of course, at least some elements of both.

For example, the United States has spent billions of dollars in trying to destroy cocaine crops in Bolivia and to convince Turkish farmers to stop growing poppies for the opium they produce and in sealing U.S. borders to the illegal importation of illegal drugs into this country. By most accounts, none of these efforts has attained the degree of success that officials had hoped.

The other approach is to develop educational programs to convince people not to begin using drugs and/or to create penalties that are severe enough to discourage people from using illegal drugs. The concept of drug testing in the workplace is another way of trying to reduce the nation's drug problem by eliminating the demand for illegal drugs.

Drug Abusers in the Workplace

During the 1980s, a number of public officials, corporate leaders, scientists, and private individuals saw that employers might serve an important function in dealing with the nation's drug problems. They pointed out that about three-quarters of all those who use illegal drugs are employed. If tests could be devised and used to detect workers who had used illegal drugs, an important tool for controlling drug use might become available.

Some of the earliest drug testing programs were developed in settings where potential subjects were not very likely to object to the practice. The U.S. military, for example, instituted a program of drug testing for all personnel in 1982. A variety of sports organizations at both amateur and professional levels also instituted drug testing programs in the 1980s. Opposition to similar programs in the private sector was vigorous, however, and such programs developed slowly at first. Eventually, however, private employers were also won over by the arguments for drug testing in the workplace. By 1994, about 85 percent of the nation's largest firms were using some form of drug testing of their employees.

Methods of Drug Testing

Testing for the use of illegal drugs by an individual is conceptually a simple process. One obtains a sample of some type of bodily fluid from a person and then tests that fluid for the presence of a drug. The bodily fluids that could, in principle, be tested include blood, urine, and perspiration. Other body parts can also be tested for drugs. For example, as hair and nails grow, they incorporate drug chemicals into their structure. One method of drug testing of some interest, for example, involves the analysis of a single strand of a person's hair. Analysis of the chemical structure of the hair is usually able to tell whether a person has used an illegal drug or not.

Testing methods usually do not detect a particular drug itself. Instead, they tend to detect the products formed when that drug breaks down within the body. These products are known as *metabolites*. For example, when cocaine is ingested into the body, it is broken down in the digestive system and in cells to produce a variety of metabolites. Since a test is usually performed some time after drug use, it is designed to detect the metabolites of the drug and not the drug itself.

One of the problems associated with drug testing is that substances other than illegal drugs may also produce the same metabolites as those that come from illegal drugs. For example, a person who has been taking the legal drug codeine will produce a positive drug test for opiates. In such cases, a person who has *not* been using illegal drugs may actually be suspected of doing so as a result of a drug test.

For this reason, positive drug tests (tests in which the metabolites of illegal drugs have been found) are usually followed by a second, confirmatory test. The most common initial drug test is called a thin-layer chromatography (TLC) test. Many high school students are familiar with this process because they have used it in chemistry classes. In a TLC test, a sample of urine, blood, or other bodily fluid is passed through a thin piece of filter paper. The presence of various components in the fluid can be detected and identified.

The most common confirmatory test is known as a gas chromatography/mass spectrometer (GC/MS) test. The GC/MS test is also a method for separating the components present in bodily fluids but is more reliable than is the TLC test. In any good drug testing program, an employee who tests positive for an illegal drug on a TLC test is also then required to take a GC/MS or other confirmatory test. Unfortunately, a large number of drug testing programs do not include this second step, largely because of the time and expense involved. In many cases, an employee is simply fired (or a job applicant denied a job) if he or she tests positive for an illegal drug.

Issues with Drug Testing Programs

Critics raise a number of objections to the use of drug testing as a preemployment screening step or as a routine testing program for existing employees. They point to possible "false positives" as one problem. A false positive is a test result that indicates that a person has been using illegal drugs when in fact he or she has not. Errors in testing methodology, the ingestion of legal drugs or foods, and other factors can result in the presence of false positive tests. Critics also express concern that false positives are not followed up with confirmatory tests as often as they should be.

Opponents also raise the issue of personal privacy. In most cases, the most common form of drug testing is urinalysis, in which a person is asked to provide a sample of his or her urine for testing. Critics say that this procedure is embarrassing and degrading and that the employee is being regarded as "guilty until proven innocent" by the practice. This presumption is, of course, just the reverse of the usual philosophy in the United States that a person is regarded as innocent until proven guilty.

Critics also point out the cost of drug testing in comparison to the results it produces. Drug testing for all job applicants and employees in a large company can cost millions of dollars a year. The number of actual drug users found in such programs, critics say, is far too small to justify such large expenses. Finally, critics argue that employees should be judged on the quality of the work they do, not on whether they use drugs or not. If a person can smoke marijuana on Saturday night and still perform his or her job on Monday morning, then that is all the employer should care about, say opponents of drug testing.

Present Status

The argument about drug testing of employees, athletes, students, and other groups has died down to a considerable extent in the last few years. Probably the main reason for this fact is that most companies and individuals have accepted drug testing in the workplace as a fact of life.

The irony of this situation is that evidence that employer testing actually contributes to a decline in drug use nationwide is not at all clear. But most people feel that drug testing is "the right thing to do." If nothing else, it sends a message to everyone who will listen that the employer, the school, the sports team, or any other group does not approve of illegal drug use and will do whatever it can to put an end to that practice. Even if researchers are not able to show definitively that drug testing actually achieves the objectives for which it is designed, most people still believe it is a worthwhile and praiseworthy effort in reducing drug abuse in the country.

References

Bach, Julie S. *Drug Abuse: Opposing Viewpoints.* St. Paul, MN: Greenhaven Press, 1988.

Black, David L., ed. *Drug Testing in Sports.* Las Vegas, NV: Preston Publishing, 1996.

Coombs, Robert H., and Louis Jolyon West, eds. *Drug Testing: Issues and Options.* New York: Oxford University Press, 1991.

Gilliom, John. *Surveillance, Privacy, and the Law: Employee Drug Testing and the Politics of Social Control.* Ann Arbor: The University of Michigan Press, 1994.

Indiana Prevention Resource Center at Indiana University: http://www.drugs.indiana.edu/druginfo/home.html# testing. 12 October 1997.

Jones, Del. "Low Jobless Rate Hinders Drug Policies." *USA Today,* 20 June 1997, B1.

Ligocki, Kenneth. *Drug Testing: What We All Need to Know.* Bellingham, WA: Scarborough Publishing, 1996.

Macdonald, Scott, and Peter Roman. *Drug-testing in the Workplace.* New York: Plenum Press, 1994.

Newton, David E. *Drug Testing in the Workplace.* Springfield, NJ: Enslow Publishers, 1997.

Swisher, Karin L. *Drug Abuse: Opposing Viewpoints.* San Diego: Greenhaven Press, 1994.

Early-Term Surgical Abortions

Timing is often an important issue in debates over abortion. The reason is that people differ as to when life begins. Some people argue that life begins at the moment that a sperm enters an egg. Others argue that life begins only when a fertilized egg is implanted into the wall of the uterus. Still others trace the beginning of life to the moment a fetus becomes viable, that is, is able to survive on its own outside the mother's body.

These differences are important because they determine for a person when abortion is equivalent to the taking of a human life. A person who believes that life does not begin until the fetus becomes viable is likely to have fewer problems in supporting abortion *prior* to that point in the life of the fetus. For someone who believes that life begins at the moment of conception, however, abortion at *any point* is equivalent to the taking of a human life.

Advances in Technology

This issue becomes more complex as medical technology improves. At one time, it was not possible to detect a pregnancy until a woman had missed at least one menstrual period. By that time, a fertilized egg would have been implanted and, in many people's minds, was already "alive."

Technology now allows medical workers to detect pregnancy as early as seven to ten days after fertilization, perhaps even before a fertilized egg has been implanted. This step has been made possible by improved methods of ultrasound technology and more sensitive tests for chemicals released when pregnancy occurs. The techniques and tests are usually not new, but they are significantly more sensitive than those used earlier.

One consequence of this improved technology is the ability to perform abortions at a much earlier stage of pregnancy than had previously been possible. For example, in late 1997, the Planned Parenthood Federation of America announced the availability of a new abortion technique developed by its medical director, Dr. Jerry Edwards. The technique is based on the fact that the earliest form of a fertilized egg can now be seen with ultrasound techniques as a small sac about the size of a match head. That sac can then be removed by techniques used traditionally in later abortions, but with smaller, simpler equipment. Removal of the embryonic sac can be accomplished, for example, with a handheld syringe rather than an electrically-powered vacuum pump.

Those who favor the availability of abortion for women applaud the new technique. It provides, they say, another safe, simple, and effective means of ending an unwanted pregnancy. Critics of abortion claim that earlier detection and intervention have no effect on the fundamental debate over abortion. As one representative of the National Right to Life Committee has said, "There's no difference between a fertilized egg and what you have three weeks later. Saying it's OK to kill it in the early stages because you're more comfortable with that is completely arbitrary" (Lewin 1997, A1).

See also: Abortion; Intact Dilation Evacuation

References

Lewin, Tamar. "A New Technique Makes Abortions Possible Earlier." *New York Times,* 21 December 1997, A1.

Echinacea

Echinacea is a genus of wildflower consisting of nine major species. The plant has been known since the seventeenth century, and Native American Indian tribes have long been aware of its medicinal benefits. In recent years, those benefits have become more widely known among the general public, and as a result, the plant is now being harvested in very large numbers to meet the demands of the consumer market.

The species of Echinacea most in demand is *Echinacea purpurea,* commonly known as the purple coneflower. A member of the daisy family, the purple coneflower is found only in North America, usually in prairies and open woodlands. Native Americans learned long ago that the plant could be used to treat a variety of medical disorders, ranging from coughs and colds to toothaches and snakebites.

In recent years, evidence has been obtained that suggests the plant may also be effective in strengthening the immune system. Many natural food stores, homeopathic physicians, natural healers, and others interested in health and herbs now recommend taking echinacea as a regular part of maintaining one's health. The herb is available in over 300 different formats, including tablets, capsules, and skin creams. It is sold by itself or in combination with another herb, goldenseal. A search for the term *Echinacea* on any web browser provides an instantaneous introduction to the widespread demand for the herb.

Profit and Extinction

Within the past five years, the demand for echinacea has skyrocketed. In 1997, it was the fourth most popular herbal preparation sold in the United States. Demand was so great that collectors were being paid up to

The use of echinacea as a medicinal product for humans raises questions as to how important it may be to save a single species of wildflower from extinction. (Corbis/D. Robert Franz)

$21 a pound for roots of the plant. After preparation, the herb was being resold to retailers at a cost of up to $150 per pound.

At these rates, many individual entrepreneurs were scouring the landscape for echinacea. In region after region, plant poachers had virtually wiped the landscape clear of the plant. It had become essentially extinct in at least two states, Nebraska and Kansas, and was seriously threatened in most other parts of its habitat in the Western states. Botanists had begun to fear that the frenzied harvesting of the plant would lead to its extinction in less than a decade.

Efforts to prevent the loss of echinacea were often thwarted by the economic factors driving poachers to steal the plants. On some Indian reservations, for example, unemployment rates reach a level of 50 percent. Poaching for echinacea is a relatively simple and very profitable alternative to unemployment. In such instances, tribal leaders may feel that the economic survival of their people is more important than the protection of a single plant.

Another practical problem in protecting echinacea is one of sheer space. Governmental agencies responsible for protecting the natural environment seldom have the staff needed to patrol hundreds of thousands of acres of prairies, watching for plant poachers. By the time poachers are detected, they may already have removed all the plants from a given area.

The future of echinacea is not a promising one. Its story is not so different from that of many other plant and animal products found useful to humans, for health of other reasons. Economic forces that encourage the massive, wholesale harvesting of the plant or animal can easily lead to the decimation or extinction of the species. Whether that fate is justified by its contribution to human needs is a question that may never be debated or answered in time for the organism's survival.

References

"Experts Worry Echinacea Boom Will Wipe Species Out." *Oregonian,* 10 August 1998, A8.

"The Herb Top 10: America's Top-selling Herbs in 1997." Business Wire, 29 December 1997.

Lantz, Gary. "Coneflower's Popularity: Prescription for Trouble?" *National Wildlife,* 16 June 1997, 12–13.

Electrical Stimulation of the Brain

Electrical stimulation of the brain (ESB) refers to any procedure employing the use of an electrical current to modify behavior in animals, including humans. The basis of modern techniques for ESB lies in the work of German physician Hans Berger. In the 1920s, Berger found that activities in the brain are accompanied by electric currents. By attaching electrodes to the scalp, Berger was able to measure the electrical currents associated with vision, speech, movement, thought, feeling, and other human activities. It was an enormously important discovery that the previously mysterious phenomenon of human behavior could be expressed in the concrete form of electrical signals.

Applications of Berger's Discovery

It did not take long for scientists to realize the consequences of this discovery. If physical, mental, and emotional functions are nothing other than electrical phenomena, it should be possible to influence human thoughts, acts, and feelings from the outside by electrical stimulation. In fact, experiments of that kind were initiated within a few years of the announcement of Berger's discoveries.

At first, the research was not very sophisticated. Electrodes were inserted into the brains of experimental animals and electrical impulses transmitted at random. Scientists were not very sure what might happen: An animal might demonstrate hunger, fear, aggression, boredom, or any other behavior or emotion. Gradually, researchers began to identify the ways in which very specific parts of the brain are responsible for very specific physical, mental, and emotional functions. Stimulation of a particular part of a cat's brain might cause it to go to sleep, while stimulation of another point only a few millimeters away might cause it to urinate.

Some of the most elegant experiments were conducted by U.S. psychologist James Olds. Olds accidentally discovered the location of certain parts of rat brains that evoked the sensations of "pain" and "pleasure." By stimulating these regions of the brain, he was able to control the behavior of rats to a very high degree. For example, he was able to get rats to perform tasks over and over again, for as much as sixteen hours a day, simply by stimulating the "pleasure" centers in their brains. They ignored food and other stimuli simply to continue performing the tasks that brought a "jolt" of electricity to their brains.

Using ESB to Control Human Behavior

Perhaps the most famous proponent of ESB was Spanish physiologist José M. R. Delgado. Delgado's research spanned nearly half a century, during which time he moved from the study of cats and monkeys to research on humans. He discovered that ESB could be effective in treating a variety of brain-related problems, such as schizophrenia, epilepsy, and Parkinson's disease. Based on these successes, Delgado began to make claims that ESB could be used to treat a much broader range of human mental problems. Neuroses, psychoses, violent behavior, and other forms of "undesirable" behavior were capable of being brought under control by ESB, he argued.

The approach might be as follows: Suppose a person has an abnormal and uncontrollable fear of cats. A possible form of treatment might be to implant a tiny transmitter-receiver in the patient's brain. The miniaturization of electronic equipment would make that procedure feasible and simple to carry out. Then, when the person encountered a cat, the brain would send out "fear" signals. These electrical impulses would be transmitted to a "controller" somewhere. The controller—a computer, for example—would read the incoming signals, interpret them, and send out a new signal to the implanted electrode. The signal would be directed to a part of the brain telling the patient to "calm down." By means of this process, nearly any "undesirable" form of human behavior could be brought under control.

The potential for using ESB to control human behavior in a variety of settings has not escaped science fiction writers. Author David Rorvik, for example, in his book *As Man Becomes Machine*, wrote about an "electoligarchy" in which ESB is used to distinguish the role of each person in a society. Everyone in the society has a certain number of electrodes—or no electrodes at all—implanted into his or her brain. Rulers of the society have no electrodes at all, while the lowest members of the society have as many as 500 electrodes. These individuals constantly receive electric shocks to the brain's pleasure center as they perform the hardest, least pleasant work required by the society. And they do so with great joy because of the stimulation received by their brain's pleasure centers.

Some research is also being conducted on the use of ESB to help relieve severe pain in patients with cancer and other diseases. Used in this context, ESB is often referred to as *neuroaugmentation*. There is some evidence that neuroaugmentation can

be used to relieve severe pain when other, more traditional forms of treatment have not been successful.

See also: Behavior Modification; Electroconvulsive Shock Therapy; Human Experimentation; Psychosurgery

References

"About Mind Control." http://infox. eunet.cz/misc/mind-control/article.html. 20 June 1998.

Delgado, Jose M. R. *Physical Control of the Mind, Toward a Psychocivilized Society.* New York: Harper and Row, 1969.

"Dr. Jose M. R. Delgado." http://earthops. org/delgado.html. 20 June 1998.

"Electrical Stimultion of the Brain Reverses Symptoms of Parkinson's Disease." http://www.thehosp.org/new/ brainstim.html. 20 June 1998.

"FDA Approves Implanted Brain Stimulator to Control Tremors." http://www. fda.gov/bbs/topics/NEWS/NEW00580. html. 20 June 1998.

Fincher, Jack. *The Brain: Mystery of Matter and Mind.* Washington, DC: U.S. News Books, 1981, 97–101.

Hooper, Judith. "Robert G. Heath." *Omni,* April 1984, 86–95+.

"Neuroaugmentation." http://www.stat. washington.edu/TALARIA/talaria0/ LS5.4.3.html. 20 June 1998.

Electroconvulsive Shock Therapy

Electroconvulsive shock therapy (ECT) is a procedure used to treat severe forms of mental disorders. In this procedure, large amounts of electricity are applied to the human body, producing a generalized seizure. The word *convulsive* refers to the violent shaking that occurs in a patient as a result of being treated by this procedure. The shaking may become so severe as to result in broken bones, bruises, and other evidence of violent agitation.

History of ECT Use for Behavior Modification

Electroconvulsive shock therapy (often called simply shock therapy) has been used for three-quarters of a century to treat a wide variety of mental disorders, emotional problems, and "abnormal" or undesirable behaviors. Peak interest in ECT occurred in the 1930s and 1940s. During that period, physicians and psychiatrists became enthusiastic about the ability of electric shock to modify many kinds of behavior. It was employed to cure every manner of disorder, from cancer and asthma to brain tumors and multiple sclerosis to mental retardation and homosexual behavior. Even today, ECT remains a popular therapeutic procedure among a large majority of psychiatrists for the treatment of at least a certain number of disorders.

ECT became a popular procedure long before scientists had any understanding as to how it achieved its results (and in spite of that fact). No one really understood how large amounts of electrical current affected the human body. But many physicians and psychiatrists were convinced that changes occurred that were, for the most part, improvements.

Questions about the Use of ECT for Behavior Modification

Over time, questions began to arise about the use of ECT. One basis for this concern was the violent and traumatic nature of the procedure. In early years, little effort was made to protect patients from the violent physical trauma that accompanies ECT. As a result, broken bones and serious bruises were common among patients treated with ECT. Today, physical restraints and the use of medications have reduced the most serious injuries experienced by patients treated with ECT.

Patients also experienced mental and emotional effects as a result of ECT treatments. Loss of memory, a reduced ability to perform intellectual tasks, and other forms of mental disruption were not uncommon among patients.

Questions were raised, also, about the nature of the "cure" that resulted from ECT treatments. In some cases, people who were declared to be "cured" emerged as little more than "human vegetables." Over the years, a number of patients and medical professionals have written articles and books

that vividly portray the horror that often accompanies ECT treatments. Organizations of ex-patients have been formed to monitor and/or protest the use of ECT therapy.

Current Status of ECT Therapy

Today, ECT is used primarily in the psychiatric units of general hospitals and in psychiatric hospitals. A survey conducted in 1980 found that 2.4 percent of all psychiatric admissions during the year (a total of 33,384 patients) were treated with ECT. Many physicians and psychiatrists are still enthusiastic about the value of ECT for treating one or more specific mental or emotional disorders.

Standards for the use of ECT have, however, become much more rigorous over the past few decades. Complaints from groups of ex-patients and others interested in the use and misuse of ECT have prompted the federal government to become much more interest in monitoring the way ECT is administered and the kinds of patients on whom it is used.

In 1985, the National Institutes of Health (NIH) published a formal statement outlining its position on ECT therapies. The statement was produced as the result of a three-day conference held on 10–12 June 1985. The NIH statement focused on five major questions:

1. What is the evidence that ECT is effective for patients with specific mental disorders?
2. What are the risks and adverse effects of ECT?
3. What factors should be considered by the physician and patient in determining if and when ECT would be an appropriate treatment?
4. How should ECT be administered to maximize benefits and minimize risks?
5. What are the directions for future research?

NIH concluded that ECT was a safe and effective treatment for a variety of mental and emotional disorders, such as severe depression, delusional depression, and acute manic episodes. It reported that evidence for the efficacy of ECT with regard to a number of other disorders was either not available or not convincing. It also concluded that careful precautions were required in the preparation, administration, and follow-up of ECT treatment to ensure patients of safety and efficacy.

ECT and the Rights of Patients

Like other forms of behavior modification, the use of ECT raises some profound questions about the rights of patients. One normally expects patients to understand the procedures that they will be asked to experience and to give their free consent to taking part in those procedures. This principle is generally known as giving "informed consent" to a procedure. Yet, physicians, psychiatrists, and others in authority sometimes have argued that people who might benefit from ECT therapy are not capable of giving informed consent. Their mental or emotional state precludes them from making such intellectual decisions. In such cases, they say, it is appropriate to force patients to experience ECT, whether they want to or not.

A similar argument has been used for individuals who demonstrate antisocial behavior. Does a rapist, for example, have the right to refuse ECT therapy if there is good reason to believe that the treatment will eliminate his desire to rape?

Such questions become very complex as decision makers compile their lists of "undesirable" or "antisocial" behaviors. For example, ECT was widely used in the 1950s and 1960s to "cure" homosexuality. Many authorities felt that those who engaged in such behavior were either mentally ill or criminal and did not deserve the opportunity to refuse ECT treatments. Relatively few professional therapists hold this view today because the image of homosexuality as a mental disorder or criminal act has declined significantly since the 1960s.

See also: Behavior Modification; Electrical Stimulation of the Brain; Psychosurgery

References
Abrams, Richard. *Electroconvulsive Therapy.* New York: Oxford University Press, 1997.
———. "Out of the Blue." *The Sciences,* November/December 1989, 24–30. For follow-up, see Letters to the Editor, March/April 1990, 10.
"Electroconvulsive Therapy: NIH Consensus Statement Online 1985 June 10–12." 5(11): 1–23. http://text.nlm.nih.gov/nih/cdc/www/51txt.html. 18 March 1998.
"Fact Sheet: Electroconvulsive Therapy (ECT)." Ask NOAH about: Mental Health. http://www.noah.cuny.edu/illness/mentalhealth/cornell/tests/ect.html. 20 June 1998.
Sackheim, Harold A. "The Case for ECT." *Psychology Today,* June 1985, 36–40.
The National Association of Electroshock Survivors is located at P.O. Box 1426, Giddings, TX 78942, (409) 542–6049.

Electromagnetic Fields, Health Effects of

See Health Hazards of Electromagnetic Fields

Endangered Species

See Bison; Echinacea; Endangered Species Act; Headwaters Forest; Masked Bobwhite Quail; Savage Rapids Dam; White Abalone; Wild Horses and Burros; Wildlife Management

Endangered Species Act

The Endangered Species Act (ESA) was passed by the U.S. Congress and signed by President Richard Nixon in December 1973. It was the fourth of four major legislative actions designed to protect plants and animals from extinction. The three earlier acts were the Endangered Species Preservation Act of 1966, the Endangered Species Conservation Act of 1969, and the Marine Mammal Protection Act of 1972.

Species Extinction

Species extinction is a normal and natural part of biological evolution. Extinction occurs when environmental conditions change in such a way that too few individuals in the species are able to reproduce efficiently enough to guarantee survival of the population. Numbers typically dwindle over a period of time until the last remaining members of the species die off.

Human activities are responsible for the extinction of many plant and animal species. In some cases, the story of the human role in extinction is well known and dramatic. The destruction of the passenger pigeon in the United States as the result of unobstructed hunting patterns is now well known to most Americans.

Today, hunting is no longer the primary reason that plants and animals are endangered. The loss of habitat for development, resource extraction, or some other reason has become the main cause of species extinction and endangerment. For example, lumber companies would like to have access to as many trees as possible to meet the demands of their customers. In cutting down trees, however, they may destroy the natural habitat in which birds and other animals live. The controversy between logging companies and environmentalists trying to protect the northern spotted owl is probably the best known of such controversies.

Species are sometimes also threatened by direct human action. For example, the white abalone is now regarded as endangered because of aggressive harvesting of the mollusk in its natural home off the California coast. Some authorities believe that the abalone population has already been reduced to a point where the animal's survival is problematic at best.

Legal Protection for Plants and Animals

The Endangered Species Act of 1973 was passed at the height of environmental consciousness in the United States. The act provided for three categories of plants and animals put at risk because of human activities:

endangered, threatened, and critical. The placing of a species into one of these categories indicated the seriousness with which its survival is regarded, ranging from most serious ("endangered") to least serious ("critical").

The two strongest sections of the ESA were Sections 7 and 9. Section 7 contained very strong provisions against any federal action that would harm an endangered, threatened, or critical species directly or that would harm that species' habitat. The section has been used over and over to retain or even improve the environmental conditions tending to contribute to the survival of a species. Section 9 dealt with the more specific issue of species removal. The section prohibited the killing or removal of any species listed as endangered, threatened, or critical.

"Son of ESA"

The original Endangered Species Act expired in 1992, twenty years after its enactment. In such cases, the Congress often adopts a new bill to extend, replace, modify, or continue the original act. A new version of the ESA had still not been adopted, however, by the end of 1998. Instead, Congress provided annual allocations of funds to keep the ESA program intact and operating in its original form.

Adoption of a new version of the ESA has proved to be particularly difficult because of changes in the national mood about endangered species and changes in the composition of the U.S. Congress. Republican control of the House and Senate, accomplished in 1994, meant that industry interests received somewhat greater attention than they might have under a Democrat-controlled Congress. Modifications of the ESA suggested by Republican representatives and senators contained enough questionable points for environmentalists that no bill was able to work its way through Congress.

On the one hand, environmentalists pointed to the successes of the original ESA. They pointed out that a number of species, such as the gray wolf, the brown pelican, the bald eagle, and the American peregrine falcon, had all been removed from the endangered list maintained by the U.S. Fish and Wildlife Service (FWS). In early 1998, in fact, the FWS announced that it was removing seventeen species of animals and twelve species of plants from the endangered list. On the other hand, critics have pointed out that nearly a thousand more species remain on the list. The original ESA therefore has done little or nothing to help remove the majority of endangered and threatened species.

The fundamental issue in almost every discussion over endangered species tends to be the needs of humans versus the survival of a plant or animal species. Typically, a group of humans (a community, a lumber company, a developer, for example) decides that it would like to make use of a piece of land that is home to an endangered, threatened, or critical species. The group tends to argue that the human needs and demands it can satisfy are more important than the survival of some little-known and unimportant plant or animal. Or the group may offer to provide some minimal level of protection for that species.

Environmentalists respond that humans do not have the right to act in such a way as to bring about the extinction of any form of life. We have to make accommodations in our activities, they say, to guarantee that other forms of life on the planet also have a chance to survive and prosper.

This debate is replayed again and again in nearly every part of the nation. In some cases, it involves a dispute between a single developer who wishes to construct a new community of a few dozen homes versus the survival of a species of butterfly that lives only on the land needed for development. In other cases, the dispute may involve thousands of acres of forest and millions of dollars of lumber resources versus the threat to one or two little-known birds or other animals.

Wherever the debate occurs and whatever the extent of its impact, the fundamental questions of human rights versus those of endangered plants and animals will be repeated over and over again. The latest version of the Endangered Species Act, when it is passed, will represent the nation's latest decision as to how those opposing needs and demands are to be met.

References

"Endangered Species Act Maintained." http://www.crossover.com/reus/nwiw12t13s1.html. 6 August 1998.

"The Endangered Species Recovery Act." http://www.defenders.org/esesra.html. 6 August 1998.

Freedman, Bill. "Endangered Species." In *Gale Encyclopedia of Science,* edited by Bridget Travers, 1341–1346. Detroit: Gale Research, 1996.

Hogan, Dave, and Jim Barnett. "Rewriting Environmental Law." *Oregonian,* 12 June 1997, A8.

Miller, G. Tyler, Jr. *Living in the Environment.* Belmont, CA: Wadsworth Publishing, 1985, 195–203.

"Science and the Endangered Species Act." http://www.nap.edu/readingroom/books/esa/executive.html. 6 August 1998.

Snape, Bill, and Mike Senatore. "S.1180 and the Endangered Species Act." http://www.defenders.org/esacomm.html. 6 August 1998.

Watson, Traci. "Intervention Brought Endangered Animals Back." *USA Today,* 6 May 1998, 3A.

Environment

See Environmental Justice; Feedlot Pollution; Global Warming; Hazardous Wastes, International Dumping; The Oceans; Slash-and-Burn

Environmental Inequity

See Environmental Justice

Environmental Justice

Environmental justice refers to policies and practices that are designed to correct or prevent harmful or deleterious environmental practices that have a disproportionate impact on communities of color. The term *communities of color* refers to districts populated primarily by people whose skin is other than white but, more important, whose culture is different from that of those with white European backgrounds.

Origins of the Environmental Justice Movement

The beginnings of the environmental justice movement are often traced to a single event that occurred in Warren County, North Carolina, on 15 September 1982. The Ward Transformer Company of Raleigh had been directed to bury 40,000 cubic yards of soil contaminated with its toxic wastes in rural Shocco Township in Warren County. Residents of the county were largely poor Black people. The county ranked 97th out of 100 on the basis of per capita income in the state.

Residents of the county had little experience in political organizing or protest, but they banded together and filed a formal protest against the dumping of wastes in their region. They also protested physically, in much the tradition of the civil rights movement of the 1960s, when trucks began to arrive with the toxic materials. The people of Warren County eventually lost that battle. A total of 7,223 truckloads of wastes, containing highly toxic polychlorobiphenyls (PCBs), were eventually buried near their homes. But the protest they organized marked the beginning of what was to become the environmental justice movement.

Environmental Inequities, Environmental Racism, and Environmental Justice

The Warren County protest alerted a number of scholars, political organizers, environmentalists, civil rights leaders, and other individuals and groups to the problems of environmental inequity in the United States and the rest of the world. The term *environmental inequity* refers to the tendency of environmental hazards to be distributed unequally in society such that communities of color are more likely than White communities to suffer their impacts.

A number of scholars began to document what they claimed were examples of environmental inequities. For example, an eighty-five-mile-long stretch of land along the Mississippi River in Louisiana has become known as "cancer alley" because of the high number of cancer cases reported in this region. Some authorities believe that the high cancer rate is a consequence of the heavy concentration of chemical and petroleum companies along the river. More than 125 plants in the area produce fertilizers, gasoline, paints, plastics, and other chemical products whose manufacture results in the formation of dozens of toxic by-products. The communities along "cancer alley" are almost entirely non-White, poor communities.

Concerns about environmental inequities first developed among Black populations. Before long, researchers pointed out that such inequities were also common among other people of color. Mexican farm workers, for example, were often exposed to pesticide risks at a far greater rate than were White populations. Native American communities were selected as the dump site for hazardous or radioactive wastes more frequently than were White communities. Poor communities, in general, tended to have toxic waste sites constructed in their areas rather than in areas occupied by middle- and upper-class families.

Causes of Environmental Inequities

Many individuals interested in the subject of environmental inequities argue that such inequities are really examples of environmental racism. *Environmental racism* was first defined in 1993 by then-executive director of the National Association for the Advancement of Colored People (NAACP) Benjamin Chavis in these terms:

> Environmental racism is defined as racial discrimination in environmental policy making and the unequal

enforcement of environmental laws and regulations. It is the deliberate targeting of people of color communities for toxic waste facilities and the official sanctioning of a life-threatening presence of poisons and pollutants in people of color communities. It is also manifested in the history of excluding people of color from the leadership of the environmental movement. (U.S. Congress 1993, 6)

Chavis and his colleagues argued that communities of color were chosen for toxic waste dumps and other environmental hazards partly because of pure racism and partly because such communities have traditionally had little or no political influence or understanding of the political process.

Growth of the Environmental Justice Movement

A number of stages occurred in the transformation of a local political protest in Warren County to a national political movement. Present at the Warren County protest was Walter E. Fauntroy, representative from the District of Columbia to the U.S. House of Representatives. Fauntroy asked the U.S. General Accounting Office (GAO) to carry out a study of the relationship between pollution and minority communities. The GAO study was a modest one, but it provided the first data to support the possible existence of environmental inequities in the United States.

The GAO's study served as the impetus for a much more ambition research program carried out by the Commission on Racial Justice (CRJ) of the United Church of Christ. The CRJ report, released in 1987, documented a widespread and pervasive national pattern in which communities of color were consistently more likely to be exposed to environmental hazards than were White communities.

The first national conference devoted to a discussion of environmental inequities was held at the University of Michigan School of Natural Resources in 1990. Representatives from a variety of communities of color attended that conference to discuss the causes of environmental inequities and the kinds of policies that could be adopted to combat these patterns. A year later, an even larger conference was held under the auspices of the CRJ. This was the First National People of Color Environmental Leadership Summit, held in Washington, D.C. More than 500 participants were present at that meeting. The major accomplishment of the conference was the framing and adoption of a document, "Principles of Environmental Justice."

Concerns about environmental justice received relatively little attention at the national political level until the election of Bill Clinton in 1992. In February 1994, President Clinton issued Executive Order 12898 establishing an office of environmental justice within the U.S. Environmental Protection Agency (EPA). The executive order also created a National Environmental Justice Advisory Council to advise the EPA on policies and practices.

Questions about the Environmental Justice Movement

Questions have been raised about a number of aspects of the environmental justice movement. Some critics have suggested that the basic data used to support claims for the existence of environmental inequities are flawed. These critics say that communities of color are no more likely to be the site of hazardous waste sites or other environmental hazards than are White communities.

Other critics admit that communities of color are more likely to be exposed to environmental hazards, but they question that racism is the reason for this pattern. They point out that, in some cases, communities have become less White *after* a hazardous waste site has been created. That is, people may actually choose to move into such an area because it costs less to live there and jobs may be more plentiful.

Finally, some observers argue that environmental inequities do exist but that they are not the result of some purposeful program on the part of business, industry, the government, or some other agency. Environmental hazards are a part of our society today, they say, and those hazards have to impact some communities more than others. It is just an unfortunate fact of life that poor people are more likely to get their share of environmental hazards than wealthier people. What we really need to work on, they say, is not environmental racism but finding ways of improving life for all Americans of all color and all economic status.

References

Alston, Dana, ed. *We Speak for Ourselves: Social Justice, Race and the Environment.* Washington, DC: Panos Institute, 1990.

Bryant, Bunyan, ed. *Environmental Justice: Issues, Policies, and Solutions,* Washington, DC: Island Press, 1995.

Bryant, Bunyan, and Paul Mohai. *Race and the Incidence of Environmental Hazards: A Time for Discourse.* Boulder, CO: Westview Press, 1992.

Bullard, Robert D., ed. *Confronting Environmental Racism: Voices from the Grassroots.* Boston: South End Press, 1993.

———. *Dumping in Dixie: Race, Class, and Environmental Quality.* Boulder, CO: Westview Press, 1994.

Center for Investigative Reporting and Bill Moyers. *Global Dumping Ground.* Cambridge: The Lutterworth Press, 1991.

Newton, David E. *Environmental Justice.* Santa Barbara: ABC-CLIO, 1996.

Petrikin, Jonathan. *Environmental Justice.* San Diego: Greenhaven Press, 1995.

U.S. Congress. *Environmental Justice.* Hearings before the Subcommittee on Civil and Constitutional Rights of the House Committee on the Judiciary, 103d Congress, 1st Session, 3–4 March 1993.

U.S. Environmental Protection Agency. Environmental Equity Workgroup. *Environmental Equity: Reducing Risks for All Communities.* Washington, DC: Environmental Protection Agency, 1992.

The Environmental Justice Information Page is maintained by the School of Natural Resources and Environment at the University of Michigan and is located at http://www-personal.umich.edu/~jrajzer/nre/index.html.

Environmental Racism

See Environmental Justice

Ethanol as a Fuel

Ethanol is an organic compound belonging to the alcohol family. It is also known as ethyl alcohol or grain alcohol. Its chemical formula is C_2H_5OH.

Production and Uses of Ethanol

In the 1930s, nearly all of the ethanol needed by chemical industries was produced by the fermentation of sugar from sugarcane. That process is a very old and very simple one by which enzymes convert a molecule of sucrose (table sugar) into two molecules of ethanol and two molecules of carbon dioxide. The chemical reaction is

In spite of controversies over the sale of gasohol, the U.S. Congress continues to encourage the practice by providing significant tax advantages to companies that develop the product. (U.S. Department of Energy)

one used by humans for thousands of years to make alcoholic beverages.

Beginning in the 1950s, chemists found a simpler and less expensive way to make ethanol. In this method, sulfuric acid (H_2SO_4) is added to ethylene (ethene; C_2H_2). By the 1980s, production trends had completely reversed, and more than 90 percent of the ethanol produced in the United States was made by this general method.

More than half of all the ethanol produced in the United States is used as a solvent in products such as detergents, household cleaners, toiletries and cosmetics, and paints and inks. The other major use of ethanol is as a chemical intermediate. That is, it is used to make other important organic compounds.

Ethanol as a Fuel

Traditionally, very small amounts of ethanol have been used as a fuel for special types of vehicles. Ethanol burns very cleanly at high temperatures, producing relatively few pollutants. In the 1970s, pressure developed for a much wider use of ethanol as a fuel. The recommendation was for a mixture of ethanol and regular gasoline, sometimes referred to as gasohol. Gasohol typically consists of about 5–10 percent ethanol mixed with 90–95 percent gasoline.

Interest in gasohol as a fuel arose for two major reasons. First, the United States was experiencing the pressure of severe petroleum shortages as members of the Organization of Petroleum Exporting Countries decided to cut their shipments to a number of countries, including the United States. Long lines at gasoline stations made consumers and politicians realize that it might be a good idea to begin developing alternatives to the nation's dependence on foreign sources of oil. Gasohol seemed to be a good response to that problem.

In addition, gasohol was thought to burn more cleanly and more completely than pure gasoline. As the nation's consciousness about problems of air pollution began to develop, a cleaner alternative to gasoline—ethanol—seemed to be attractive.

Corn, Ethanol, and Politics

But other factors were at work in the movement to promote gasohol as a gasoline substitute. A commitment to the use of gasohol would mean a bonanza for corn farmers in the United States. The plan proposed by those most interested in gasohol was to obtain the additional ethanol needed for gasohol production from increased production of corn crops.

Many environmentalists, politicians, scientists, and members of the general public were convinced about the future role of gasohol in the nation's energy equation. The U.S. Congress passed legislation to encourage development of gasohol production facilities and use of the product in cars and trucks. The legislation called for two tax benefits for gasohol use. First, a tax credit was given to companies that produce gasohol. Second, gasohol was exempted from a certain portion of the federal gasoline taxes charged the consumer.

By some estimates, the gasohol program cost the U.S. treasury more than $7 billion between the mid-1970s and the mid-1990s. At least half of that amount went to a single company, Archer Daniels Midland, a firm that calls itself "supermarket to the world."

Reassessments and Reconsiderations

In 1997, the U.S. Congress reconsidered its position on its support of gasohol as a gasoline substitute. Factual data about the gasohol program were somewhat discouraging. A report prepared by the General Accounting Office (GAO) on gasohol found that the use of gasohol would have virtually no effect, positive or negative, on the environment. Elimination of gasohol might result in slightly higher release of carbon monoxide but slightly lower release of chemicals from which ozone is produced. The report also said the substitution of gasohol for gasoline would have "minimal" effects on the release of greenhouse gases. Finally, the GAO pointed out that the gasohol program had had virtually no effect on the nation's dependence on foreign oil sources. The independent think tank Cato Institute observed that the ethanol program had been "a costly boondoggle with almost no public benefit."

When Congress reconsidered the gasohol support program, however, it drew a different conclusion. It continued to see the program as an important element in reducing air pollution and the nation's dependence on foreign petroleum supplies. It was also convinced that the production of ethanol was "a vote for American jobs" that would contribute to a better economy. In May 1998, the House of Representatives voted 297 to 86 and the Senate 88 to 5 to extend the ethanol support program through the year 2007. Reductions of 0.2 cents in the ethanol tax incentive were built into the program reauthorization, but the rest of its elements remained essentially the same.

References

"Congress Votes to Extend Ethanol Tax Incentive." http://www.ethanolrfa. org/pr052298.html. 19 July 1998.

"Ethanol as a Fuel." http://www.mct.gov. br/GABIN/CPMG/CLIMATE/ PROGRAMA/ingl/alcohol3.htm. 19 July 1998.

Fumento, Michael. "ADM's Ethanol Gets by with Help from Its Friends." http:// www.consumeralert.org/fumento/ adm.htm. 19 July 1998.

"Getting the Facts on Fuel Ethanol." http:// www.greenfuels.org/ethaques.html. 19 July 1998.

Moore, Stephen. "Push Ethanol off the Dole." http://www.cato.org/dailys/ 7–10–97.html. 19 July 1998.

Peaff, George. "Ethanol in Fuel Debated Again." *Chemical and Engineering News,* 4 August 1997, 20–22.

"Politics-Ethanol Mix Keeps Bad Tax Break Humming." *USA Today,* 19 June 1997, 12A.

Eugenics

The term *eugenics* was coined in 1883 by Sir Francis Galton. Galton was a cousin of Charles Darwin and himself a child prodigy. He studied to be a physician but then became interested in anthropology and meteorology and, later, in genetics.

Galton chose the term *eugenics* to express his views as to the future direction the human race should take. The word comes from two Greek words meaning "good birth." Galton believed that the discoveries made by Darwin and other biologists of the time showed that humans had more control over their future than they had ever realized. He thought that "more fit" members of society should be encouraged to breed, while those who are "less fit" should be discouraged from breeding. In this way, Galton thought, human society would continue to improve over many generations.

Galton was certainly not the first person to hold these views. As early as the fourth century B.C., Plato had written in *The Republic* that "the best of both sexes ought to be brought together as often as possible, the worst as seldom as possible." He thought this dictum would permit "our herd . . . to reach the highest perfection."

Positive and Negative Eugenics

Eugenic programs are often described as either *positive* or *negative.* In positive eugenics programs, people who are valued by a society are encouraged to mate with others of the same kind. The goal of such programs is to bring about an overall improvement in the quality of the gene pool in the society. A program that currently exists in the United States and Western Europe is known as the Artificial Insemination by Donor (AID) program. In this program, women are impregnated with sperm donated by men of unusually high attributes, such as intelligence or physical characteristics. A similar project was initiated in 1979 when Nobel Laureates were encouraged to donate sperm to an ongoing sperm bank to be used to father children in future years.

Negative eugenics programs are those in which people who are regarded as less valuable by a society are discouraged or prevented from having children. The most common way to carry out programs of negative eugenics is compulsory sterilization. Women who are regarded as "substandard" by some measure are required to be sterilized, thus preventing the chance that they will pass on their "less desirable" genes to future generations.

By far the most famous and most horrible example of a eugenics program was that carried out by the Nazi Party in Germany after they came to power in the 1930s. Adolf Hitler held that there was a certain "Aryan" personality that should be encouraged in the "new Germany," while other, less desirable personalities should be discouraged. This philosophy made use of both positive and negative eugenics. For example, men and women with "Aryan" traits, such as blond hair and blue eyes, were encouraged to have as many children as possible. The national government called attention to and publicly praised those "breeders" who were most successful in helping to perpetuate the "Aryan" race.

By contrast, residents of Germany who were not "Aryans" and, therefore, presumably inferior were to be eliminated by a variety of negative eugenic techniques. These included forced sterilization, of course, but also the outright murder of millions of Jews, gypsies, homosexuals, and other "unsuitables."

Eugenics in the United States

Eugenics became widely popular in Great Britain as a result of Galton's writings and teachings. National associations of those interested in eugenics were organized in Great Britain, the United States, and a number of other countries. The goal of these societies was expressed on the inside cover of *Eugenics Review,* a publication of the British Eugenics Society. That goal was "the study of agencies under social control that may improve or impair the racial qualities of future generations, whether physically or mentally."

The American Eugenics Society was founded in 1926. It carried on the efforts of its British counterpart to protect "better" Americans from the hordes of immigrants who had been arriving on these shores as well as from the insane, retarded, crimi-

nally inclined, and other miscreants within society. The movement was remarkably successful for a period of time, with more than half of the states passing forced sterilization laws for those who were regarded as "unfit."

Even the U.S. Supreme Court was sympathetic to the principles of eugenics. In 1927, the Court was asked to rule on the forced sterilization of a "feeble-minded" woman, Carrie Buck. The Court ruled, eight to one, that the sterilization could proceed. Justice Oliver Wendell Holmes ruled that "it is better for the world, if instead of waiting to execute offspring for crime, or to let them starve for their imbecility, society can prevent those who are manifestly unfit from continuing their kind. The principle that sustains compulsory vaccination is broad enough to cover cutting the Fallopian tubes."

By some estimates, forced sterilizations of the kind to which Carrie Buck had to submit were carried out on more than 50,000 Americans.

Eugenics Today

One might imagine that the eugenics program devised by the Nazi Party for Germany might have signaled the death knell for efforts to "improve" humanity by scientific means. It certainly clarified what are probably the fundamental issues surrounding eugenics.

First, who is to decide which individuals are "fit" and which "unfit"? The Nazis were by no means the first or the last people to suggest that Jews, gypsies, and homosexuals are "unfit" and should not be a part of future human generations. One might imagine, however, that a Jewish state could hold similar views about other groups of people, such as Arabs.

Second, does any society have the right to violate the most important privacy a person has, her or his own body, in order to build a "better" future for that society? Or do individuals have some rights, such as the right of procreation, that a society cannot violate?

Some critics point out that elements of eugenic thinking remain today. For example, one writer has argued that population control advocates, such as those who belong to Planned Parenthood, may embrace eugenic principles, even if they do so unconsciously. This writer points out that Margaret Sanger, the "mother" of birth control programs in the United States, was a member of both the English and American Eugenics Societies. She actually wrote that "those least fit to carry on the race are increasing most rapidly. . . . Funds that should be used to raise the standard of our civilization are diverted to maintenance of those who should never have been born."

An important underlying fact about such views is that they tend to be directed at minorities in society, such as people of color. The same writer points out that the majority of birth control programs are directed at Blacks, the poor, and other groups with little political power. In that regard, birth and population control programs may be regarded as remnants of a eugenic philosophy first defined by Galton and later carried out to its most complete degree by the Nazis.

Some observers are also concerned as to how our rapidly expanding knowledge about the human genome might be used by those with eugenic beliefs. As scientists learn more and more about the genes that determine human traits, will attempts be made to encourage the transmission of some genes and to discourage the transmission of others? To the extent that such attempts are made, the historic effort to implement eugenic principles will be achieved.

See also: Genetic Testing; Human Genome Project; IQ; Sterilization, Human

References

"American Eugenics." http://www.africa2000.com/ENDX/aepage.htm. 5 July 1998.

Duster, T. Backdoor to Eugenics. New York: Routledge, Chapman, and Hall, 1990.

Galton, Francis. Essays in Eugenics. London: Scott Townsend Publishers, 1996.

Kevles, Daniel J. In the Name of Eugenics: Genetics and the Uses of Human Heredity. Cambridge: Harvard University Press, 1995.

Kuhl, Stefan. The Nazi Connection: Eugenics, American Racism, and German National Socialism. New York: Oxford University Press, 1994. Also see the review of this book at http://www.execpc.com/~jfish/fff/aug94-f5.txt. 5 July 1998.

Larson, Edward J. Sex, Race, and Science: Eugenics in the Deep South. Baltimore: Johns Hopkins University Press, 1995.

Smith, J. David. The Eugenic Assault on America: Scenes in Red, White, and Black. Washington, DC: Georgetown University Press, 1993.

Wooldridge, Adrian. "Eugenics a Trait Found in Many Nations' Makeup." Sunday Oregonian, 14 September 1997, G1+.

Euthanasia

Euthanasia is the act of killing a person or allowing a person to die in order to provide a less painful but certain death. The word comes originally from two Greek terms meaning "good" or "easy death."

Active and Passive Euthanasia

Euthanasia is usually classified as either active euthanasia or passive euthanasia. In active euthanasia, some person actually causes the death of the seriously injured or ill person, such as by shooting the person or giving the person poison. In passive euthanasia, the injured or ill patient is simply allowed to die, with no effort made to provide food, water, drugs, or other substances that would otherwise prolong his or her life.

All human societies have had to deal with euthanasia. People do become terminally ill. They do suffer injuries that incapacitate them or will almost certainly result in their deaths. And people do get old and die, a process that is sometimes long and filled with suffering. Societies have to decide how they will deal with such individuals. Will they allow them to experience a prolonged and unpleasant experience of dying? Will they make every conceivable effort to keep them alive, no matter how painful that experience may be? Or will they take some positive action that hastens the individual's death?

These questions have been made much more difficult in recent decades by advances in modern medical technology. Untold numbers of individuals are now being kept alive who would almost certainly have died much sooner in earlier years. Medical tools and techniques such as drugs, organ transplants, artificial respiration, and artificial organs can often keep people alive almost indefinitely. Indeed, it is not totally beyond reason to imagine a day in the not-too-distant future when human life will be maintained artificially and indefinitely by medical means.

In such cases, the question of deciding when and how a person is to die is no longer one that concerns a relatively small part of the population. Conceivably most families will have to decide at what point they will have to "let a loved one go" or even "hasten that person on his or her way." The recent interest in assisted suicide in the Right to Die movement reflects this growing concern as to how such issues are to be handled.

One, the Other, Neither, or Both?

Moral philosophers, physicians, governments, and individuals have often given different moral weight to the two forms of euthanasia. In general, active euthanasia, the act of killing a person, has been regarded as "bad" while passive euthanasia has often been regarded as "good." There are almost no instances in which active euthanasia has received official or legal approval in any group or society. In fact, the passage in Oregon in 1994 of a Right to Die initiative was one of the very few instances in U.S. history when active euthanasia was given legal approval.

Passive euthanasia, however, has been received with growing approval over time, as medical advances have made natural death more uncertain and ambiguous. It is no longer unusual for physicians to allow terminally ill patients to "slip away" by withholding food, water, and medication. Currently, this form of euthanasia, with the patient's approval, is the only form permitted by law in the United States.

Some scholars have questioned the moral distinction, however, between active and passive euthanasia. After all, the end result is the same: A sick or injured person in pain is allowed or caused to die. Is not active euthanasia actually the kinder act when the person's suffering is ended immediately, rather than being drawn out, as in passive euthanasia? People who experience passive euthanasia actually suffer more, in some cases, than if medical treatment and feeding were actually continued. In this context, it might be more fair to say that active euthanasia is the "good" or kinder act and passive euthanasia the "bad" or less compassionate act.

Those who oppose active euthanasia often invoke the "slippery slope" argument in their presentations. The "slippery slope" argument suggests that once a society starts in one direction (such as trying to walk down a slippery slope), it may not be able to keep from moving farther in that direction (sliding down the slippery slope) than it had originally intended. In the case of active euthanasia, opponents ask, what is to prevent the principle of killing sick people from being extended to other groups. What about people who are old, but not ill; very young children with congenital disorders; other people who may be severely disabled but not terminally ill; or people with whose race, ethnicity, sexual orientation, or other attributes we may not be comfortable. Opponents of active euthanasia, in fact, often point to the Nazi attempt to eliminate Jews, gypsies, homosexuals, and other "undesirables" as examples of what can happen if active euthanasia becomes acceptable.

Debates about euthanasia are likely to become even more widespread in coming decades. Polls continually indicate that an overwhelming number of Americans—

usually at least eight out of ten—do not want to have their lives sustained when all hope of recovery had passed. Yet, legislation and initiatives to permit active euthanasia have, with few exceptions, failed, usually because of concerns about the proper safeguards for dying patients and the medical establishment.

See also: Right to Die

References

Campbell, Courtney S., and Bette-Jane Criggen, eds. "Mercy, Murder, and Morality: Perspectives on Euthanasia." *Hastings Center Review,* January/February 1989, special supplement (whole).

Kohl, Marvin, ed. *Beneficent Euthanasia.* Buffalo: Prometheus Books, 1970.

Rachels, James. "Active and Passive Euthanasia." *New England Journal of Medicine,* 9 January 1975, 78–80. See replies to this article in Tom L. Beauchamp, "A Reply to Rachels on Active and Passive Euthanasia," in *Social Ethics,* edited by Thomas A. Mappes and Jane S. Zembaty, 67–76 (New York: McGraw-Hill, 1977); and Thomas D. Sullivan, "Active and Passive Euthanasia: An Impertinent Distinction?" in *Social Ethics,* 4th ed., edited by Thomas A. Mappes and Jane S. Zembaty, 115–121 (New York: McGraw-Hill, 1992). Rachel's reply to Sullivan follows on 121–131. Also see "An Evaluation of James Rachel's Active and Passive Euthanasia." *Philosophika, The Internet Journal of Philosophy.* http://www.philosophika.com/Campbell.html. 27 June 1998.

"Right to Die: Euthanasia and End-of-life Decisions." at http://ethics.acusd.edu/euthanasia.html. 27 June 1998.

"The Rights of Euthanasia." http://www.death-dying.com/survey/paper.html. 27 June 1998.

Schneiderman, L. J. "Euthanasia: Can We Keep It a Special Case?" *The Humanist,* May/June 1990, 15–17+.

Steinbock, Bonnie, and Alastair Norcross. *Killing and Letting Die.* 2d ed. New York: Fordham University Press, 1994.

"The VE [voluntary euthanasia] bulletin." http://www.on.net/clients/saves/bull.htm. 27 June 1998.

"Voluntary Euthanasia—the Basic." http://dialspace.dial.pipex.com/town/parade/ga46/basic.htm. 27 June 1998.

Wekesser, Carol, ed. *Euthanasia: Opposing Viewpoints.* San Diego: Greenhaven Press, 1995.

The home page for the Voluntary Euthanasia Society is at http://www.ves.org.uk.

Experimentation on Humans

See Drug Testing; Human Experimentation

Faith Healing

Religion and science have historically come into conflict over many issues. In modern society, one of those issues has been the treatment of illness. People of religious faith tend to believe that God, in whatever form he or she is conceived, causes or allows disease to occur. A view of an omnipotent Being could hardly allow any other belief.

Religions differ, however, as to the extent to which they rely on communication between humans and God—usually in the form of prayer—for the cure of physical illness. In many religions, believers accept the fact that medical practitioners can work as an adjunct to or agent of God and, therefore, possess the ability to effect cures. They believe that prayer alone may be insufficient to bring about a cure and are willing to seek medical care for illness and injury.

Complete Dependence on Supernatural Cure

A few religions reject that view. They accept the absolute power of prayer in all realms, including physical illness. They believe that a person's health is absolutely in the hands of a supernatural being, such as God, and that the intervention of human medical practices is never warranted. Indeed, it may even be regarded as an insult to the Supreme Being for an individual *not* to place his or her life in that Being's hands. Christian Scientists may be one of the best known religious denominations to hold a view of this kind.

There has traditionally been relatively little debate about the practice of allowing adults to choose prayer over medical sci-

ence for the treatment of disease and injury. Such has not been the case when children are involved. Children do not have the same opportunity to make choices in health care issues as do adults. Therefore, the question has arisen as to whether adults have the right to withhold medical treatment for religious reasons from a child who would almost certainly benefit from medical care.

There are now numerous cases in which children have died from such decisions. The question raised in such cases is usually a conflict of rights. On the one hand, do parents have the right to follow their own consciences about the best treatment for their offspring and withhold medical care from seriously ill children? Do they have the right to impose on those children their own beliefs that prayer alone is sufficient to deal with illness? Some individuals argue that this right is granted to parents by the First Amendment right of religious freedom.

On the other hand, some people argue that sick children also have a constitutional right, the right to life. They say that when children are suffering from a disorder that can be treated by medical science, they have a right to receive that treatment. For example, a child with diabetes or treatable forms of leukemia should have the opportunity to undergo standard, well-known, and generally successful methods of medical treatment for those disorders.

Legal Precedence

Historically, courts have generally ruled in favor of parents in such cases. When chil-

Many faith healers believe that they have the right to decline the use of medical services for their children. Do they? (Corbis/Adrian Arbib)

dren have died as the result of parents' refusal to seek medical treatment, the courts have almost always accepted the religious freedom argument in exonerating them from blame. In a relatively few cases, judges have ordered emergency medical care for children who were in clear and imminent danger of dying. In 1991, for example, the Oregon Supreme Court ordered immediate brain surgery for the infant child of a member of the state's Church of the First Born. The church does not allow its members to seek medical care for its children who become ill. The child survived and is now healthy.

Some observers think that courts may begin to adopt a somewhat different view about the death of sick children for whom no medical care is sought. Prosecutions of parents for homicide are still largely unsuccessful. However, the U.S. Supreme Court in January 1996 upheld a wrongful death decision in the amount of $1.5 million against four Christian Scientists for the death of an eleven-year-old boy. The boy had died in 1989 after falling into a diabetic coma.

References

Barrett, Stephen. "Some Thoughts about Faith Healing." http://www.quackwatch. com/01QuackeryRelatedTopics/faith. html. 30 May 1998.

Dossey, Larry. *Healing Words: The Power of Prayer and the Practice of Medicine.* San Francisco: Harper San Francisco, 1995.

Finley, Carmel. "Parents Go on Trial for Denying Son Medical Care." *Oregonian,* 17 April 1996, A1+.

Larabee, Mark, and Peter D. Sleeth. "Faith Healing Raises Questions of Law's Duty—Belief or Life?" *Sunday Oregonian,* 7 June 1998, A1+.

Munns, Roger. "Christian Science's No. 1 Enemy." *San Francisco Chronicle,* 26 April 1996, A10.

Pearson, Mark A. *Christian Healing: A Practical and Comprehensive Guide.* Chicago: Fleming H. Revell, 1995.

"Prayer and Faith Healing Instead of Medical Treatment." http://www. religioustolerance.org/medical.htm. 30 May 1998.

Walker, James. "*Christian Science:* Couple found guilty of manslaughter." http:// www.watchman.org/csguilty.htm. 30 May 1998.

Feedlot Pollution

One of the most significant changes in modern agricultural practices has been the introduction of feedlots, enclosed areas in which animals spend most of their lives. Food is delivered to the animals in a feedlot rather than, as in traditional systems, animals having to forage for their food. Feedlots offer a significant economic advantage to farmers because they can raise a much larger number of animals on a much smaller land area than is possible by allowing animals to roam across the countryside.

Animal Wastes on Feedlots

Probably the most important new problem created by feedlots, however, is that of animal wastes. A single dairy cow, for example, regularly produces about twenty-five times as much waste as does a single adult human. In traditional farming and ranching techniques, those animal wastes are spread out over the many acres on which foraging animals live. In feedlots, the wastes are concentrated on a smaller area to which the animals are confined.

An example of the scope of the modern feedlot problem is Circle 4 Farms in Milford, Utah. Circle 4 Farms raises hogs on a 50,000-acre plot 300 kilometers (190 miles) southwest of Salt Lake City. The hogs are housed 120,000 to a barn. They live, eat, and sleep in the barns, leaving only to go to the slaughterhouse. In one year, Circle 4 Farms will turn over 2.5 million hogs. The wastes produced on this one farm annually exceed the total amount of human wastes generated by the city of Los Angeles.

Agricultural experts have developed environmentally sensitive techniques for handling this level of concentration of animal wastes. The usual technique is to construct an artificial lagoon into which manure and urine are dumped. The lagoon may be

Legislative actions to reduce the problems posed by feedlot pollution are of little value unless adequate funding is provided to make sure such actions are enforced. (Corbis/Phil Schermeister)

lined with plastic or some other impermeable material to prevent wastes from leaching into the ground. Within the lagoon, wastes are partially degraded naturally by the action of bacteria. The products of this process are then pumped out of the lagoon and used to fertilize nearby crops.

Problems with Waste Disposal Systems

This system is a safe, effective, and profitable way of dealing with feedlot wastes ... when it works properly. The problem is that in many instances, it does not do so. In most cases, the reason the system fails is that the ratio of animals to crops is too large. Wastes are produced faster than they can be used up as fertilizer.

In such cases, storage lagoons may simply overflow. Wastes escape into nearby streams and are then carried into larger rivers and eventually into the oceans. It is no longer unusual to hear about streams, rivers, lakes, and even parts of the ocean that can no longer support aquatic life because of such runoffs. For example, 10 million fish were killed off the coast of North Carolina in 1995 following a heavy rain that caused the overflow of 130 million liters (35 million gallons) of animal wastes from storage lagoons.

Health Effects

Animal wastes pose a threat to the environment and human health primarily for two reasons. First, they are a rich nutrient that encourages the growth of algae in lakes and the oceans. Algae deplete oxygen from such waters, however, and make it difficult or impossible for fish and other aquatic organisms to survive. Some coast regions of the Gulf of Mexico, for example, are now regarded as "dead zones" because of the flow of animal wastes into the area. Second, compounds present in animals wastes, such as nitrates, may present a threat to human health. The young, elderly, and ill are especially vulnerable to this danger.

The loss of animal wastes from storage lagoons is a problem, however, for another reason. These wastes are simply a valuable natural resource, rich in potassium, nitrogen, and phosphorus, that should be used as fertilizer and not allowed to be lost.

Solving the Problem of Feedlot Wastes

The question as to how to deal with animal wastes is one that is largely unsolved and, to a large extent, essentially ignored. The state of California, which now produces more milk than any other state, is an example. Between 1987 and 1996, there was an increase of 46 percent in the number of dairy cattle in the eight counties that make up the state's Central Valley. More than 1,600 dairies produce a volume of animal wastes equal to that of a city of 21 million people. Yet, the state has only one inspector for the entire industry. A majority of the state's dairies have never been seen or inspected.

Officials concede that the animal waste problem is simply out of control. They estimate that less than half of all dairies make any serious effort to control waste runoffs. The majority simply allow wastes to escape into the ground or into nearby streams or rivers.

The effects of unsatisfactory waste treatment practices have already begun to show up. Test wells drilled in the Central Valley have found nitrate levels in groundwater five times higher than that permitted by government standards. One expert has predicted that there are "plumes of high-salt, high-nitrate water under dozens, if not hundreds, of these sites." "What concerns me," he continues, "is there are a lot of rural residences that still have old wells that don't go down so deep. I suspect a lot of those people are drinking water exceeding the nitrate standard" (Diringer 1997, A6).

How the problem of animal wastes will be resolved is not clear. Many farmers point out that their job is already a difficult twelve-hour-a-day, seven-day-a-week job and that they do not have time to worry about "what's coming off the back end of the dairy." The drive to increase profitability can do nothing but increase the pressures they already feel.

At the same time, farm and dairy interests often have powerful influences on state legislatures. Those interests are seldom matched by individuals and organizations speaking out for environmental and health concerns. As a result, the tendency in many instances is for states to retain the status quo and not overemphasize the problems raised by feedlot pollution.

References

Diringer, Elliot. "In Central Valley, Defiant Dairies Foul the Water." *San Francisco Chronicle.* 7 July 1997, A1+.

"Feedlot Permits." http://www.mda.state. mn.us/docs/agdev/MANURE.HTM. 20 June 1998.

"Groundwater Quality Protection for Livestock Feeding Operations." http:// leviathan. Tamu.edu:70/01pubs/ animalsc/b-1700.abs. 21May 1999.

"Minnesota's Approach to Feedlot Pollution Control." http://www.epa.gov/OWOW/ info/NewsNotes/issue12/nps12agr.html. 20 June 1998.

U.S. Department of Agriculture. National Resources Conservation Service. *Agricultural Waste Management Field Handbook.* Springfield, VA: National Technical Information Services Division, U.S. Department of Commerce.

U.S. Environmental Protection Agency. *Water Pollution from Feedlot Waste: An Analysis of Its Magnitude and Geographic Distribution.* Washington, DC: Environmental Protection Agency, 1993.

Fetal Tissue Research

Fetal tissue research (FTR) refers to the use of tissue taken from a fetus for the treatment of animal (usually human) disorders and for research on such treatments. FTR is to be distinguished from a similar term, *fetal research,* that refers to research on a fetus itself while it is still in the womb.

Why Use Fetal Tissue?

FTR has been the subject of considerable controversy in recent years. However, this kind of research goes back at least half a century. Medical researchers have long been interested in the use of fetal tissue because it has special qualities that are potentially useful in the treatment of a number of human disorders.

In the first place, fetal tissue tends to be largely undifferentiated. It is still at a stage of growth in which it can adapt and adjust to its surrounding environment. It has the ability to grow and change that older tissue found in postbirth humans, particularly older individuals, does not have.

Fetal tissue also has a tendency to grow rapidly. This trait makes it useful in the development of cell lines. A cell line is a series of cell generations that develop over time from a single set of cells. Because of their common origin, the cells in a cell line are all very similar to each other, a property of importance in some forms of biological research. Fetal cell lines were used during the 1950s, for example, in research that led to the development of polio vaccines.

Examples of FTR Therapies

Researchers have explored the use of fetal tissue cells for the treatment of a number of diseases, including Alzheimer's disease, Parkinson's, Huntington's chorea, diabetes, multiple sclerosis, epilepsy, leukemia, hemophilia, sickle-cell anemia, spinal cord injuries, birth defects, and disorders of the eye. One of the most promising of these lines of research involves the treatment of Parkinson's disease.

Parkinson's is a disorder characterized by the loss of nerve function, with the result that people gradually lose the ability to walk, speak clearly, and perform other motor activities. Parkinson's is thought to be caused by the loss by certain brain cells with the ability to produce a chemical neurotransmitter ("nerve messenger") known as dopamine. As levels of dopamine drop in a person's brain, nerve function becomes more erratic.

Scientists have found that the transplantation of fetal tissue cells into the brains of Parkinson's patients can result in some relief from the symptoms of the disorder. Apparently, the fetal cells grow and adapt to the surroundings in which they have been transplanted. They begin to act like new brain cells, some of which produce the dopamine that older cells around it can no longer produce.

Another promising line of research involves the use of fetal tissue cells in the treatment of an eye condition known as macular degeneration of the retina. The disorder is very common among older people and ultimately leads to blindness. In 1997, researchers reported on studies in which they transplanted fetal tissue cells into the retina of individuals with advanced cases of macular degeneration. Short-term results indicated that the cells grew and adapted to their surroundings and took over the function of native cells that had ceased to function properly.

Issues Surrounding FTR

Fetal tissue research has had its critics. Some people have argued that the practice might lead to an increase in the number of abortions performed. Aborted fetuses are the primary source of tissue used in FTR. If FTR becomes a common and approved procedure, critics say, women who might not otherwise choose to have an abortion might agree to the procedure. They might feel that their own fetus might be able to serve a worthwhile function in the treatment or cure of some other person's disease.

Legal Status of FTR

Legislators have been very conscious of the issues involved in FTR and have taken different, and quite opposing, views on the subject. During the administration of President George Bush, for example, all funding for FTR by the federal government was banned. Shortly after Bill Clinton became president in 1993, however, he rescinded that ban. Clinton's action was supported by a 93 to 4 vote in the U.S. Senate.

In 1997, the Senate considered the issue once again. Senator Dan Coats (R-Ind.) offered an amendment to a funding bill banning the release of any money for FTR. That amendment was defeated by a vote of 60 to 38.

Fetal tissue research in the United States, then, is very much alive but very highly controlled. The 1993 Senate bill contained a number of provisions designed to make sure that tissue was removed only from fetuses that would have been aborted anyway and that no woman would feel encouraged in any way at all to have an abortion for the purpose of donating its tissue to researchers.

Fundamentally, decisions as to how tissue from aborted fetuses is to be used are regulated by general laws and regulations dealing with the donation of body parts from dead people. All fifty states and the District of Columbia have laws known as Uniform Anatomical Gift Acts that tightly control the donation, removal, and use of body parts for medical or research purposes. The use of tissues obtained from aborted fetuses is regulated by these same acts.

References

Boer, G. J. "Ethical Guidelines for the Use of Human Embryonic or Fetal Tissue for Experimental and Clinical Neurotransplantation and Research." at http://www.bm.lu.se/~nectar/eth.1.html. 14 July 1998.

DoctorNET Online Fetal Tissue Research Links. http://www.comedserv.com/fetal.htm. 14 July 1998.

"Fetal Research Discussed in IOM Conference Report." http://www2.nas.edu/whatsnew/24a6.html. 14 July 1998.

Focus on Family. "When an Embryo Becomes Alive." http://user.mc.net/dougp/ftrnews7.htm. 14 July 1998.

Gorner, Peter. "New Hope for Old Eyes." *Chicago Tribune*, 31 January 1997.

"The News on Fetal Tissue Research." http://shrike.depaul.edu/~dmiller3/fetal.html. 14 July 1998.

"Position Statement: Fetal Tissue Research." American College of Rheumatology.

http://user.mc.net/dougp/ftrnews2.htm. 14 July 1998.

Fluoridation

Fluoridation is the process of adding certain chemicals known as fluorides to public water supplies. This process has become popular in the United States and other parts of the world as a method for reducing the rate of dental caries (cavities) among adults and, especially, children.

History of Fluoridation

Interest in the use of fluorides to improve dental health originated in the 1920s. A series of research studies indicated that people living in certain communities with natural fluoride in their waters tended to have fewer cavities than those who lived in areas with little or no natural fluoride. Some of the communities studied were Galesburg and Monmouth, Illinois, and Glassboro, Pitman, and Woodstown, New Jersey.

Dental researchers were excited about these studies because they held the promise for solving one of the nation's most common health problems: tooth decay. Those researchers instituted a number of additional studies to see whether decay could be reduced through the artificial addition of fluorides to a community's water supply.

The first of these studies was conducted in Grand Rapids, Michigan, beginning in June 1944. Enough sodium fluoride (NaF) was added at the city's water filtration plant to raise the fluoride content from a natural level of 0.05 ppm (parts per million) to an experimental level of 1.0 ppm. Over the next ten years, the number of cavities reported for schoolchildren in Grand Rapids was noted and compared with the number for children in Muskegon, Michigan. Muskegon is less than fifty miles from Grand Rapids, but it had no artificial or natural fluoride in its water supply. Results from Grand Rapids were also compared to those obtained for children from Aurora, Illinois. Aurora is demographically similar to Grand Rapids and Muskegon, but it has a natural fluoride level of 1.2 ppm in its water supply.

The results of this study showed a dramatic decrease in tooth decay among children in Grand Rapids. This and similar studies performed throughout the country convinced many doctors, dentists, government officials, and lay citizens that fluoridation of water supplies was an effective means of reducing dental caries. By 1965, virtually every professional organization, both national and international, had endorsed the practice of fluoridation. Since the late 1940s, many communities of every size and in every part of the nation have decided to start fluoridating their public water supplies.

Biochemical Basis for the Effects of Fluoridation

Today, scientists understand the chemical reasons that fluoridation would reduce the rate of dental caries. Tooth enamel contains a stonelike chemical compound known as calcium hydroxyphosphate (apatite; $Ca_{10}[PO_4]_6[OH]_2$). Acids found normally in the mouth can attack this compound and cause it to break apart. The loss of apatite from tooth enamel eventually leads to the formation of tiny holes in the enamel known as dental caries.

When fluoride is added to a person's diet, it reacts with apatite to form a compound known as fluorapatite. Fluorapatite is identical to apatite except that the two hydroxide groups (OH) in apatite have been replaced by fluorine atoms. The formula for fluorapatite, $Ca_{10}(PO_4)_6F_2$, illustrates this change. Fluorapatite is also a stonelike material, but it is much stronger and more resistant to acids than is apatite. Enamel that incorporates fluorapatite into its structure is more resistant to mouth acids and, thus, less likely to form dental caries.

Objections to Fluoridation

Not every community, however, has been convinced about the practice of fluoridation. In spite of the enthusiastic recommendations of professional and governmental authorities, a large number of communities still do not have fluoridated water. It is obvious when the question of fluoridation is put to the voters that a great many people—both professionals and laypersons—still have strong reservations about the desirability of adding fluorides to public water supplies.

One problem that sometimes arises in dealing with the fluoridation issue is the confusion over fluorides and fluorine. *Fluorine* is a chemical element, a greenish-yellow gas that is highly toxic and extremely dangerous to work with. By contrast, *fluorides* are compounds that are made from fluorine and one other element. In small concentrations, fluorides tend to be relatively harmless, although they can have health effects in larger doses.

It is the health effects of large doses of fluorides that trouble some people. They point out that studies have shown that fluorides can cause damage to teeth and other parts of the bodies. Some opponents of fluoridation have argued that the chemicals can increase the rate of diseases such as diabetes, heart disease, stroke, cancer, and mental disorders. They have also been accused of causing miscarriages, stillbirths, and crippling diseases in children.

It is true that serious health problems can be caused by high levels of fluorides, defenders of fluoridation agree. But the levels needed to cause these problems are much higher than those used in the fluoridation of water supplies.

Critics of fluoridation also claim that the practice is a form of forced medication of the general public. They agree that many public health practices are important and essential to the maintenance of good health in a community. Vaccinations may be necessary, for example, to prevent the spread of epidemics like typhoid and typhus. But tooth decay can hardly be described as a "horrible epidemic." Therefore, people should have the right to choose whether or not they want to use fluorides. For example, they can choose to use toothpastes that contain fluorides or they can take fluoride pills. Once fluorides are added to the public water supply, however, individual citizens no longer have those options, critics say. Another argument that has often been used is that certain groups, such as members of the Communist Party, support flouridation as a way of gaining control of public water supplies. In this way, the argument goes, such groups hope to be able to threaten a community by having the ability to add other substances, such as toxic chemicals, to the water supply.

The Debate Today

A half century after the Grand Rapids experiments, controversy still rages over the fluoridation of public water supplies. Some people who once supported the practice have even changed their minds and begun to question the use of fluoridation. They cite statistics that indicate dramatic decreases in the number of dental caries among young children, even in regions where there is no fluoridation of public water supplies. The reason for these decreases is not always clear, although improvements in nutrition and overall health practices are thought by some to account for the change.

At the beginning of the twenty-first century, most professional health and dental groups still strongly support the practice of fluoridation. Yet, in most case where the practice is put to a public vote, it tends to lose. Those who study this subject are intrigued by the confrontation that occurs between "science" and "propaganda" in such cases. They have tried to determine how so many communities reject, often by lop-sided margins, the arguments of prestigious leaders of the community's scientific communities and defeat proposals for fluoridation. The practice of fluoridation of public water supplies presents, indeed, a classical case study in the way nonscientists think about and act on scientific issues that affect their everyday lives.

References

Colquhoun, John. "Why I Changed My Mind about Water Fluoridation." *Perspectives in Biology and Medicine,* Autumn 1997; also at http://www.cadvision.com/fluoride/colquh.htm. 22 June 1998.

Easley, Michael W. "Fluoridation: A Triumph of Science over Propaganda." http://www.asch.org/publications/priorities/0804/fluoridation.html. 22 June 1998.

Exner, F. B. "Fluoride: A Protected Pollutant." http://www.cadvision.com/fluoride/exner.htm. 22 June 1998.

"Fluoridation: The Cancer Scare" and "Six Ways to Mislead the Public." *Consumer Reports,* July and August 1978, 392–396 and 480–482.

"The Fluoridation Controversy in Calgary, Alberta, Canada." http://www.cadvision.com/fluoride/calgary.htm. 22 June 1998.

"Fluorides and Fluoridation." http://www.ada.org/consumer/fluoride/fl-menu.html. 22 June 1998.

Marshall, Eliot. "The Fluoride Debate: One More Time." *Science,* 19 January 1990, 276–277.

McClure, Frank J. *Water Fluoridation: The Search and the Victory.* Washington, DC: National Institutes of Health, U.S. Department of Health, Education, and Welfare, 1970.

Oldenburg, Don. "Warning Labels Caution Fluoride Toothpaste Unsafe for Kids." *San Francisco Chronicle,* 18 June 1997, A5.

Preventive Dental Health Association. http://emporium.turnpike.net/P/PDHA/health.htm. 22 June 1998.

Sears, Cathy. "Fluoridation: Friends and Foes." *American Health,* October 1989, 36–38.

Two articles dealing with the question as to why voters so consistently reject fluoridation are the following:

Mausner, Bernard, and Judith Mausner. "A Study of the Anti-scientific Attitudes." *Scientific American,* February 1955, 35–39; For responses, see Letters to the Editor, April 1955, 2.

Sapolsky, Harvey M. "Science, Voters, and the Fluoridation Controversy." *Science,* October 1968, 427–433; For responses see Letters to the Editor, January 1969, 17.

Food Additives

See Vitamins and Minerals

Foods

See Delaney Clause; Irradiation of Food; Olestra; Vitamins and Minerals

Forests

See Clearcutting; Logging Roads; Northwest Forest Plan; Prescribed Burn; Rain Forests; Slash-and-Burn

Formaldehyde

Formaldehyde is the simplest member of the organic family of compounds known as the aldehydes. It is a gas with the chemical formula HCHO. For many years, the most important uses of formaldehyde were as a preservative for biological specimens and as an embalming fluid. In such applications, formaldehyde is typically used in the form of a 37 percent aqueous (water) solution known as *formalin*.

In these applications, formaldehyde can pose a serious health hazard in the specific area in which it is being used. But it presents little or no widespread threat to the general population. Formaldehyde is also released to the atmosphere by other human activities. For example, it is formed in smog, and it is produced during the combustion of methanol (methyl alcohol) and ethanol (ethyl alcohol).

Formaldehyde and Indoor Air Pollution

By far the most serious environmental effects of formaldehyde today are as a component of indoor air pollution because a number of products used in construction are made from formaldehyde. Once these products are installed in a structure, they may continue to release trace amounts of the gas over long periods of time.

Insulation made with formaldehyde may continue to pose a health threat for many years after its installation. (Corbis/James L. Amos)

For example, formaldehyde is used in the manufacture of particle board, plywood, and imitation wood paneling. Most of the formaldehyde is changed chemically during the manufacturing process and poses no health or environmental threat in its new form. A small amount of formaldehyde remains trapped within the new product, however. It is this formaldehyde that can escape into the atmosphere and cause problems for humans and other organisms.

For example, various kinds of artificial wood may release anywhere from 0.03 to 9.2 micrograms of formaldehyde per gram of material per day. While these amounts may seem small, the total mass of building materials used can be large, and ventilation systems may be insufficient to provide adequate removal of the gas. Carpeting, fiberglass insulation, and clothing are other common sources of formaldehyde.

The National Institute for Occupational Safety and Health has set a standard for exposure to formaldehyde of 0.1 ppmv (parts per million by volume) for indoor air and a standard of 1.0 ppmv over a thirty-minute period for air in factories where formaldehyde is used or released. Studies have shown that the level of formaldehyde in mobile homes (usually made with many formaldehyde-based materials) can range as high as 2.9 ppmv and may average 0.38 ppmv. Homes with urea-formaldehyde insulation may have concentrations ranging from a low of 0.01 ppmv to a high of 3.4 ppmv (with an average of 0.12 ppmv). By comparison, homes with insulation other than urea-formaldehyde products have formaldehyde levels ranging from 0.01 to 0.08 ppmv (with an average of 0.03 ppmv). For comparison, formaldehyde concentrations in a plywood factory tend to range from 1.0 to 2.5 ppmv.

The health effects of formaldehyde depend on the concentration of the gas and length of exposure. The gas can typically not be detected, for example, at exposures of less than 0.05 ppmv for one minute. When a level of 0.08 ppmv has been reached, the cerebral cortex may be affected. Throat, eye, and nose irritation occurs at a concentration of 0.2 ppmv for sixty minutes, and damage to the central nervous system begins at a concentration of 0.8 ppmv for a period of ten minutes.

See also: Indoor Air Pollution

References

Gammage, R. B., and K. C. Gupta. "Formaldehyde." In *Indoor Air Quality,* edited by P. J. Walsh, C. S. Dudney, and E. D. Copenhaver. Boca Raton, FL: CRC Press, 1984.

Hileman, B. "Formaldehyde: How Did EPA Develop Its Formaldehyde Policy?" *Environmental Science and Technology* 16 (1982): 543A.

Indoor Air Pollution/Formaldehyde. Washington, DC: American Lung Association, 1995.

The Inside Story—A Guide to Indoor Air Quality. Washington, DC: U.S. Environmental Protection Agency, 1995.

"Instructional Module: Formaldehyde in the Home." http://www.montana.edu/wwwcxair/formald.html. 26 May 1998.

U.S. Consumer Product Safety Commission. *An Update on Formaldehyde.* Washington, DC: U.S. Consumer Product Safety Commission, 1990; also at http://www.epa.gov/iaq/pubs/formald2.html. 21 June 1998.

Freedom of Speech

See Privacy and the Internet

Gasohol

See Ethanol as a Fuel

Genetic Engineering

See Bovine Somatotropin; Cloning; Genetically Manipulated Foods; Human Gene Therapy

Genetic Testing

Scientists have learned a great deal about the chemical basis of heredity in the last half century. The breakthrough in that field of research came in 1953 when U.S. biologist James Watson and English chemist Francis Crick discovered the structure of deoxyribonucleic acid (DNA). DNA is the molecule in cells that determines the characteristics of an organism, such as human hair and eye color, ear shape, hair texture, and height.

Each physical characteristic is determined by one or more sections of a DNA molecule known as a *gene*. In some cases, a specific gene is responsible exclusively and entirely for one particular physical characteristic. In other cases, two or more genes operate in combination to determine a given physical characteristic. In still other cases, a physical characteristic is dependent on the existence of two or more genes as well as environmental factors to which the organism is exposed after birth.

Since the Watson/Crick discovery, scientists have been aware that it is at least theoretically possible to produce a "map" of all the genes in human DNA (or the DNA of any other organism). In the case of humans, such a map would consist of about 100,000 genes. A "gene map" would be useful for a number of purposes, one of which would be to identify abnormalities in genes and the ultimate consequences that those abnormalities might have for an organism.

For example, suppose that scientists discover that gene #48,371 (according to some imaginary system of coding) is responsible for the production of a chemical needed to digest proteins in the body. Scientists will, at the same time, have discovered the chemical structure of that gene.

Now, suppose that it is possible to extract from a fetus a sample of blood that can be tested for gene #48,371 (and, presumably, at least some other genes). And further suppose that tests indicate an abnormality in that gene, an abnormality that would prevent its normal function of producing the needed protein-digesting chemical. In such a case, it would be possible to predict that the fetus would eventually be born with a genetic abnormality resulting from an error in gene #48,371. That genetic error could have no major health consequences for the individual, it could result in relatively minor health problems, or it could have effects serious enough to cause the death of the individual.

Methods of Gene Testing

The scenario described here is that involved in gene testing. Techniques are now available for the two steps required in such a procedure: (1) the extraction of blood or

101

The possibility of wide distribution of information obtained from genetic testing increases as information technologies, such as the Internet, continue to grow and become more sophisticated. (Corbis/Jim Sugar Photography)

other bodily fluid from a fetus, and (2) testing of those fluids for the presence of healthy and/or abnormal genes.

The two most common procedures for collecting fetal fluids are *amniocentesis* and *chorionic villus sampling* (CVS). Amniocentesis involves the removal of amniotic fluid surrounding the fetus in a woman's uterus. That fluid contains chemicals excreted by the fetus and containing DNA that can be tested for genetic disorders. Amniocentesis is typically conducted when a fetus is sixteen to twenty weeks old. It is a relatively safe procedure, although it causes miscarriages in 0.25 to 0.50 percent of all cases.

Chorionic villi are tiny hairlike projections that appear early in the development of a fetus. These villi eventually develop into the placenta. In CVS, a thin needle is inserted into the uterus and a small sample of villi is removed. Those villi contain fetal DNA and can be used to look for genetic abnormalities. CVS can be conducted at a

somewhat earlier stage than amniocentesis, usually when the fetus is nine to twelve weeks old. The risk of miscarriage is somewhat higher with CVS than with amniocentesis, about 1 percent.

Current technologies are being tested that are even more sensitive than either amniocentesis or CVS. One currently in use can be done on a three-day fetus, while research on an eight-cell test is now under way.

At the present time, tests for more than 400 genetic disorders are available. These tests detect a variety of conditions that range from the relatively rare to the relatively common, including Down syndrome, Tay-Sachs disease, and anencephaly (absence of a brain). They make it possible for parents to learn as early as sixteen weeks (with amniocentesis), nine weeks (with CVS), or earlier (with new technologies) about the existence of fetal genetic disorders.

Issues

Questions always arise as to the relative value of such information compared to the medical risks posed to the pregnant woman. Perhaps more difficult, however, are questions as to how such information is to be used by parents of the fetus. Suppose, for example, that a pregnant woman learns that her child will be born with anencephaly (usually fatal within a week) or with Down syndrome (not fatal, but associated with physical and mental retardation). What action should she (and her mate) take when confronted with this information? In most instances, that question can be rephrased to ask whether the fetus should be aborted, since that procedure is essentially the only alternative to bringing the fetus to term.

The dramatic impact of this question is one reason that some women choose not to have genetic tests done at all. Of those who do, the decision to abort or not abort is related significantly to the severity of the genetic disorder. For disorders that are likely to lead to a child's early and/or painful death, close to 100 percent of all pregnant women choose abortion. For nonfatal disorders, such as Down syndrome, between 50 and 90 percent of pregnant women choose abortion.

Choices of this kind are likely to become even more difficult for women (and their mates) in the future. As technology advances, a greater number of disease-causing genes are being discovered. For example, a gene thought to be responsible for the development of breast cancer has now been identified. Are there justifications for aborting a fetus found to be carrying that gene?

Other Uses of Genetic Testing

Questions have begun to arise about other possible uses of gene testing. For example, suppose a physician draws a sample of blood from an individual as part of a routine application for health insurance. And suppose that tests performed on the blood show the applicant to have a gene related to heart disease. Is the insurance company justified in denying the individual's application on the basis of this information? Would a company be justified in denying a person a job based on the same information?

On the one hand, one might argue that companies have a right to protect themselves from risks that are likely to pose a financial hardship for them in the future. For example, should an insurance company really be expected to offer a policy to someone known to be at risk for a serious health disorder?

On the other hand, one might ask whether a person should be denied health coverage, a job, or other benefits simply by reason of inheriting a particular gene that may or may not be manifested later in life. Decisions such as these are very complex, of course, because of the many different genetic abnormalities that might be discovered by gene testing and the variety of uses to which that information might be put.

Currently, the law generally comes down on the side of corporations in such disputes. As of 1998, only nineteen states had laws that prohibited discrimination on the basis of genetic tests. Most of those laws tend to be relatively weak, however, and tend to exclude a significant number of employees, employers, and insurance programs.

On a federal level, bills extending that kind of protection to all citizens have been pending in Congress since 1995, although no action has been taken on any of them. Those bills would prohibit health care companies from denying, canceling, or refusing to renew policies to anyone based on genetic testing. They would apply similar restrictions to the use of genetic testing in employment and firing.

References

Boyd, Robert S. "Genetic Progress Stirs Fears of Bias in Jobs, Insurance." *Oregonian,* 25 October 1996, A23.

Cranor, C., ed., *Are Genes Us? The Social Consequences of the New Genetics.* Brunswick, NJ: Rutgers University Press, 1994.

Findlay, Steven, and Mimi Hall. "Clinton Backs Bill to Block Genetic Bias." *USA Today,* 15 July 1997, 1A; and follow-up report on page 2A.

Hendren, John. "Gene Testing Avoided over Fears of Bias." *San Francisco Chronicle,* 16 April 1998, A4.

Holtzman, Neil A., and Michael S. Watson, eds. *Promoting Safe and Effective Genetic Testing in the United States: Final Report of the Task Force on Genetic Testing.* Baltimore: Johns Hopkins University Press, 1998.

Palmer, Kim. "Doctors Have Prenatal Test for 450 Genetic Diseases." *USA Today,* 15–17 August 1997, 1A+.

Rothenberg, Karen H., and Elizabeth J. Thomson, eds. *Women and Prenatal Testing: Facing the Challenges of Genetic Technology.* Columbus: Ohio State University Press, 1994.

Schwartz, Robert. "Genetic Knowledge: Some Legal and Ethical Questions." In *Birth to Death: Science and Bioethics,* edited by David C. Thomasma and Thomasine Kushner, 21–34. Cambridge: Cambridge University Press, 1996.

Genetically Manipulated Foods

Genetically manipulated (GM) (or altered) foods are foods that have been changed by adding to them one or more genes from another organism. The purpose of adding the genes is to change the character of the host organism in some desirable way. For example, scientists at the University of California at Davis announced in 1992 that they had developed a method for protecting plants that are normally attacked by caterpillars. The scientists extracted a gene from scorpions that produce a toxin that kills caterpillars. They inserted that gene into a variety of plants normally eaten by caterpillars. When caterpillars attacked those plants, they also ingested the transplanted scorpion gene. The gene killed the caterpillars.

Why Modify Plants Genetically?

The reasons for wanting to modify plants are legion. For example, one of the earliest experiments in genetic engineering involved the insertion of a gene into tomato plants that protects them from freezing. Plants carrying this gene are able to survive at much colder temperatures than plants without the gene.

One of the most popular forms of genetically manipulated plants is one that carries a gene providing resistance to herbicides. These plants can be grown at far less expense than can native plants lacking that gene. The reason is that farmers normally protect their crops from weeds by spraying with herbicides designed to kill the weeds. But most plants are either killed or damaged by those herbicides. Spraying must be done very carefully in order to avoid harming the crop itself. Crops that contain the herbicide-resistant gene, however, do not need such care. The herbicide can simply be sprayed everywhere. It will kill weeds but will not affect the crop itself.

Plants have also been developed that are resistant to a variety of pests. A swarm of pests may attack and begin eating such plants. But as they do so, they ingest the gene that produces a chemical toxic to them. The remaining portion of the crop is left unharmed.

Another purpose for altering plants genetically is to have better control over their harvesting and storage. For example, a new kind of tomato called the Calgene Flavr Savr tomato first appeared in stores in May 1994. This tomato contains a gene that is not natural to the plant, one that delays the ripening of the fruit. Since the tomato remains on the vine longer, it develops more flavor.

Some GM plants also have properties that protect them during storage. For example, a group of researchers announced in 1994 that they had introduced into peas a gene that produces a chemical toxic to weevils even after the plants had been harvested. As they lie in storage, ready for packaging and sale, the peas are protected from attack by weevils by the transplanted gene.

Increasing production is another goal of genetically modifying organisms. Perhaps the best-known example involves the hormone known as bovine growth hormone (BGH) or bovine somatotropine (BST). Cows treated with BST produce significantly larger amounts of milk than do cows that do not receive the hormone. At one time, the hormone could be obtained only from slaughtered cows and was, thus, in relatively short supply. In the 1980s, the Monsanto Corporation developed a method for producing the hormone from genetically-altered bacteria. As a consequence, the hormone is now readily and inexpensively available to dairy farmers.

A similar technique can be used with animals. In 1994, for example, researchers announced that they had introduced a growth hormone gene from sockeye salmon into coho salmon. The inserted gene caused the cohos to grow to unusually large sizes. One fish in particular was thirty-seven times as large as the average coho salmon.

Risks of GM Products

In the United States, authority to regulate GM plants and animals lies with the U.S. Food and Drug Administration (FDA). FDA makes its decisions on a case-by-case basis when approving such organisms for commercial use and sale. In general, its policy is that the item will be approved if it is indistinguishable from the natural form of the item. The problems for which the FDA looks are the presence of a toxic material or allergen within the GM food, changes in nutritive value of the food, or the use of antibiotic genes in the food. As of late 1998, more than forty different GM foods had been approved for commercial sale by the FDA. Those already on the market include GM forms of canola (rapeseed oil), cotton and cottonseed oil, soybeans, squash, and tomatoes. The remaining foods are currently in a state of further research or development.

A number of questions have been raised about the sale and consumption of GM foods, however. Some of these questions relate to their effects on human health and some on environmental effects. For example, studies have shown that some people are allergic to genetically modified forms of a food but not to natural forms of the same food. In 1996, for example, researchers reported on such a reaction among people who had eaten soybeans that carried a gene transplanted from a Brazil nut.

One of the major concerns expressed by critics is the presence of antibiotic genes in GM foods. The presence of these genes may reduce the effectiveness of antibiotics now in use and make humans more susceptible to a number of diseases.

Other observers emphasize the environmental dangers posed by GM plants. Agricultural crops and weeds do cross-pollinate, and it may happen that herbicide-resistant genes may be transferred from crops to weeds. If that were to happen, we might see the growth of very resistant "superweeds" in future years. Perhaps more to the point, it is always difficult to know exactly what will happen when new organisms are released into the environment. And GM plants do qualify as "new organisms" whose future effects no one can really specify at the moment.

Laws and Regulations

Regulatory and legislative bodies have not yet decided exactly how to deal with the issues of GM foods. Some critics have called for a ban on all such foods, citing too many unknown risks associated with them. Other authorities have asked for more rigorous controls on the approval of such foods. They claim that the FDA and other regulatory bodies have not taken seriously enough the potential threats these foods may pose to humans and the environment.

In 1998, the European Union (EU) decided to require the labeling of all GM foods provided that the transplanted genetic material can be detected in the final food product. This decision was applauded by some observers as a step in the right direction, although it left many issues unresolved. For example, the "genetic fingerprint" of transplanted material may not remain after a food has been processed. According to some experts, the EU decision would result in the labeling of only a small fraction of all products actually made from GM plants.

In the United States, the FDA still tends to take a somewhat laissez faire attitude toward the issue, arguing that a GM food that is chemically indistinguishable from a natural food should not have to be labeled as being "different" from it. As the number and variety of GM foods expand, however, one can imagine that the discussion over labeling and regulation might become more intense.

References

Benson, Susan, Mark Arax, and Rachel Burstein. "A Growing Concern." *Mother Jones,* January/February 1997, 36–43+.

Burros, Marian. "Trying to Get Labels on Genetically Altered Food." *New York Times,* 21 May 1997; also at http://www.genetic-id.com/nyt.htm. 8 July 1998.

Fagan, John. "Genetically Altered Food: Buyer Beware." *USA Today,* 6 March 1997, 13A.

"First Commercial Growing of a GE-crop in Europe." http://www.greenpeace.org/~geneng/main.html. 8 July 1998.

"Gene Genie." *The Guardian,* 4 June 1998; also at http://www.netlink.de/gen/Zeitung/1998/980604.htm. 8 July 1999.

"Genetically Engineered Food—A Serious Health Risk." http://www.netlink.de/gen/fagan.html. 8 July 1998.

"Manipulating the Genetic Debate." http://www.netlink.de/gen/Zeitung/1998/980609.htm. 8 July 1998.

Global Warming

Global warming refers to an anticipated increase in the Earth's annual average temperature expected to occur as the result of certain human activities. This change is caused by an increase in the amount of carbon dioxide and other so-called greenhouse gases released by activities such as the combustion of fossil fuels and the burning of tropical rain forests. An increase in the Earth's average annual temperature could conceivably have profound effects on the Earth, producing significant changes in regional climates and agricultural patterns and the loss of land masses adjacent to the oceans. Considerable scientific and political differences of opinion exist as to the causes and possible effects of global warming.

The term *climate change* is sometimes used. It provides a somewhat broader perspective on possible planetary changes than does the more restrictive term *global warming.* Climate change emphasizes that a number of major environmental changes may take place beyond a simple increase in the average annual temperature of the planet.

Greenhouse Effect

Sunlight that strikes the top of the Earth's atmosphere consists of radiation with a wide range of wavelengths, from about 0.1 μm (micrometer) to about 2.0 μm. Various forms of radiation experience different fates as they enter the atmosphere. Some are reflected back into space, some are absorbed by gas molecules in the atmosphere, and some penetrate the atmosphere to the Earth's surface. This pattern is determined by the ability of certain molecules in the atmosphere to absorb radiation of certain wavelengths. Water molecules, for example, tend to absorb radiation with wavelengths between about 0.8 and 2.0 μm. They do not absorb—and, therefore, are transparent to—radiation of other wavelengths.

Radiation that reaches the Earth's surface may be absorbed or reflected back into the atmosphere. During the process of reflection, the wavelength of the radiation is changed. For example, ultraviolet radiation that strikes the Earth's surface may be reflected in the form of infrared radiation (heat). This reflected radiation may pass through the atmosphere into outer space, or it may be absorbed by gas molecules in the atmosphere. One of the substances most efficient in absorbing reflected radiation is carbon dioxide. The capture of infrared radiation by carbon dioxide in the atmosphere is an important factor in maintaining the relatively high temperature of our planet compared to that of other similar planets.

Because of its general similarity to the process by which horticultural greenhouses are thought to trap heat, the above process is known as the *greenhouse effect.* The greenhouse effect is an entirely natural phenomenon in the Earth's atmosphere and differs in intensity only as a result of human activity.

Changes in Carbon Dioxide Concentrations in the Atmosphere

One might reasonably predict that human activities would have an effect on the intensity of the natural greenhouse effect. For more than two centuries, humans have been burning fossil fuels (coal, oil, and natural gas) at a prodigious rate. The two most important products of fossil fuel combustion are carbon dioxide and water. As fossil fuel combustion continues to grow, therefore, one might expect to find increasing concentrations of carbon dioxide in the atmosphere and a corresponding increase in the Earth's annual average temperature.

Such a scenario, however, is far too simplistic. The Earth's atmosphere and climate are influenced by many other factors and, as a result, are highly complex systems. Humans have barely begun to understand the operation of these systems.

For example, carbon dioxide produced by fossil fuel combustion does not simply escape into the air and remain there. Instead, some of the gas is captured in natural sinks, where it is removed from the atmosphere. A sink is a region where some natural resource is collected and stored. Some carbon dioxide may dissolve in the oceans, for example, and some is used by growing plants in the process of photosynthesis.

One of the most fundamental questions in studies on global arming, however, *is* a simple one: What trends are there, if any, in the amount of carbon dioxide in the atmosphere? Fortunately, scientists have a very nice set of data on this question obtained from an observatory on top of Mauna Loa in Hawaii. These data date back to the late 1950s and show a very clear and steady increase in the level of carbon dioxide over the observatory. That level has increased from about 315 ppm (parts per million) in the late 1950s to more than 355 ppm in the late 1990s.

Data from other sources are also available. For example, it is possible to obtain ice cores (long, cylindrical sections) from glacial ice sheets, such as those that cover Greenland. Analysis of air bubbles trapped in these ice cores shows that levels of carbon dioxide in the atmosphere have been increasing since the late 1700s. This finding confirms the simple-minded prediction that increasing use of fossil fuels on Earth corresponds to increasing levels of carbon dioxide in the atmosphere.

Other Greenhouse Gases

While the best known substance capable of trapping heat in the atmosphere, carbon dioxide is certainly not the only substance known to do so. Other gases capable of absorbing radiation in the infrared region are methane, nitrous oxide, chlorofluorocarbons (CFCs), and ozone. Collectively, these gases are known as *greenhouse gases.* All except the CFCs are produced by both natural and anthropogenic (human) means. The concentration of greenhouse gases in the Earth's atmosphere, like that of carbon dioxide, is increasing as a result of human activities such as fossil fuel burning and agricultural practices.

Temperature Increases

It is possible to predict temperature increases in the Earth's atmosphere based on increasing levels of carbon dioxide. For example, according to the best estimates now available, a doubling in the concentration of carbon dioxide in the atmosphere should produce an increase in the Earth's average annual temperature of 1.5 to 4.5 degrees C. This range of uncertainty is rather large and reflects our currently poor understanding of the effects of clouds and other natural phenomena on the global warming process.

With regard to the cloud problem, an increase in the Earth's average annual temperature results in an increase in the rate of evaporation from the world's oceans and lakes. As more water evaporates, the formation of clouds is likely to increase. A larger cloud covering for the planet, however, may result in an increase in the amount of sunlight reflected back into space. The overall effect of increased cloud formation, then, might actually be a decrease in annual average temperatures.

Measurements of temperature changes over the past fifty years or more are much more problematic than are those of carbon dioxide concentrations. These changes are relatively small (fractions of a degree), differ from place to place around the globe, and are difficult to determine with precision. Also, precise long-term measurements of temperature changes, such as those possible for carbon dioxide in ice cores, are generally not available.

Nonetheless, atmospheric scientists are now virtually convinced that the Earth is in a period of warming that has seen the atmospheric temperature rise by about 0.3–0.7 degrees C over the last century. The period from 1987 through 1997, for example, included nine of the warmest years ever measured. In addition, 1997 was the warmest year ever recorded scientifically.

Potential Effects

What are the possible effects of an overall, long-term increase in the Earth's annual average temperature? Perhaps the most obvious prediction is that large ice masses, such as those at the poles and in continental glaciers, would begin to melt. A significant fraction of the Earth's water now stored as ice would be converted to liquid form, raising the overall level of the oceans. By some estimates, an increase of only 1 degree C would result in a sea-level rise of about 25 centimeters.

Melting of polar and glacial ice is actually a less significant factor in ocean-level rise than is simple expansion due to heating. That is, as the Earth's temperature rises

and its oceans warm, they expand to take up more space. It is expansion, rather than melting, that is expected to produce the most important amount of sea-level rise as a result of global warming. According to the best predictions available, melting and expansion may result in a sea-level rise of 25–75 centimeters by the year 2100 and by 1–3 meters over a period of 500 years.

An increase of that magnitude might not be significant if it happened only once. If it were to continue over time, however, it might pose a serious problem for human settlements. Some of those most concerned about global warming are inhabitants of island nations whose land area is very flat. The highest point on some Pacific Ocean islands, for example, is no more than a few meters. Should global warming become an ongoing reality, those nations might actually disappear beneath the rising oceans.

Nations with more extensive land masses would also have reason for concern as a result of global warming. Many of the world's largest cities—New York City, London, Sydney, Honolulu, Rio de Janeiro, Miami, and Los Angeles, for example—would have large sections engulfed by rising sea levels. Untold millions of people would lose homes, agricultural lands, and infrastructure built along the ocean's coastlines.

Changes in agricultural patterns might also result from global warming. The effects of global temperature increases would probably not be uniform worldwide. They would be ameliorated by factors such as proximity to oceans, landforms (such as mountains or prairies), and soil type. Regions that are now agriculturally productive might become less so, while those now poorly adapted for agriculture might become more productive.

Proposed Solutions

Programs for dealing with global warming can be categorized into one of two general types: technological "fixes" and political solutions. A technological fix involves the use of some scientific technique for dealing with global warming. For example, some scientists have proposed the release of large amounts of powdered iron on the surface of the oceans. Iron appears to promote the growth of small, one-celled organisms called phytoplankton. Phytoplankton are a major consumer of carbon dioxide in the oceans. The spreading of iron might be one way, therefore, of reducing the amount of carbon dioxide present in the oceans and, ultimately, in the atmosphere.

Thus far, no technological fix has been proposed that appears to be economically feasible and to have earned widespread acceptance among scientists and politicians. Instead, most proposals for dealing with global warming have involved political solutions. The most obvious of these solutions is to get the nations of the world to agree to reduce the amount of carbon dioxide released to the atmosphere.

Such agreements have been very difficult to negotiate, however, given two major stumbling blocks. First, substantial disagreements remain among some scientists, politicians, and business leaders with regard to many aspects of the global warming debate, disagreements about both scientific data and the interpretation of those data. Second, the economic costs of such agreements could be very large indeed. Cutting back on carbon dioxide emissions means cutting back on fossil fuel emissions and, given the technological basis of modern society, that move could mean reducing industrial output. Neither businesses nor governments look forward to international agreements that call for cutbacks in their economies.

Nonetheless, concerns about global warming have grown to the point where two worldwide conferences have been held to discuss ways of dealing with the problem. The first of these conferences, the United Nations Conference on Environment and Development, was from June 3 to 14, 1992, in Rio de Janeiro. This conference became better known as the Earth Summit. Representatives from 178 nations attended the conference. One major result was an agreement among the world's industrial-ized nations to a voluntary program of industrial cutbacks designed to reduce carbon dioxide emissions by the year 2000 to their 1990 levels.

An important debate at the Rio conference focused on the role of developing nations in industrial cutbacks. Those nations argued that they should not be penalized for trying to attain a standard of living comparable to those of industrialized nations and should be subject to less severe restrictions on industrial development than were developed nations. Ultimately, developing nations were given a longer period of time to reduce their carbon dioxide emissions and were promised technological and financial help from industrialized nations to meet these goals.

The Kyoto Conference of 1997

It rapidly became clear that the voluntary restraints adopted at the Rio summit were going to be ineffective. The majority of industrialized nations—especially the United States—made no progress toward reducing carbon dioxide emissions. In fact, those emissions continued to increase in the years immediately following the Rio agreement. Concerned scientists and politicians soon began planning a second international meeting, one that they envisioned as producing more concrete and enforceable carbon dioxide emission standards. That meeting was scheduled for Kyoto, Japan, in December 1997.

At preliminary planning sessions for the Kyoto conference, four major issues evolved. The first concerned the greenhouse gases that were to be included in any final treaty. Some nations called for the listing of six gases—carbon dioxide, methane, nitrous oxide, hydrochlorofluorocarbons (HCFCs), perfluorocarbons, and sulfur hexafluoride—while other nations wanted only the first three to be included.

Debate on the second issue centered on the possibility of "emissions trading," the practice by which one nation could exceed its assigned carbon dioxide reduction limits provided it worked out an agreement with a second nation that produced less carbon dioxide by a comparable amount.

The third issue was once again that of limitations for developing nations, with those nations demanding less restrictive regulations than the ones placed on industrialized nations. Fourth, a crucial conflict arose over the specific size of the reductions to be set for individual nations, with decreases in carbon dioxide emissions of as much as 15 percent requested by the European Union compared to increases of as much as 20 percent called for by Australia.

When 159 nations met in Kyoto, the possibility of reaching an agreement on these issues seemed remote. Only a day before the meeting was to adjourn, various nations were still espousing positions so remote from each other that some observers were predicting that the meeting would end without a treaty.

Over the last twenty-four hours of the conference, however, chair of the conference Raúl Estrada-Oyuela, of Argentina, kept delegates in session until they reached compromises on all remaining issues. The final document included all six greenhouse gases, provided for the trading of emissions credits, and required no immediate commitments from developing nations for greenhouse gas reductions. Most important, it established specific reduction targets for thirty-eight industrialized nations, ranging from 6 percent below 1990 levels for Japan to 8 percent below those levels for members of the European Union.

Differences of Opinion

The preceding discussion does not begin to reflect the level of disagreement and controversy surrounding the issue of global warming. In fact, there are relatively few points in this whole question about which everyone is in agreement. One of these points of agreement is that almost no one doubts that human activities have substantially added to the concentration of greenhouse gases in the Earth's atmosphere or that the effect of those gases is to increase the average annual temperature of the

planet. Nor is there much doubt that the average annual temperature of the troposphere (the lowest layer of the atmosphere) has increased by a fraction of a degree over the last century.

Beyond these basic "facts," however, most statements about global warming need to be categorized as "probably true" or "possibly true." With regard to those statements, critics of various global warming scenarios have been able to suggest alternative scenarios and explanations that are also "probable" or "possible."

For example, critics point out that long-term variations in the Earth's average annual temperature are a natural phenomenon caused by changes in the Earth's orbit around the sun, its angle of inclination to the plane of the Earth's orbit, and other factors. It is entirely likely, they say, that temperatures changes observed over the last century are caused by nothing other than these natural fluctuations.

Also, critics argue that humans know far too little about major Earth systems, such as movement of air masses in the atmosphere, cloud formation, and ocean circulation, to predict the effects of very small temperature increases. They point out that most projections about the effects of global warming are based on computer models, models that are still based on questionable assumptions, that contain very large errors and uncertainties and make use of inadequate and sometimes incorrect data. Given these limitations, they warn, computer-generated models of the future can, in some circumstances, give dramatically different predictions with equal levels of reliability.

Uncertainties of Information and Political Decisions

The bottom line for many scientists and politicians interested in global warming is the question as to what policies to adopt and what actions to take given the level of uncertainties inherent in the process of global warming. Such policies and actions usually involve very substantial economic decisions and transformations. Agreements adopted at the Kyoto conference, for example, could conceivably require industries throughout the developed world to cut back in their operations or, at the least, to make huge investments in alternative technologies. Gail McDonald, president of the industry-based Global Climate Coalition, announced in 1997 that reducing greenhouse gas emissions would mean "canceling an expected 26 increase in America's economic growth."

Other agencies have drawn other conclusions about the effects of cutting back on greenhouse gas emissions. A study conducted by the World Resources Institute, for example, predicted an economy in 2020 that would be 75 percent larger than that of 1997 with emission controls versus one that would be 77.4 percent larger without such controls.

The fundamental question is how much, if at all, economic growth should be limited given the many uncertainties in our current understanding of the global warming process. On the one hand, nations might invest billions of dollars in order to reduce economic growth with the possibility that global warming was never a real problem produced by human activities. On the other hand, the failure to take political and economic action could conceivably result in an environmental catastrophe of planetary proportions if human activities *are* an important factor.

Finally, the factor that makes decisions among such options so difficult is the time frame of possible global warming effects. Left undisturbed, any anthropogenic effects on global climate probably could not be detected with any level of assurance for decades, if not centuries. By the time a degree of certainty satisfactory to all observers had been obtained, global changes would already have been set in motion that would be impossible to stop or reverse for very long periods of time.

References

Cushman, John H., Jr.. "Industrial Groups Plan to Battle Climate Treaty." *New York Times,* 26 April 1998, A1.

Gerstenzang, James. "Campaign to Halt Global Warming Moves Glacially." *Sunday Oregonian,* 14 June 1998, A6.

Hileman, Bette. "Global Climate Change." *Chemical and Engineering News,* 17 November 1997, 8–16.

———. "Global Climate Treaty Reached." *Chemical and Engineering News,* 15 December 1997, 9.

Mahlman, J. D. "Uncertainties in Projections of Human-caused Climate Warming." *Science,* 21 November 1997, 1416–1417.

Manning, Anita. "'90s Contain Hottest Years of Hot Century." *USA Today,* 23 April 1998, D1.

Miller, G. Tyler, Jr. *Environmental Science: Sustaining the Earth.* Belmont, CA: Wadsworth Publishing, 1991, Chapter 10.

Newton, David E. *Global Warming: A Reference Handbook.* Santa Barbara, CA: ABC-CLIO, 1993.

Pickering, Kevin T., and Lewis A. Owen. *An Introduction to Global Environmental Issues.* London: Routledge, 1994, Chapters 2 and 3.

Schneider, Stephen. *Global Warming: Are We Entering the Greenhouse Century?* New York: Random House, 1989.

Warrick, Joby. "5 Years Later, Leaders Gather for Ssecond Earth Summit." *Oregonian,* 23 June 1997, A4.

Grazing Legislation

Grazing is the primary mechanism by which cows, sheep, and other ruminants feed. These animals spend most of the day eating grass and small shrubs.

Grazing and North American History

One of the great natural resources possessed by the North American continent was a huge expanse of grassland covering the land between the Mississippi River and the Rocky Mountains. When pioneers first began to move into this region in the 1800s, they found an abundant supply of natural food for the cattle and sheep they brought with them. "Raising cattle," for those pioneers, was to a large extent a process of letting their animals loose on the open rangeland, collecting them only when it came time to bring them to market.

Issues of how best to use grazing lands are often resolved not on the basis of the best scientific information available but on the basis of social and political priorities. (Corbis/Wolfgang Kaehler)

For some time, it appeared that grazing was an ecologically harmless activity. There seemed to be far more grassland than any amount of cattle could ever consume. But by the 1930s the falsity of that argument became clear. Cattle were grazing some areas so aggressively that serious problems were appearing. When cattle are allowed to spend too much time in an area, they eat grass plants down to their bare roots. The grass dies and is replaced by weeds, bushes, and other plants inedible to cattle. As grass cover is removed, rains and wind begin to remove top soil, soil erosion begins, and lakes and rivers are contaminated with mud and silt.

In 1934, Representative Edward T. Taylor of Colorado introduced a bill to create certain rules and regulations for grazing on public lands. Taylor had originally been a fervent advocate of unlimited grazing rights for cattle ranchers, but he had seen firsthand the disastrous effects this policy was having on Western lands. The bill he sponsored was designed to permit the continuation of traditional grazing patterns but with certain restrictions designed to protect the environment from further destruction.

One of the most important features of the Taylor Grazing Act was the establishment of permits valid for up to ten years to be given to those who wished to graze cattle on public lands. In addition, cattlemen were required to pay a certain fee based on the kind and number of animals they had in a given area.

The Public Rangeland Improvement Act of 1978

The general principles established by the Taylor Grazing Act remained in effect for four decades. By the 1970s, however, environmentalists began to attack these principles aggressively. They pointed out a number of problems with traditional grazing policies. In the first place, the grazing fees charged by the Bureau of Land Management were very low. Those fees were usually about one-third to one-fifth of the fees charged for grazing on state and private lands.

In addition, critics argued, cattlemen were inclined to ignore sound environmental practices on the land they used. Various studies by governmental agencies showed that less than 10 percent of all public rangeland was in "good" condition and as much as a quarter was in "poor" condition.

In response to these concerns, the U.S. Congress passed the Public Rangelands Improvement Act of 1978 (PRIA), which called for an increase in grazing fees and a thorough study of federal grazing policies. The act was designed to expire in 1985, after which time, it was assumed, a new policy would be adopted by the Congress.

Post-PRIA Developments

A decade after PRIA expired, the anticipated new grazing policy had not yet been adopted. Congress made a number of attempts to increase grazing fees, but all such efforts failed. Then Secretary of the Interior Bruce Babbitt also tried to increase grazing fees by administrative order, but he was blocked from doing so by a suit brought by the National Cattlemen's Beef Association. Grazing reform went nowhere until the late 1990s.

The latest efforts to develop a new grazing policy occurred with bills offered by Senator Pete Domenici (R-N.Mex.) in 1996 and by Representative Bob Smith (R-Oreg.) in 1997. Domenici's bill was criticized by environmentalists for being an "industry" bill that weakened public control over grazing practices and maintained grazing fees at unrealistically low levels. It was not approved by the Senate.

Smith's original bill contained many of the features thought to be objectionable by environmentalists, such as the assignment of increased rights and responsibilities to ranchers and decreased involvement of the general public in grazing decisions. Even-

tually, Smith removed most of these provisions from his bill, and it passed the House on 30 October 1997.

The primary feature of the Smith bill was an increase in grazing fees from $1.35 per animal unit month (AUM) to $1.82 over a five-year period. An animal unit month is the amount of forage needed to sustain a single cow and her calf, one horse, or five sheep or goats for a month. The Smith bill included a number of other administrative provisions, such as regulations for payments to individuals who do not actually own cattle, but own their grazing rights, and for methods to be used to assess land quality on rangelands.

Many environmentalists were very dissatisfied with Smith's bill. They pointed out that the proposed grazing fee of $1.82 was about a fifth of the highest rate of $10.92 charged by some states. At such low rates, cattlemen had little incentive to improve grazing practices that had caused serious environmental damage in the past, they argued, and that would continue to cause such damage in the future.

References

"Action Alert: Domenici Bill Would Hurt Rangelands, Waste Money." http://www.foe.org/act/grazing1.html. 25 July 1998.

Batt, Tony. "Bill Aims to Stabilize Grazing Fees." *Las Vegas Review Journal*, 18 September 1997; also at http://www.lvrj.com/lvrj_home/1997/Sep–18–Thu–1997/business/6081423.html. 25 July 1998.

"Federal Grazing Legislation." http://www.beef.org/librfacts/fs_grazing_legislation.htm. 25 July 1998.

"Grazing Bill and Proposed Legislation Bills." http://www.defenders.org/rn091896.html. 25 July 1998.

Kupchella, Charles E., and Margaret C. Hyland. *Environmental Science*. Boston: Allyn and Bacon, 1986, 532–533.

"Livestock; Grazing." http://www.nfu.org/html/issue_analysis_24.html. 25 July 1998.

Petulla, Joseph M. *American Environmental History*. San Francisco: Boyd and Fraser, 1977, 323–324.

Halons

Halon is a commercial name for a group of chemical compounds with molecules that contain carbon and one or more halogen atoms. The halogens are fluorine, chlorine, bromine, and iodine. The name Halon™ itself, for example, is the trademark for a polymer made from tetrafluoroethylene by the Allied-Signal Chemical Company. The compound known as Halon 1211™ is the trademark for bromochlorodifluoromethane ($CBrClF_2$) produced by the Great Lakes Chemical Company. The numbers that are used as suffixes following the word *halon* are part of a system developed by the chemical industry for identifying specific compounds in the family.

In the 1970s and 1980s, the halons were implicated in the destruction of ozone in the Earth's atmosphere. Their production was at first limited and later banned by the Montreal Protocol on Substances that Deplete the Ozone Layer of 1987 and its amendments.

See also: Chlorofluorocarbons; Ozone Depletion

References
"Questions and Answers on Halons and Their Substitutes." http://www.epa.gov/docs/ozone/title6/snap/hal.html. 23 June 1998.

Hazardous Wastes, International Dumping

A hazardous waste is any material left over from some commercial, industrial, agricultural, or other process that poses a substantial threat to human health or the environment when improperly handled. Some examples of hazardous wastes, as defined by the U.S. Environmental Protection Agency (EPA), are heavy metals, such as arsenic, barium, cadmium, lead, and mercury; certain pesticides; wastes from chemical processing operations, such as petroleum refining and the manufacture of chlorine; and by-products of the manufacture of many synthetic chemicals, including those used in pharmaceuticals, automotive products, and cosmetics.

At one time, hazardous wastes were treated in much the way as nonhazardous wastes. That is, they were buried in landfills, incinerated, or dumped into the ocean. Those practices long ago became unsatisfactory for two reasons. First, the materials disposed of in this way soon migrated into groundwater, drinking water sources, and other parts of the environment from which they caused damage to human health and/or other organisms in the environment. Second, the volume of hazardous wastes became so great that, even under the very best circumstances, traditional disposal methods were insufficient to deal with the growing problem they presented.

Overseas Disposal
By the 1980s, the problem of hazardous waste disposal became a serious issue for many companies and municipalities. The problem was that ocean dumping was prohibited, very little space remained for the burial of these wastes, and very severe con-

Deciding what to do with hazardous wastes becomes more and more of a problem as increasing quantities of such materials are produced by industrial processes and by our culture's "use-it-up, throw-it-out" philosophy. (Corbis/Michael S. Yamashita)

ditions had to be satisfied in order for burial to occur when space *was* available.

At that point, the possibility of shipping hazardous wastes out of the United States to other nations began to loom as a solution to the problem. One rationale was that many countries in the world had large amounts of land under which these wastes could be buried. Also, many countries were delighted with the possibility of earning money by accepting hazardous wastes from other nations.

This viewpoint was expressed somewhat bluntly in an internal memo by Lawrence Summers, then chief economist of the World Bank. The memo is reputed to have said: "I think the economic logic behind dumping a load of toxic waste in the lowest wage country is impeccable and we should face up to the fact that . . . underpopulated countries in Africa are vastly under-polluted." "Just between you and me," Summers went on to say, "shouldn't the World Bank be encouraging *more* migration of the dirty industries to the LDCs (less developed countries)?" (as cited in Foster 1993, 10).

Examples

Many examples could be cited of the dumping of hazardous materials in developing countries by industries or municipalities from the developed world. For example, the city of Philadelphia hired a Norwegian company, Bulkhandling, Inc., in 1988 to transfer 15,000 metric tons of toxic incinerator ash to Kassa Island in the African nation of Guinea. At first, the Guineans were delighted at the profit they made in accepting Philadelphia's toxic wastes. But they soon changed their mind. Both plants and animals on Kassa Island began to die off as a result of exposure to the toxic materials deposited there. Eventually, the Guinean government changed its mind and ordered the toxic wastes returned to the United States.

The government of Benin has had a particularly long history of dealing with those wishing to dispose of their hazardous waste materials. The nation is very poor and has long struggled to find ways of balancing its budget and making payments on its enormous foreign debt. As early as 1984, Gen. Mathieu Kerekou, ruler of Benin, negotiated a deal with the Soviet Union in which several tons of that nation's radioactive wastes would be deposited in Benin. It appears that some of that waste was used as fill under a new runway near Abomey, Benin's third largest city.

External pressures forced the Benin government to reconsider this arrangement with the Soviet Union. But the lack of any progress in strengthening its economy led to additional deals later. In 1988, for example, the country negotiated an arrangement with the government of France, agreeing to accept radioactive and industrial wastes in return for a cash payment of $1.6 million and an extended program of economic aid. In the same year, the Benin government signed a similar arrangement with the Anglo-American company, Sesco-Gibraltar. In this deal, Sesco-Gibraltar received permission to stockpile 50 million metric tons of hazardous wastes in Benin over a ten-year period.

The Basel Convention

By the late 1980s, individuals, environmental groups, and national governments were expressing outrage over the practice of dumping hazardous wastes in developing nations. The result of that outrage was a meeting held in Basel, Switzerland, attended by more than 100 nations. In March 1989, these nations agreed to the Basel Convention on the Control of Transboundary Movements of Hazardous Wastes and Their Disposal. As of 1998, 117 nations, including the United States, had ratified that treaty.

The major feature of the Basel Convention was an agreement that anyone (a company or municipality) wishing to ship hazardous wastes to some other country must first receive written approval from that country. Any country that chooses not to accept hazardous wastes is allowed to refuse shipment of such materials.

On the surface, the Basel Convention would appear to be a satisfactory solution to the problem of hazardous waste dumping. Not all observers felt that way, however. Some environmental groups and certain nations felt, for example, that the Basel Convention legitimized a practice that should, instead, be banned. They argued that there are no circumstances under which the producers of hazardous wastes should be allowed to dump those materials on some innocent party's land. Some African nations refused to ratify the Basel Convention for this very reason.

Later efforts were made, therefore, to modify the Basel Convention to express more strongly the objection of developing countries and other interested parties to the practice of hazardous waste dumping. In December 1992, for example, certain members of the convention adopted an amendment to the original document asking industrialized countries to prohibit the export of hazardous wastes to developing countries. This statement was stronger than the original convention because it did not provide conditions under which dumping could occur, suggesting instead that the dumping not occur at all.

Two years later, an additional amendment was passed banning the export of all hazardous wastes from the Organization for Economic Cooperation and Development (OECD) to non-OECD nations. A year later, the same ban was extended to all nations that had signed the original convention. As of 1998, however, only seventeen nations had signed the later amendments to the convention. Of those, sixteen were members of the OECD. The United States has not signed the later amendments.

Supporters of the Basel Convention and its amendments are hopeful that much wider support for these changes will soon develop. That support has been held up as nations argue over the exact meaning of "hazardous wastes" and the nations to which the amendments will apply. As decisions on those issues are reached, the amendments are likely to be ratified by more nations.

Another stumbling block, however, is opposition from a number of businesses and business groups (such as the U.S. Chamber of Commerce) and from some major countries (such as the United States, Canada, and Australia). These groups and nations are still faced with the very realistic issue as to what they are to do with the hazardous wastes being produced in industrialized nations. Land is still lacking on which to dispose of those wastes, and the option of using land in less developed nations is still appealing.

See also: Environmental Justice

References

Alter, Harvey. "Halting the Trade in Recyclable Wastes Will Hurt Developing Countries." Chapter 9 in *Environmental Justice*, edited by J. S. Petrikin. San Diego: Greenhaven Press, 1995.

"The Basel Ban—A Triumph for Global Environmental Justice." http://www.ban.org/Library/briefing1.html. 22 July 1998.

"Benin Hazardous Waste." http://gurukul.ucc.american.edu/TED/BENIN.HTM. 24 July 1998.

"Earthlife Africa Stops Waste Ship." http://www.gem.co.za/ELA/cupric.html. 24 July 1998.

Foster, John Bellamy. "Let Them Eat Pollution." *Monthly Review*, January 1993, 10–20.

"The Global Waste Trade." http://www.zerowasteamerica.com/WasteTrade.htm. 22 July 1998.

The Greenpeace International Toxics Campaign. http://xs2.greenpeace.org/ctox.html. 21 May 1999.

Mpanya, Mutombo. "The Dumping of Toxic Waste in African Countries: A Case of Poverty and Racism." Chap. 15 in *Race and the Incidence of Environmental Hazards: A Time for Discourse*, edited by Bunyan Bryant and Paul Mohai. Boulder, CO: Westview Press, 1992.

Norris, Ruth, ed. *Pills, Pesticide and Profits: The International Trade in Toxic Substances*. Croton-on-Hudson, NY: North River Press, 1982.

U.S. House Committee on Energy and Commerce. *Basel Convention on the Export of Waste: Hearing before the Subcommittee on Transportation and Hazardous Materials*. 102d Cong., 1st sess., 10 October 1991.

Headwaters Forest

The Headwaters Forest is a region of land near Eureka, in Humboldt County, northern California. The region covers an area of about 30,000 hectares (76,000 acres) and is owned by the Pacific Lumber Company (Palco). The name Headwaters Forest was originally given to the region in 1986 by a group of environmentalists who had made the decision to save the land from further development by Palco. That decision led to a decade-long controversy between lumbering interests and environmentalists that at times grew ugly and violent.

The Headwaters Forest as a Natural Resource

The Headwaters Forest is part of a long strip of forests running parallel to the Pacific Coast shoreline from southern Oregon to southern California. The region covers an estimated 680,000 hectares (1.7 million acres) and is an ideal habitat for the giant coastal redwood (*Sequoia sempervirens*). These trees grow very rapidly, often adding up to a meter or more of height per year. They also tend to have very long lives, reaching more than 2,000 years during their lifetime. The redwoods are among the oldest living objects on the Earth's surface. The Pacific redwoods make up a range of trees that forms a very stable ecosystem known as an *old-growth forest*.

Mature redwoods are a highly desirable commodity. Their wood is strong and close grained, superior to younger trees of the same species. Under the best circumstances, a mature redwood can bring up to $150,000 per tree on the lumber market.

The Headwaters Forest controversy in California illustrates how difficult environmental issues can be made even more complex by business decisions unrelated to the environment itself. (Corbis/Galen Rowell)

Economic Profit versus Environmental Protection

For many years, Palco was widely respected for its moderate approach to the taking of redwoods on its lands. It carried out selective cutting, in which only certain individual trees are removed from an area, followed by the planting of young seedlings. By using this practice, the company ensured that forested regions would not be wiped out as a result of lumbering practices.

That pattern changed in 1986, however, when Palco was purchased by Houston financier Charles Hurwitz. Hurwitz gained a reputation as a "corporate raider" for his hostile takeovers of a number of companies, one of which was the United Savings and Loan of Texas (USAT). When USAT went bankrupt in the savings and loans debacle of the late 1980s, Hurwitz was accused by the Federal Deposit Insurance Corporation of financial mismanagement. He was threatened with fines of more than $750 million.

The purchase of Palco by Hurwitz dramatically altered the financial status of both parties. Palco, renamed Maxxam after the takeover, went from a debt-free company to one with more than $700 million in high-interest debt. Hurwitz became owner of a company with timber assets estimated in the billions of dollars. Hurwitz decided to convert some of those assets into cash as quickly as possible. Logging in old-growth forests by Maxxam was accelerated significantly.

Concerned about the possible loss of the old-growth forests in northern California, environmentalists pursued every form of protest they could imagine. These ranged from lawsuits and proposed new legislation to tree-sitting and physical civil disobedience on logging sites.

Environmentalists were aided in their efforts by the existence of a number of threatened and endangered species in the region. One of these, the marbled murrelet, makes its nests only in old-growth forests along the Pacific Coast between Monterey Bay and Puget Sound. It had been declared an endangered species and was, therefore, protected by federal law. Coho salmon, a threatened species, also spawns in rivers and streams that pass through the redwood forests. Environmentalists argued that species such as these would be wiped out if Maxxam proceeded with its plans to log the old-growth redwoods.

Compromise and Agreement

For nearly five years, the federal government, the state of California, Maxxam, and environmentalists attempted to find common ground over the Headlands Forest issue. A number of near agreements fell through before the controversy was finally resolved in 1996. The agreement reached on September 28 of that year called for the state and federal government to purchase 2,200 hecatres (5,600 acres) of Maxxam-owned land in the Headwaters region for $380 million dollars. The land includes both the Headwaters Forest itself, the nearby Elk Head Springs Forest, and adjacent "buffer lands" between the protected areas and forests that Maxxam plans to log. The agreement was to take effect on 1 March 1999 and last for fifty years.

As with any agreement of this type, a number of fine points had still not been resolved when it was signed. For example, the state was required to approve a Sustained Yield Plan prepared by Maxxam outlining its future logging practices. The federal government was also expected to approve a long-term Habitat Conservation Plan. Two years after signing the original agreement, however, very little progress had been made in obtaining either of these agreements.

In apparent frustration, Maxxam announced that it would begin *salvage logging* inside the protected area. Salvage logging is a process by which fallen trees are removed from an area without disturbing standing trees. Maxxam expressed the view that this plan did not thwart the intent of the original agreement. Environmentalists, however, questioned whether salvage logging could be carried out without causing serious damage to standing trees and disruption of the forest floor with new logging roads. As Maxxam actually began work, it also appeared to be violating some specific conditions of the September 1996 agreement, such as the distance from streams within which it would remove trees. Critics said that the company's logging practices were resulting in erosion of stream banks with the consequent muddying of stream waters and damage to salmon spawning.

The September 1996 agreement proved to be a shaky one. Over the next two years, first one side and then the other threatened to withdraw from the deal. In December 1998, for example, the California Senate announced that it would require additional environmental protections from Maxxam. It said it would not authorize any state money for the purchase of Headwaters without such protections.

Two months later, Maxxam weighed in with its own ultimatum. It had decided that the $480 million price to be paid for its lands was inadequate. Unless the price was increased by an unspecified amount, it said, the sale was off.

Only 24 hours before the agreement was to take effect it was still unclear as to whether the two sides could agree on a final compromise. Were that not to happen, the federal government's share of the purchase price—$250 million—would expire, and most observers felt that Washington's offer would not be approved again.

At the last moment, negotiators from the state and federal government and from Maxxam hammered out an agreement. The final plan retained the $480 million purchase price for 3,020 hectares (7,470 acres) of land and imposed timber-cutting restrictions on an additional 330 square miles of old-growth forests owned by the company.

Some environmentalists were still unhappy with the deal. They thought it did not go far enough in protecting other parts of the redwood forests. For many involved in the decade-long controversy, however, the agreement was good enough to permit a long sigh of relief.

References

Cockburn, Alexander. "The Feds Blinked on Redwood Swap." *Los Angeles Times,* 2 August 1996, B9.

Diringer, Elliot. "Cutting a Deal on Redwoods." *San Francisco Chronicle,* 4 September 1996, A1+.

Gerstenzang, James. "Logging Deal Puts Headwaters Forest off Limits." *Los Angeles Times,* 28 February 1998, A1.

"Headwaters Forest and the Pacific Lumber Company." http://www.palco.com/hforest.htm. 29 June 1998.

The Headwaters Forest Website. http://www.envirolink.org/orgs/headwaters/index.html. 29 June 1998.

Headwaters Interactive News and Actions. http://www.envirolink.org/orgs/headwaters-ef/. 29 June 1998.

Headwaters Murrelet Project Page. http://sei.org/headwaters.html. 29 June 1998.

Meehan, Brian T, "Scientists Seek to Understand the Murrelet." *Sunday Oregonian,* 4 August 1996, A1+.

"Miller Questions Impacts of Headwaters Deal." http://www.house.gov/resources/105cong/democrat/press/relea306.htm. 29 June 1998.

"Redwood Deal Is Just a Start." *Los Angeles Times,* 15 May 1998, B8.

Health Hazards of Electromagnetic Fields

Electromagnetic (EM) radiation is a form of energy consisting of pulsating electrical and magnetic fields traveling through space at right angles to each other. Many forms of electromagnetic radiation exist, including visible light, ultraviolet radiation, infrared radiation, X-rays, gamma rays, radio waves, radar, and microwaves.

These forms of EM radiation differ from each other on the basis of their wavelength and frequency. The wavelength of an EM wave is the distance between two adjacent troughs or crests of the wave. The frequency of an EM wave is the number of wave crests or troughs that pass a given point in space per second. Wavelengths of EM radiation range from about 0.01 nanometers, or billionth of a meter, for gamma rays to more than 10,000 meters for radar waves. The frequencies of various forms of EM radiation range from about 100,000 hertz, or cycles per second, to about 10^{19} (a hundred trillion trillion) hertz for gamma rays.

Possible Health Effects from EM Radiation

Most forms of high-frequency EM radiation have long been known to be harmful to living cells. Both X-rays and gamma rays, for example, destroy cells and cause illness or death in both plants and animals. Until somewhat recently, however, there was little or no evidence that low-frequency EM radiation had any such effects on the health of organisms.

That situation changed in the 1960s. Men and women in the Soviet Union who worked around high voltage electrical equipment began to complain of fatigue, headaches, appetite loss, insomnia, and reduced sexual drive. Scientific studies designed to find out if electrical or magnetic fields actually caused these problems were not able to answer that question. However, they did create interest about the issue in other nations of the world, including the United States.

By the 1970s, residents of some parts of the United States were beginning to wonder if their health was endangered by overhead power lines. In some cases, citizens expressed their fears before regulatory agencies. In other instances, people took matters into their own hands through protests against companies who owned or used power lines.

In one case, judges in New York State granted approval for the construction of two new power lines about which nearby residents had complained. But the judges also acknowledged that evidence about possible health effects from the lines was too strong to ignore completely. In another case, people living near a proposed new power line in Minnesota and North Dakota tried to sabotage the project. They caused

Report after report shows no health effects from exposure to low levels of electromagnetic radiation, but such reports do not reassure critics who believe that such effects really do exist. (U.S. Bureau of Reclamation)

more than $9 million in damage before the line was actually completed in 1978.

Results of Scientific Studies

At first, most scientists did not take very seriously the idea that low-frequency electrical and magnetic fields (EMF) could affect the health of humans and other animals. No existing theory or body of research existed to support that idea. But evidence gradually began to accumulate. In 1979, for example, two scientists reported that children living near high-voltage power lines were more likely to develop cancer than those who did not.

By the end of the 1990s, dozens of studies had been completed on this question. In fact, a bibliography on electromagnetism

and health compiled by Richard W. Woodley for the Bridlewood Residents Hydro Line Committee included 754 items on the subject. Some of these studies examined the effect of EMF on cancer rates. Others studied the relationship between EMF and reproductive problems. In some cases, researchers reviewed the health records of people who lived or worked near high levels of low-frequency EMF. Others examined the direct effects of EMF on individual cells, tissues, or experimental animals.

For many years, the results of these studies were inconclusive. Some showed a correlation between high levels of low-frequency EMF and cancer rates or levels of reproductive problems. But scientists know that a high correlation does not necessarily mean a cause-and-effect relationship. The fact that people who live near power lines have a higher-than-average rate of cancer does not automatically mean that the power lines *caused* the cancers. Some third factor, of which scientists are not aware, may also have been involved.

Other studies showed no correlation between EMF and health conditions. They suggested that people who live near high-intensity EMF sources are as healthy as those who do not.

Between 1996 and 1998, three major studies of the health effects of EMF were reported. They were a review of the literature by a research team from the National Research Council of the National Academy of Sciences of 1996, the National Cancer Institute's Linet Study of 1997, and the Royal Adelaide Hospital (Australia) ELF Mice Study of 1998. All three of these studies concluded that there was no basis for concern about the health effects of EMF on humans or other organisms.

Even these supposedly "conclusive" studies, however, have not put to rest concerns about the health effects of low-frequency electrical and/or magnetic fields. One review of these three studies, for example, concluded that "When one considers the totality of current evidence now available, there are indications that a risk [to human health] may indeed exist. That risk may be small when compared to other risks in our modern society, but important due to its pervasive nature and possibility of serious consequences for affected individuals" (Maisch and Rapley 1998). Among the "affected individuals" who should still be concerned about this issue, according to this review, are physicians, patients, and other medical workers who are continually exposed to high-intensity low-frequency electrical and/or magnetic fields.

It appears that even the most recent and supposedly definitive studies on this issue have not totally put to rest the concerns of at least some individuals about the health effects of EMF.

References

Beardsley, Tim. "Guessing Game." *Scientific American,* March 1991, 30–32.

Black, Pam. "Rising Tension over High-tension Lines." *Business Week,* 30 October 1989, 158–160.

Brodeur, Paul. "The Annals of Radiation." *The New Yorker,* 12, 19, and 26 June 1989, 51–88, 47–77, and 39–68.

"The Electromagnetic Radiation Health Threat, Parts I and II." http://www.nzine.co.nz/features/neilcherry.html. 8 and 15 May 1997, 25 June 1998.

LaMacchia, Diane. "Study of Biological Effects of Electromagnetic Radiation Inconclusive." http://www.lbl.gov/LBL-Science-Articles/Archive/electromagnetic-radiation-study.html. 25 June 1998.

Maisch, D., and B. Rapley. "Powerline Frequency Electromagnetic Fields and Human Health—Is It the Time to End Further Research?" http://www.tassie.net.au/emfacts/3studies.html. 25 June 1998.

Marshall, Jonathan. "Cell Phones Linked to Lymphoma in Mice." *San Francisco Chronicle,* 9 May 1997, A1+.

Morgan, M. D., et al. "Power-frequency Fields: The Regulatory Dilemma." *Issues in Science and Technology,* Summer 1987, 81–100.

Noland, David. "Power Play." *Discover,* December 1989, 62–68.

Overbeck, Wayne. "Electromagnetic Fields and Your Health." http://www.arrl.org/news/rfsafety/wo9404.html. 06/25/98.

U.S. Congress, Office of Technology Assessment. *Biological Effects of Power Frequency Electric and Magnetic Fields: Background Paper.* 1989.

Woodley, Richard W. "Bibliography on Electromagnetic Radiation and Health." http://www.envirolink.org/seel/emf.html. 25 June 1998.

One of the best single sources of information about this topic is the Electric Power Research Institute. Its journal, *EPRI Journal,* has carried at least three major articles on this topic. They are:

Douglas, John. "Electromagnetic Fields and Human Health." *EPRI Journal,* July-August 1984, 14–21.

Moore, Taylor. "Pursuing the Science of EMF." *EPRI Journal*, January-February 1990, 4–17.

Shepard, Michael. "EMF: The Debate on Health Effects." *EPRI Journal*, October-November 1987, 4–15.

ELECTROMAGNETICS Forum is a quarterly publication dealing specifically with the environmental effects of electromagnetic radiation. It is available from EMFacts Information Service, P.O. Box 96, North Hobart 7002, Tasmania, Australia, or on-line at http://www.tassie.net.au/emfacts/.

HIV

See HIV and AIDS; Needle Exchange Programs

HIV and AIDS

HIV is the abbreviation for the human immunodeficiency virus, the agent that causes the disease known as AIDS. AIDS, in turn, is the abbreviation for the full name of the disease, acquired immunodeficiency syndrome.

AIDS first appeared in the United States in about 1981. Doctors in New York City began to see a few patients with symptoms of a disease known as Kaposi's sarcoma (KS). KS was a well-known but rare disease. Typically, it occurred in older men of Mediterranean ancestry.

At about the same time, doctors in Los Angeles had begun to see a few patients with a rare form of pneumonia known as Pneumocystis carinii pneumonia (PCP). This disease was typically seen in patients who were recovering from chemotherapy treatments with damaged immune systems. It was very rare in patients with healthy immune systems.

The striking fact about these two sets of observations was that, in all cases, the patients presenting with KS or PCP were young homosexual (gay) men. All had been healthy until a few months before they saw their doctors. All lived a relatively short time before dying.

Learning about AIDS

For about four years, scientists were mystified as to the cause of AIDS and its method of transmission. This uncertainty was reflected in the disagreement among medical workers as to what even to call this new disease. One suggestion was gay-related immune deficiency syndrome (GRID). Another was community-acquired immune deficiency syndrome (CAIDS). Both terms reflected the fact that AIDS appeared to occur most commonly among gay men.

However, these terms were unsatisfactory for two reasons. First, scientists discovered early on in the AIDS epidemic that individuals other than gay men can contract the disease. Drug abusers who share needles, for example, can also become infected. Also, nongay men and women who have sexual contact with those infected can also contract the disease.

In addition, the terms GRID and CAIDS both seemed to suggest that the new dis-

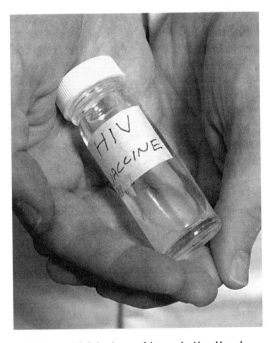

An HIV vaccine might help solve one of the worst health problems the world has ever seen, but what new social, educational, and political issues would it raise among men and women who engage in unsafe sexual or drug practices? (Corbis/Bryn Colton; Assignments Photography)

ease was somehow associated with a particular group of people, specifically gay men. While it was certainly true that gay men were the first individuals to be diagnosed with the disease, it was misleading to call the new condition a "gay disease." That terminology seemed to suggest that the disease somehow sought out and infected only certain groups of individuals. Most people understand that infectious diseases are not respecters of one's gender, age, ethnic background, sexual orientation, or other trait.

Eventually the name acquired immunodeficiency syndrome was selected for the disease. The name suggests that the primary problem associated with the disease is that one's immune system becomes severely compromised, and one's body is unable to fight off infections that would be of little or no consequence to a person with a healthy immune system.

In 1984, the long search for the causative agent for AIDS came to an end. A French research team led by Dr. Luc Montagnier discovered that AIDS is caused by a virus, which the team named the human immunodeficiency virus. Researchers soon confirmed that HIV is transmitted between two individuals almost exclusively as the result of the exchange of body fluids, blood, or semen.

Today, physicians try to distinguish among the various stages that constitute an HIV infection. Those who have been infected with the virus are said to be HIV positive. At one time, it was thought that everyone infected with HIV was destined to die. The first stage of the anticipated sequence was a period of time during which a person was infected but showed no outward signs of the disease. Eventually, as the disease progressed, a person became more and more ill and eventually died of KS, PCP, or some other disease.

Today, it is clear that a person who is HIV positive is not living under a death sentence. Some people have now been infected for nearly two decades and are living reasonably healthy and normal lives. Other people are living similar lives because of the availability of a number of drugs that have been developed to treat HIV infection.

Public Reaction to People with AIDS

The appearance of AIDS in the 1980s set into motion one of the most dramatic and instructive episodes in public health in recent human history. To begin with, the disease itself was very frightening. For nearly five years, the medical community was not entirely certain as to how it was transmitted. For a period of time, some experts thought that the virus could be transmitted by mosquitoes, by sitting near an infected person, or by talking with such a person. Even today, some scientists dispute the generally held view that AIDS is caused by HIV and believe that some other factor is involved.

The fear of the general public instilled by AIDS was augmented by the population of individuals most affected by the diseases: gay men and drug abusers. One would be hard-pressed to find two groups within society in the United States who are more disliked and feared. As terrified as they might be of their own security, many nongay, non–drug user Americans were inclined to feel that AIDS was somehow a less serious health problem because it affected individuals with low social status.

In fact, it was not uncommon early in the AIDS crisis for religious leaders to point out that the disease was "God's punishment" for the immoral lifestyle led by those who became ill. One minister in Nebraska developed a certain notoriety for attending funerals of men who had died of AIDS, preaching to the assembled family that the man's death was punishment for his sin. Signs reading "Thank God for AIDS" were not uncommonly seen along gay rights marches and other public assemblies of gay men and lesbians.

Issues Surrounding HIV Infection

Under these circumstances, it is hardly surprising that the progress of HIV infection in the United States and the rest of the world over two decades has been fraught with controversy. At the outset, medical researchers found it very difficult to obtain adequate financial funding for the studies of HIV that were necessary to find a cure or treatment for the disease. In fact, the story as to how gay men and lesbians organized to exert pressure on the federal government to obtain funds for AIDS has become something of a classic in the annals of community organizing. Groups concerned with specialized health issues, such as breast cancer, say that they have learned a great deal as to how to get the government's attention as the result of the work of AIDS activists.

Some of the most difficult issues surrounding HIV infection have now been settled or at least significantly resolved. For example, there was a vigorous debate at one point as to whether children infected with HIV should be allowed to attend public schools. Many parents and educators were terrified that the presence of such a child in the schoolroom would pose a threat to the health of those around him or her. As an understanding of HIV infection has improved, those fears have largely receded, although they have not entirely disappeared.

Similar concerns were expressed about the possibility of an HIV-positive individual's participating in a competitive sport. Some people worried that an injury and loss of blood would increase the likelihood that HIV would be transmitted from the infected person to teammates or opponents. Largely as the result of the efforts of "Magic" Johnson, who announced in 1991 that he was HIV positive, that fear has also been significantly reduced.

However, other issues surrounding HIV and AIDS remain. Some are still contentious points whose final resolution is still not clear. Others are still the subject of some debate but appear to be on their way to settlement.

Testing for HIV

Within a year of the discovery of HIV, a test had been developed to detect the presence of HIV infection in a person. The test was no sooner available than questions began to arise as to who should take the test and what should be done with the information.

At first, many gay men and drug abusers saw no point in taking the HIV test. Even if they discovered that they were positive, there was essentially no help available. They could expect, in most cases, to die a long, lingering, and unpleasant death no different from that of the person who had never been tested.

Today, that argument no longer holds true. Researchers have developed—and are continuing to develop—drugs and drug combinations that can make it possible for a person infected with HIV to live a relatively long and healthy life. Testing, therefore, is an essential early step for those who engage in behaviors that place one at risk for HIV infection, such as homosexual acts and sharing of needles.

Testing involves an even more profound social issue, however. Many medical experts believe that the results of HIV tests should be reported to state health departments. This practice is followed with every other major communicable disease, they say. It is important for public health workers to be able to know which persons have which communicable diseases. It is only with this knowledge that they can prevent the spread of the disease.

Other medical experts disagree. They acknowledge that knowledge about HIV-infected individuals can be helpful to the public health community. But they point out the risks associated with collecting, storing, and disseminating these data. The majority of people infected with HIV, after all, are not "just any old citizen." They are people despised by society who, in many cases, spend much of their lives hiding who they are, what their sexual orientation may be, what kind of lives they live, and so on. The risk that this information would become generally available is too high to permit the sharing of HIV status information, these experts say.

As of early 1999, more than half of all states had laws requiring that the names of HIV-positive individuals be reported to state health departments. The two states with the largest number of HIV-infected individuals—California and New York—had still not passed such a law, however.

Dangers to Medical Workers

One group that has been particularly concerned about the issue of HIV testing has been the medical profession and others in health-related fields. Suppose that an emergency room doctor is confronted with a young man bleeding profusely from a car accident. What if that man should be HIV positive? Should the physician have the right to test the patient to determine his HIV status?

In the past, many medical workers have argued that the answer to this question is yes. They say that they deserve to take all necessary precautions from becoming infected themselves while treating this patient. In some cases, physicians, dentists, and other medical workers have actually refused to treat some patients whom they knew or believed to have been infected with HIV or, at the very least, at risk for the disease.

Most medical authorities now believe that this position is indefensible. Medical workers already are expected to use reasonable and prudent methods of protection whenever they treat a patient. They have learned to assume that a patient has some deadly disease and to take every possible precaution to protect themselves from contracting that disease. Health workers who simply practice good medical technique, these authorities say, will protect themselves against HIV infection.

Dangers from Medical Workers

The other side of the coin is that medical workers may themselves be HIV positive. Should such individuals be allowed to continue their usual practices? Or should they be identified and then banned from coming into contact with patients?

This question was debated vigorously in the early 1990s. One reason for the debate was a report that a dentist in Florida had infected about a half dozen of his patients with HIV in the late 1980s. That dentist died of an AIDS-related condition in 1990.

This debate has been resolved to some extent because no confirmed reports have yet appeared as to any other instance in which HIV has been transmitted from a medical worker to a patient. Given the millions of cases in which patients are treated by physicians, dentists, nurses, and other health personnel, the fears of worker-to-patient transmission have been greatly reduced.

HIV Issues in the Twenty-first Century

Most of the issues discussed above are those with which Americans and people living in other developed nations have struggled for two decades. Many of those issues remain controversial, within one context or another. And other issues continue to develop. For example, some health experts are now suggesting that all pregnant women be tested for HIV. A fetus can contract HIV infection from its mother while still in the womb. By testing a pregnant woman, transmission of the virus may be prevented and/or the baby can be treated for the disease as early as possible.

Critics argue that counseling for all pregnant women and voluntary testing is a more efficient way to stop the spread of HIV infection. If women know that they will be tested for HIV, they may refuse the prenatal care that is so necessary for their own health and that of their child.

The American Medical Association reflected this difference of opinion in 1996 when it voted to require HIV testing of pregnant women. The affirmative side on this issue prevailed by only four votes and came in opposition to the association's own Council on Scientific Affairs, the American College of Obstetricians and Gynecologists, and the American Academy of Pediatrics.

But some of the most difficult debates about HIV have shifted from the United

States and other developed countries to the developing nations of the world. It is in these nations that HIV infection is extracting its greatest toll and presents the greatest challenges for the twenty-first century.

For example, more than 21 million people are believed to be infected with HIV in sub-Saharan Africa. There may be another 6 million infected people in South and Southeast Asia and as many as 1.5 million HIV-positive individuals in Latin America. By some estimates, 40 million people around the world will have died from AIDS-related diseases by 2000, and more than 9 million children will have been orphaned. It is hardly surprising, then, that many HIV-related issues in the future are likely to be centered on ways of dealing with the disease in other nations.

One of the first questions, of course, has to do with treatment of the disease in poor countries. Many HIV-positive individuals in the United States and other developed countries have access to drugs that will allow them to live out most of their natural lives. But those drugs are enormously expensive. They all cost a few thousands of dollars per year. There is virtually no country in Africa, Asia, or Latin America that can afford to provide these drugs to its HIV-infected citizens.

For that reason, many scientists place their greatest hope in the development of a vaccine that can be used to prevent the further spread of HIV infection. But that route also holds its risks and problems. For example, complaints were raised in 1997 about the testing of a new combination of drugs for use against HIV in parts of Africa. The problem was that some people were being given drugs known to be effective against HIV, while others in the same study were being given a placebo. That is, some people involved in the study were actually being prevented from having access to a drug that was known to be of help to them.

One of the issues involving vaccine testing in developing countries involves the nature of the anticipated HIV vaccine itself. Most experts think that the vaccine will not

be able to prevent *infection*, although it should be able to prevent the *disease* itself. Suppose such a vaccine is successfully developed and tested, say, in Africa. Subjects on whom that vaccine is tested will become infected, by the nature of the vaccine. It then becomes a moral obligation of the testers to provide drug therapy to those individuals to prevent AIDS' actually developing. But by providing that therapy, the researchers also make it impossible to determine the long-range effectiveness of the vaccine.

On the other side of the coin, many health officials in developing nations are asking for a change in the way vaccines are tested. At one time, drug manufacturers sometimes tested new products in poor countries; when the products were found successful, they were marketed only in developed countries, where they could be sold at a high price. In response to that practice, drug manufacturers are now ethically required to test their new drugs only in the nations where the company is headquartered or in comparable nations where the drugs are likely to be used.

But representatives of developing nations are asking for a reversal of that procedure. They *want* experimental vaccines to be tested in their own countries, hoping that that practice will become part of the solution for their own HIV crises. Their citizens are no longer as uninformed as they once were about experimental tests, and they should be permitted to make their own informed decisions as to whether or not they want to participate in such studies.

The issue of HIV transmission from mother to child has also arisen in developing nations. For many years, international health agencies have worked hard to encourage mothers to breast-feed their children. This position is based on a number of factors. In general, breast milk is better for a child than artificial formula, which most parents cannot afford anyway. Also, breast-feeding is likely to delay the birth of another child to the woman.

It has become clear, however, that HIV can be transmitted from mother to child through breast milk. International health agencies have now decided to *discourage* mothers infected with HIV from breast-feeding their children. They see this step as an important factor in reducing the spread and the total number of HIV cases in developing nations. The new policy is likely to be a step backward, however, if it is also adopted by women who have only recently been convinced of the value of breast-feeding.

See also: Needle Exchange Programs

References

"ACLU Position Statement on Prenatal and Newborn HIV Testing." http://www.aclu.org/issues/aids/newborn.html. 25 July 1998.

Altman, Lawrence K. "Ethics Panel Urges Easing of Curbs on AIDS Vaccine Tests." *New York Times*, 28 June 1998, A6.

———. "U.N. Discourages Nursing by Mothers with H.I.V." *New York Times*, 26 July 1998, A2.

Bayles, Fred. "Disability Law Covers People with AIDS Virus." *USA Today*, 26 June 1998, 3A.

Bloom, Barry R. "The Highest Attainable Standard: Ethical Issues in AIDS Vaccines." *Science*, 9 January 1998, 186–188.

Coles, Matthew. "The Law and Health Care Workers: Confidentiality, Testing, and Treatment." http://hivinsite.ucsf.edu/akb/1994/9–1/index.html. 24 July 1998.

Krieger, Lisa M. "Bill Fuels Battle over AIDS Test Reporting." *San Francisco Examiner*, 27 April 1997, A1+.

"Mandatory Reporting of HIV Urged." *San Francisco Chronicle*, 11 September 1997, A3.

Meier, Barry. "In War over AIDS, Battle over Baby Formula Reignites." *New York Times*, 8 June 1997, A1.

Newton, David E. *AIDS Issues: A Handbook.* Hillside, NJ: Enslow Publishers, 1992.

Perlman, David. "Poor Nations Losing Battle against AIDS." *San Francisco Chronicle*, 24 June 1998, A1+.

"Project on Legal and Ethical Issues Raised by HIV/AIDS." http://www.aidslaw.ca/elements/projectE.html. 21 May 1999.

Sandrick, Karen M. "AMA Supports Mandatory HIV Testing of Pregnant Women, Semen Donors." *Chicago Mail Tribune*, 27 June 1996.

Sternberg, Steve. "Scientists Disturbed by Slow Pace." *USA Today*, 23 July 1998, D1+.

Homosexual Behavior

Homosexual behavior is any sexual act that takes place between two people of the same sex. Some authorities also include thoughts, feelings, emotions, and other opinions of physical or intellectual attraction between two people of the same sex that may or may not be expressed outwardly.

Homosexual behavior has traditionally been regarded very differently by different cultures. In some cultures, for example, the attraction between two people of the same sex is regarded as a special gift that may qualify the individuals to become priests, shaman, or other people of authority and respect. In other cultures, homosexual behavior is regarded as normal and natural at least between some individuals under some circumstances. And in other cultures, homosexual behavior is regarded negatively and may be considered worthy of reproach or punishment.

The difference between the way individuals and cultures *feel* and the way they *act* regarding homosexual behavior can be quite striking. This reality tends to hold true for many kinds of sexual behavior, when a culture holds one belief in general about that behavior but individuals may hold different individual beliefs. The Roman Catholic Church, for example, has long expressed strong negative attitudes toward homosexual behavior. At the same time, however, there has long been a record of such behavior's occurring commonly among people who proclaim themselves to be Catholics and within the priesthood itself.

Sources of Homosexual Behavior

In the history of Western civilization, one question about homosexual behavior that has frequently been asked is how it originates. That is, what are the factors that tend to make a person feel greater attraction to a member of his or her own sex than to someone of the opposite sex?

This question is actually a fairly new one historically. It appears that through most of Western history, the existence of homosexual behavior was acknowledged between certain individuals under certain circumstances. While such behavior may not have been condoned, it was hardly the subject of scholarly analysis.

The past two centuries, however, have seen a growing interest in the question as to "how people get that way." The question is not, in and of itself, value free. That is, the question of how people develop an attraction to someone of the opposite sex is seldom, if ever, asked. Some observers point out that the main reasons for asking about the origins of homosexual behavior are that the answers to that question will provide a way of dealing with that behavior.

For example, suppose that it was found that people attracted to someone of the same sex made a conscious choice to have that feeling. They might be found to have made that choice even if they knew that the society in general had strong negative views against such feelings. In such a case, a person might legitimately be punished by society for ignoring or flaunting its moral values. Indeed, one could define homosexual behavior under such circumstances as a criminal act.

It might also be found that people who engage in homosexual acts are sinners. That is, they may have made the choice to engage in behaviors that are strictly forbidden by some dominant religious group. Under such circumstances, those individuals would probably not be subject to criminal punishment. But they might well be regarded as pariahs, especially by those who belong to the dominant religious group.

A third possibility is that people who engage in homosexual behaviors are not mentally competent. That is, they may have developed some kind of antisocial behavior as the result of early upbringing, a brain disorder, or some other psychological prob-

lem. They might, for example, be considered as maladapted individuals similar to kleptomaniacs or neurotics. In such cases, those individuals might be provided with some form of psychological treatment that would cure them of their desire to perform homosexual acts.

Finally, it might be that an attraction to someone of the same sex is a genetic trait. That is, researchers might find a gene or set of genes that predisposes a person to same sex feelings rather than opposite sex feelings. The only cure for one who engages in homosexual behavior in such cases would probably be genetic surgery, in which the "aberrant" gene would be modified to produce a behavior that is more "socially appropriate."

Diagnosis and Cure

In fact, all four of the above explanations for the cause of homosexual behavior have been proposed at one time or another. Over the past two centuries, one or the other of these explanations has tended to hold sway, leading to the use of some particular approach to the treatment of individuals attracted to someone of the same sex.

For example, the dominant psychological, psychiatric, and psychoanalytical explanation for homosexual behavior for many decades was that people who engaged in such behaviors were mentally ill. Homosexual behavior was actually defined as a mental disorder in the official "Bible" of the psychological profession, the *Statistical and Diagnostic Manual*.

In 1975, however, the psychological profession changed its mind about that diagnosis. By a vote of the members of the American Psychological Association (APA), homosexual behavior was removed from the list of mental disorders. It was replaced by a new definition in which an *unreasonable and debilitating* concern about one's sexual orientation was classified as a mental disorder. Activists in the gay rights movement have sometimes pointed out the irony of millions of Americans' having been cured of their "mental disorder" essentially overnight as the result of this action.

Since the APA's action, most critics of homosexual behavior have reverted to a religious argument for that condemnation. They suggest that homosexual behavior is prohibited by the Bible and that, therefore, those who engage in such behavior are sinners. Ironically, perhaps, some religious leaders, such as Pope John Paul II, have incorporated both views of homosexual behavior into their condemnations, calling such behavior both sinful and mentally disordered. This position raises the difficult question as to the sense in which a person can be a sinner if he or she has some debilitating mental disorder.

Genetic Origins of Sexual Orientation

In any case, the most recent development in the long debate over homosexual behavior has been the discovery of clues that such behavior may have a genetic origin. For example, in 1991, Simon LeVay, then at the Salk Institute for Molecular Biology in San Diego, found differences in the hypothalamus among gay men, heterosexual men, and heterosexual women. LeVay considered this difference to be of some significance since the hypothalamus is involved in some sexual responses.

Two years later, Dean Hamer, at the National Institutes of Health, reported on a study of the families of 114 gay men. He reported that sexual orientation appeared to be passed down through the mother's sides of those families. Hamer hypothesized that the tendency toward a homosexual orientation might be traceable to a particular segment of a gene on the X chromosome.

In 1998, a researcher at the University of Texas, Dennis McFadden, announced that he had found a small, but important, difference in the structure of the inner ears of lesbian and nongay women. McFadden hypothesized that the differences he discovered occurred at an early stage of development and were caused by modest differences in hormone levels between the two groups. These differences may also have resulted in differences in sexual orientation, he suggested.

This research, like all of the debate over homosexual behavior, is not a purely scientific or philosophical issue. The scientific information produced and the final decisions people reach about the origins of homosexual behavior have profound effects on social attitudes, legislation, political attitudes, and everyday feelings about such behavior.

That is, suppose that scientific evidence can be adduced to prove beyond doubt that homosexual feelings and acts are free will choices made by people who, if they chose to do so, could engage in heterosexual feelings and acts instead. In such a case, those individuals would clearly be making a conscious choice to violate social mores. It would be difficult to argue that they should be guaranteed certain kinds of civil liberties because they have willingly and knowingly chosen to violate social norms.

On the other hand, suppose that scientific evidence continues to show that homosexual behavior is somehow predetermined by one's heredity. In such a case, it might be more difficult to continue enforcing social restrictions on those who engage in such behaviors, restrictions such as lack of protection in employment, housing, access to public accommodations, and other areas. One, then, could no more withhold civil rights from a gay man or lesbian because of his or her sexual orientation than one could withhold civil rights from someone because of his or her skin color, sex, or other inborn characteristic.

The research on the genetic basis of sexual orientation has had its share of critics. Many observers continue to be concerned about the possibility of ascribing social behaviors of any kind to genetic factors. Geneticists Paul Billings and Jonathan Beckwith have written, for example, about "the current unbridled enthusiasm for studies relating genes with human behavior." They were particularly critical of LeVay's study as being both "inaccurate and inconsistent" ("Born Gay," 1993).

Some people who regard homosexual behavior as sinful are also unimpressed by the new data from genetics studies. They say that even if genes *are* responsible for homosexual behavior, that behavior is still not acceptable in the eyes of God.

References

Bahnsen, Greg L. *Homosexuality: A Biblical View*. Grand Rapids, MI: Baker Book House, 1978.

Bailey, Derrick Sherwin. *Homosexuality and the Western Christian Tradition*. London: Longman, Green, 1955.

"Born Gay." *Technology Review,* July 1993; also at http://www.cs.cmu.edu/afs/cs.cmu.edu/user/scotts/bulgarian/billings-beckwith.html. 12 July 1998.

Boswell, John. *Christianity, Social Tolerance, and Homosexuality*. Chicago: University of Chicago Press, 1980.

Conner, Gary L., ed. *Symposium: Sexual Preference and Gender Identity*. A special issue of the *Hasting Law Journal,* March 1979.

Editors of the *Harvard Law Review. Sexual Orientation and the Law*. Cambridge: Harvard University Press, 1990.

Gramick, Jeannine, and Pat Furey. *The Vatican and Homosexuality*. New York: Crossroad, 1988.

LeVay, Simon, and D. H. Hamer. "Evidence for a Biological Influence in Male Homosexuality." *Scientific American,* May 1995, 44–55.

McKenzie, Aline. "The Science of Sexual Orientation." *Dallas Morning News,* 31 May 1998, J1+.

Pool, Robert. "Portrait of a Gene Guy." *Discover,* October 1997, 50–55.

Vreeland, Carolyn N., Bernard J. Gallagher III, and Joseph A. McFalls Jr. "The Beliefs of Members of the American Psyciatric Association on the Etiology of Male Homosexuality: A National Survey." *Journal of Psychology,* September 1995, 507–517.

Weinberg, Martin S., and Alan P. Bell. *Homosexuality: An Annotated Bibliography.* New York: Harper and Row, 1972.

Human Experimentation

Human experimentation is the process of using human beings to test the safety and effectiveness of new physical, chemical, biological, or other procedures. For example, drugs designed for use with human beings can be tested in early stages with experimental animals, such as laboratory rats and mice, chimpanzees, or other species. These early tests are designed for two purposes: (1) to make sure that the new drug does not cause illness or death in the experimental animals, and (2) to see if the drug brings about the desired medical effect, such as curing some disease.

Using Humans as Subjects in an Experiment

At some point, however, the new drug must also be tried out on humans. Animal tests often give very accurate predictions as to how humans will respond to a particular drug. But those predictions are seldom totally accurate. It is still necessary to test the drug on humans before it is approved by government agencies, such as the U.S. Food and Drug Administration (FDA), and released to the public.

The comments that apply to this drug-testing scenario also apply to other kinds of experiments. For example, psychologists are often interested in the way humans respond to various types of reinforcements, such as receiving rewards or punishments for expressing certain kinds of behaviors.

Experiments in which humans are used as subjects, however, are fraught with a number of profound issues and questions. In general, most people would not be extraordinarily upset if a group of laboratory mice died during the testing of a new drug. But the consequences of a similar result with a group of human subjects would be very serious.

Historically, a majority of the subjects used in human experiments have come from four populations: prisoners, mental patients, students, and military personnel. One characteristic of the individuals in these four populations is that they tend to have somewhat less freedom of choice and action than may be the case in the general population. It may, therefore, be somewhat easier to enroll them in experimental protocols.

This is not to say, however, that members of the general public are not also subject to human experimentation. One of the most famous human experiments of all time, for example, was the Tuskegee Syphilis Study that was initiated in 1932. That study involved a group of southern Black men who, although members of the general population, were also members of a somewhat select group because of their minority status.

Informed Consent

Over time, a number of general principles and safeguards have been developed for experiments involving human subjects. Perhaps the most fundamental of those principles is *informed consent*. Briefly, informed consent means that a subject understands the experiment in which he or she is about to participate and that the subject accepts the possible risks incurred by taking part in the study. In practice, a conscientious researcher will spend a considerable amount of time and effort explaining to prospective subjects what an experiment is about. The researcher will point out all possible consequences of which he or she is aware that may result from the study. The potential subject must make it clear that he or she understands the risks and is willing to accept those risks in participating in the study.

In addition, certain safeguards must be observed as the study progresses. For example, any subject should be allowed to drop out of the research at any time for any reason. Also, a researcher is obligated to end a study early if it becomes obvious that subjects will be harmed by continuing the study. The researcher may also decide to end a study when partial results indicate strongly that members of the experimental group (those who are being subjected to the experimental treatment) are benefiting to an extent that the same treatment should not be withheld from the control group (those who are not receiving the experimental treatment).

Researchers are also permitted to offer pay or other benefits to people who enroll in their experiments. For example, prison-

ers might be offered a reduction in their sentences for taking part in a study of a new medicine. But the payment should not be so great that it is the sole reason for an individual's taking part in the experiment.

Today, nearly all institutions in which human experimentation is conducted have review boards for this kind of research. Any scientists who wants to use human subjects in an experiment must first submit an outline of the research to the review board. If the board determines that no individual's rights are being infringed, the research may be approved.

Human Experimentation Gone Bad

Most scientists who use human subjects are extremely careful to follow the letter and the spirit of the general principles outlined above. From time to time, however, humans are used in experiments about which they are not informed and which become highly dangerous to them. One of the most famous of these experiments was the Tuskegee Syphilis Study.

The Tuskegee Syphilis Study was sponsored by the U.S. Public Health Service (PHS) and conducted in Macon County, Alabama, beginning in 1932. Six hundred poor Black males were enrolled in the study. Of this number, two-thirds were infected with syphilis (the experimental group) and one-third were not (the control group). Members of the experimental group were told that they would be receiving free treatment for their syphilis.

In fact, the purpose of the study was not to provide any kind of treatment at all but to follow the course of syphilis over time. Subjects were given saline or sugar water "treatments" when they reported to the local clinic, where researchers were interested primarily in any changes they were able to observe during each visit.

By the early 1950s, a cure for syphilis had been developed. According to any ethical standards at all, the Tuskegee Study should have been brought to a conclusion at that point and the subjects provided with treatment to cure their disease. In fact, treatments were not offered even at that point in the experiment.

Eventually, a woman who had been familiar with the study, Jean Heller, released news of the research to the general public. The PHS admitted that no formal protocol (plan) for the Tuskegee Study had ever been written and that subjects had never been aware of the purpose of the study. It claimed that the research had been necessary and important and that the study had been an appropriate piece of research. By the time Heller released news of the study, however, more than 100 of the subjects had died as a result of their syphilitic condition, and many more suffered from the debilitating long-term effects that result from the disease.

Human Experimentation Today

One might assume that the use of uninformed human subjects in research studies is a thing of the past. Stories of the Tuskegee fiasco, Nazi experiments on human subjects, U.S. Army use of soldiers for tests of mind-altering drugs, and the like would have made it impossible for improperly designed human studies to occur any more.

In fact, such is probably not the case. Over the period 1977 to 1997, for example, the FDA conducted routine investigations of 4,154 studies on drug testing with human subjects. The agency found that 53 percent of those studies failed to meet the basic standards of informed consent expected of such studies. That is, humans were often used as subjects in those experiments without fully understanding the study in which they were participating.

As an example, the U.S. Centers for Disease Control and Prevention (CDCP) sponsored a 1991 study of a new measles vaccine in inner-city Los Angeles. Mothers were told that they were taking part in a study to determine the best measles vaccine and the best age at which to administer the vaccine. They were not told, however, of any possible side effects or other risks of taking part in the experiment. One of the pieces of information mothers might have been told was that similar tests in Africa of the new vaccine had actually resulted in the deaths of a number of young children.

Decisions as to how far researchers need to go—exactly what they must tell potential subjects—are often difficult to determine. Most researchers probably do not set out to place their human subjects at risk for health problems. The CDCP certainly had no intention, in 1991, of attempting to deceive prospective subjects in the measles vaccine experiment, nor did it intend to put them at risk. However, the Los Angeles case does illustrate the need for researchers to be as forthright as possible with humans that they intend to enroll in their research.

See also: Drug Testing

References

"America's Dirty Little Secret" [the Tuskegee Study]. http://www.aabhs.org/tusk.htm. 29 June 1998.

Annas, George J. *Informed Consent to Human Experimentation: The Subject's Dilemma*. Cambridge, MA: Ballinger Publishing, 1977.

Dowd, S. B., and B. Wilson. "Informed Patient Consent: A Historical Perspective." *Radiology Technology*, November/December 1995, 119–124.

Epstein, Keith C., and Bill Sloat. "After Tuskegee: New Experiments." *Oregonian*, 16 May 1997, A20.

French, Howard W. "AIDS Research in Africa: Juggling Risks and Hopes." *New York Times*, 9 October 1997, A1.

Freund, Paul Abraham, ed. *Experimentation with Human Subjects*. New York: G. Braziller, 1970.

"HIV Placebo Studies' Ethical Dilemma." *San Francisco Chronicle*, 23 April 1997, A3.

Jones, James H. *Bad Blood: The Tuskegee Syphilis Experiment*. New York: Collier Macmillan, 1981.

Kinney, David. "Experiments Come to Light over Prisoners as Guinea Pigs." *San Francisco Chronicle*, 18 June 1998, A11+.

"NHMRC [Australian National Health and Medical Research Council] Statement on Human Experimentation." http://www.

macarthur.uws.au/research/ethics/
nhmrc.html. 21 May 1999.

"The 1994 Rockefeller Report" [testing of
military personnel]. http://www.trufax.
org/trans/roc1.html. 29 June 1998.

"1932–1972: The Tuskegee Syphilis Study."
http://showme.missouri.edu/~socbrent/
tuskegee.htm. 29 June 1998.

Sobel, Dana. "So that Others May Live."
Omni, December 1979, 52–58+.

The home page for the Advocacy Committee
for Human Experimentation Survivors—
Mind Control is at http://www.aches-
mc.org.

Human Gene Therapy

Human gene therapy (HGT) is a method
for treating certain diseases by changing
the genes that occur normally in a person's cells. HGT can be of two major
types: somatic or germline HGT. In somatic HGT, the changes made in a person's genes will not be passed down to future generations. In that therapy only
those genes found in somatic cells—that
is, those not involved in reproduction—
are modified. In germline HGT, genetic
changes are made in cells that take part in
reproduction—the egg and sperm. Those
changes, therefore, will be transmitted to
future generations.

The Chemical Basis of Genetic Disorders

The diseases subject to treatment by HGT
are those caused by errors in a person's
genes. A gene is a segment of a deoxyribonucleic acid (DNA) molecule consisting
of thousands of distinctive groupings
known as base pairs. A particular sequence
of bases in the DNA carries instructions
that tell a cell what function it is supposed
to perform.

For example, suppose that a gene consists of the base sequence shown below:

ATCGGCTCTAGAATCGCGCGCTATATCCTCGATTTTCCCTAACTAAA ...
TAGCCGAGATCTTAGCGCGCGATATAGGAGCTAAAAGGGATTGATTT ...

The complete gene cannot be shown here
because it would take up too much space.

Each specific sequence of base pairs in a
gene carries a unique and distinctive message for a cell. For example, imagine that
the base sequence shown above tells a cell
to make the chemical ornithine transcarbamylase, an enzyme needed for the diges-
tion of proteins. In the absence of this enzyme, proteins are not digested properly
but break down instead to produce substances toxic to an individual. A baby
whose cells lack the gene for ornithine
transcarbamylase falls into a coma soon
after birth and dies shortly thereafter.

Genetic disorders typically do not result
from the complete loss of a gene but from
seemingly minor changes in a base sequence. For example, suppose that a particular individual has the base sequence
shown below instead of the one shown
above.

ATCGGCTC**T**GAATCGCGCGCTATATCCTCGATTTTCCCTAACTAAA ...
TAGCCGAGATCTTAGCGCGCGATATAGGAGCTAAAAGGGATTGATTT ...

Notice that only one base in the entire sequence is incorrect, the one shown in bold
italic type. Yet, an error as simple is this is
enough to prevent the gene from providing
a cell with correct instructions. This base sequence has no meaning to a cell because of
the error, and the cell will be unable to
make the necessary ornithine transcarbamylase.

HGT Techniques

The principle behind HGT is relatively simple. The plan is to insert into a person's
cells correct copies of a defective gene. That
is, to overcome the problem of ornithine
transcarbamylase deficiency, the plan
would be to insert a number of copies of
the *correct* gene into the body of a person
whose cells lack that gene or have only an
incorrect copy of the gene.

In practice, this insertion is accomplished by means of a *vector,* a chemical
compound with the capability of delivering
the correct gene to cells. The most common
vectors used are viruses. Viruses are very
small particles that are living by some defi-

nitions but nonliving by other definitions. Viruses are used as delivery vehicles in HGT because they have a natural tendency to "break into" cells. Viral diseases occur, in fact, when viral particles attach themselves to the outside of a cell membrane and inject their own DNA or ribonucleic acid (RNA) into the cell. It is this natural tendency of viruses that makes them so useful as deliverers of new genetic material in HGT.

Two steps are involved in HGT. First, a particular virus is selected as the delivery vehicle for the gene that is to be inserted into a patient. The portion of the virus that is responsible for any harmful effect on cells is removed. In its place is inserted a base sequence that corresponds to the correct base sequence present in the gene to be inserted. This correct base sequence can be removed from another healthy cell, or it can be produced synthetically in the laboratory.

Next, the altered virus is injected into the patient. The expectation is that the virus will "infect" the patient's cells and inject its altered DNA into those cells. Included in that altered DNA is the correct base sequence needed by the patient's cells. Researchers hope that the patient's cells will incorporate the injected base sequence into its own DNA. If that happens, the cell will have regained the ability to manufacture an essential chemical that it did not previously have.

Research and Results

Scientists have been familiar with the general techniques for HGT for more than twenty years. Yet, they have had a number of problems in getting the process to work the way they would like. For example, the human body has an elaborate defense system to protect itself against the invasion by disease-causing viruses. This system is not able to tell the difference between a flu virus, for example, and one carrying new genetic information to a patient lacking ornithine transcarbamylase. That is, the virus may never even make it into cells because it is defeated by the body's immune system.

Even if the virus does make it to a cell, it may not be able to deliver its genetic "message" to the cell. The cell has its own defense systems, including a tough membrane through which viruses may not be able to pass.

Finally, a successful insertion of the correct DNA into a cell may still not guarantee success. The DNA may be incorporated by the host cell into the wrong parts of its own DNA such that the added DNA is never expressed at all or, worse, actually interferes with the normal operation of the host DNA.

Given all of these problems, it should not be much of a surprise to realize that early tests with HGT were not very successful. Scientists were unable, in an overwhelming number of cases, to get the inserted gene to express itself in a way that would overcome the genetic disorder for which a person was being treated.

The one exception to that statement appears to have been one of the very first experiments ever attempted. In 1989, geneticists French Anderson and Michael Blaese injected engineered DNA into the bloodstreams of two young girls suffering from adenosine deaminase deficiency. The enzyme adenosine deaminase plays a role in the development of a person's immune system. Without the enzyme, the immune system does not develop correctly, and a person is subject to all kinds of infections against which a normal body can protect itself. As of 1999, the two girls appear to have responded positively to the treatment and have begun to live a normal life with a normal immune system.

Issues with HGT

The genetic changes made possible by HGT have long troubled many people. Some people take a relatively extreme view that humans should never, under any circumstances, alter a person's genetic makeup. To do so, they say, would be for humans to begin acting "like God." The genetic makeups with which humans are born are the ones that God intended, and attempts to treat diseases must operate within that general constraint, they say.

Other people see some value in HGT for the treatment of certain genetic disorders. But they worry about the use of that technique for making other than therapeutic changes. That is, suppose that a couple decided that they wanted their unborn child to be born with blue eyes and blond hair. HGT techniques provide the technical means to make the genetic changes necessary to produce that result. But is the use of HGT to produce cosmetic changes a legitimate application of the technology?

The issue becomes even more difficult if one thinks about traits that are less superficial, such as hair and eye color, and more important to one's success in the world, such as intelligence. If parents were able to have HGT performed on their unborn child in order to improve its overall intelligence, should they be allowed to do so?

Part of the problem is that the "technological genie" is now out of the bottle. That is, there is no longer a great deal of doubt that scientists have or will soon have the technological capability of changing any number of human characteristics with HGT, not just curing diseases. As long as that technology exists, what can be done to be sure that it is not used for inappropriate purposes? And who decides what "inappropriate" use of the technology will be?

Finally, questions remain about the use of HGT for germline cells. It is one thing to make genetic changes in a person that will die out when the person dies. It is not difficult to understand an argument for the use of HGT to cure a person of terrible genetic disorders. But what about making genetic changes that will be transmitted to future generations? How should a researcher respond to a couple who want to have a gene for increased intelligence inserted into the germline cells of their yet unborn child? Such a step would alter the child's own intelligence and, perhaps, that of his or her own descendants.

Some people have constructed science fiction stories based on such probabilities.

One of the most famous of these stories was Aldous Huxley's *Brave New World,* in which individuals were "engineered" by HGT-like technologies to assume certain given places in society. The 1997 movie *Gattaca* presented a similar story. In this film, members of a futuristic society were expected to select from a genetic library the traits they wanted expressed in their own children . . . and their children after them. The society that evolved from this concept was a highly structured order in which people were assigned various roles on the basis of the genes they carried.

Some critics have dismissed *Gattaca* and similar science fiction stories as too absurd to be given attention. Other critics point out that some of the most absurd science fiction tales from the past have actually come to pass and that *Gattaca* might pose issues in its fictional tale that humans might some day have to deal with in their real lives.

References

Baskin, Yvonne. "Doctoring Genes." *Science 84,* December 1984, 52–60.

"Clinical Trials." The Institute for Human Gene Therapy. http://www.med.upenn.edu/ihgt/clintri.html. 14 July 1998.

Friend, Tim. "'Gattaca' Could Foretell Reality." *USA Today,* 28 October 1997; also at http://www.usatoday.com/life/science/genetics/lsg020.htm.

Gibbs, W. Wayt. "Gene Therapy." *Scientific American,* October 1996.

Judson, Horace Freeland. "Who Shall Play God?" *Science Digest,* May 1985, 52–54+.

Petechuk, David. "Gene Therapy." In *The Gale Encyclopedia of Science,* edited by Bridget Travers, 1607–1613. Detroit: Gale Research, 1996.

Thompson, Larry. *Correcting the Code: Inventing the Genetic Cure for the Human Body.* New York: Simon and Schuster, 1994.

Weiss, Rick. "Engineering the Unborn." *Washington Post,* 22 March 1998, A1+.

———. "Gene Therapy at a Crossroads." *Washington Post,* 19 October 1994, 12–15.

Wilson, Jim. "What Is Gene Therapy?" http://www.med.upenn.edu/ihgt/info/whatisgt.html. 14 July 1998.

Human Genome Project

The Human Genome Project (HGP) is one of the largest research projects in biology— or, for that matter, in science—ever to have been undertaken. Its goal is to determine completely the human genome.

The genome of a species consists of all the genes that occur within that species. By specifying the complete set of genes within any species—its genome—the species has been completely defined from a biological standpoint.

Genes and DNA

One of the great discoveries in the history of science occurred in 1953 when the English biochemist Francis Crick and the American biologist James Watson discovered the biological role of deoxyribonucleic acid (DNA). DNA is an abbreviation for an organic molecule that occurs in living cells.

A DNA molecule consists of three major parts: a sugar (ribose), a group of phosphorus and oxygen atoms known as a phosphate group, and a set of four nitrogen bases. The sugar and phosphate groups are attached to each other in an alternating sequence that makes up a very long chain consisting of many thousands of units. Attached to each sugar group on the chain is one of the four possible nitrogen bases: adenine (A), cytosine (C), guanine (G), and thymine (T).

The key discovery made by Watson and Crick was the way in which a DNA molecule can store genetic information. They found that the sequence in which nitrogen bases occur in DNA is a system for encoding genetic information in the molecule. For example, the sequence:

$$\text{-S-P-S-P-S-P-S-P-S-P-S-P-S-P-S-P-S-P-S-P-S-P-}$$
$$| \; | \; | \; | \; | \; | \; | \; | \; | \; | \; | \; | \; | \; | \; | \; | \; |$$
$$\text{A T T C G A C T A G C G G A A T}$$

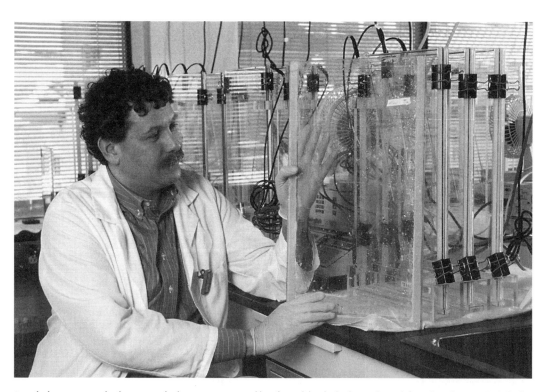

Once the human genome has been mapped, what are some potential benefits and drawbacks that might result from that information? (Corbis/ Roger Ressmeyer)

might tell a cell to perform one type of function. A somewhat different sequence, such as

might tell a cell to perform a very different type of function.

A DNA molecule is only slightly more complicated than indicated here. In fact, it actually consists of two chains of the kinds shown above. The two chains are wrapped around each other in a spiral configuration called a *double helix.* Nitrogen bases have chemical properties such that only certain pairs can line up next to each other in a DNA molecule. That is, an A on one chain always pairs with a T on the other chain; and a C on one chain always pairs with a G on the other chain.

A gene is a certain segment of a DNA molecule that has some biological meaning or significance. Genes typically consist of many thousands of nitrogen bases. For example, one might say that "gene 1" in a human DNA molecule consists of the base sequence: TCCGATCTAGCTATATATGC-GACTCTTACCCCTTATATCGGGTACGAT . . . This sequence cannot be shown completely here because it is so long. It would take a number of pages to write down the base sequence of a typical gene.

One can be said to have discovered or defined the complete genome of an organism when the precise sequence of base pairs on all DNA molecules in the organism has been identified. For the human genome, that task involves the analysis of about 80,000 genes. Those 80,000 genes consist of a total of about 3 billion nitrogen bases.

Analyzing the Genome

The Human Genome Project was initiated in 1990. The project is expected to take fifteen years. The program is administered by the U.S. Department of Energy and the National Institutes of Health. HGP is a complex and expensive project. Work on specific parts of the genome has been assigned to scientists throughout the United States. The total cost of the project has been estimated at about $3 billion.

In principle, the decoding of a gene is simple. One uses physical and/or chemical means to break a DNA molecule apart. The portion of that molecule that makes up the gene to be studied is identified. The bases present in that gene and the sequence in which they are arranged is then determined. Finally, the structure of the gene is announced and published in a master "gene map" stored at an HGP library.

An ancillary goal of HGP is the decoding of genomes of species other than that of humans. The fruit fly, laboratory mouse, and a common bacterium, *Escherichia coli,* are included among these species. Scientists hope to gain a better understanding of gene function in general by analyzing these organisms. They also hope to find ways in which this knowledge can be used to solve human health problems.

By the end of 1998, HGP was well on its way to its primary goals. The location of nearly all genes had been discovered, and the structure of many had already been identified. At the same time, however, competition for the federally funded HGP had begun to appear. Other scientists in other laboratories had decided to strike out on their own in the search to explicate the human genome. Some claimed to have developed faster, more efficient techniques than were being used by HGP scientists. Others reported that they were not interested in discovering the complete genome but only sections that might have medical or commercial value.

One major difference between the genome "newcomers" and the HGP is that the latter is a federally funded project whose results will be freely available to anyone who wishes to have the information. In fact, the results of genome research are currently available on the HGP web site for anyone who wants to look at and/or use them.

Non-HGP researchers, however, see knowledge of the human genome as a potentially profitable package of information. They hope to patent the information they find and then sell it to drug companies or others who might find use for the information.

Value of HGP Research

One of the most important reasons for studying the human genome is simply for the sake of learning more about the human species and about other species. In this regard, the HGP is fundamentally a massive example of basic research. But some very important practical information will evolve from this research also. For example, scientists will eventually be able to discover what the "normal" structure of a gene is. By *normal*, we mean the chemical structure of the gene when it functions the way it is supposed to function in a healthy body. With this information, it should be possible to identify the chemical changes that occur in a gene that prevent it from acting "normally." This knowledge, in turn, creates the possibility for curing certain health disorders by making corrections in "abnormal" genes.

For example, suppose that the gene responsible for making an essential protein in the human body has the structure:

TCCGACTGGCTAGCTAGTCTAGCTATATATGCGACTCTTACCCCT-
TATATCGGGTACGAT.

But then suppose that some individual is found whose gene has a single error within it, as:

TCCGACTGGCTAGCTAGTA*TAGCTATATATGCGACTCTTACCCCT-
TATATCGGGTACGAT.

The A* portion of this gene indicates a nitrogen base that differs from that in the "normal" gene. With this knowledge, a scientist might be able to perform genetic surgery, in which the incorrect A base is removed and replaced by the correct C base. This kind of genetic surgery might be used to relieve or cure human diseases that cannot now be cured by any existing method.

Ethical, Legal, and Social Issues (ELSI) of HGP

HGP is different from most other large research projects in one important way. Researchers have always known that a number of social, political, legal, ethical, and other issues will arise as scientists find out more about the human genome. The possible commercial uses of such information is an example. In order to deal with such issues, HGP budgets about 3 percent of its annual budget to research on these so-called ELSI questions.

Some of the questions related to research on the human genome are as follows:

1. Who owns information about the human genome? Is it possible or desirable to allow this information to fall under the control of a single company or individual? Can such an entity "own" knowledge as to what any particular individual's gene makeup is?
2. Who should have access to information about an individual person's genome? Should that knowledge be available only to researchers? to physicians? to insurance companies? to the military? to the government? to the person himself or herself?
3. What use should be made of genome information? For example, suppose that a company finds out that a prospective employee has a "defective" gene, that is, one that might be responsible for a medical disorder now or later in life. Should the employee be allowed to use that information to deny a person a job or to fire a person from a job?
4. Under what circumstances should genetic information be obtained or made available for reproductive purposes. For example, should a couple be *required* to find out any

genetic disorders their future child might have? Should the couple be *allowed* to obtain that information?

Questions like these trouble a great many biological researchers, government officials, and ordinary people. After all, unraveling the human genome will eventually provide the most complete and most intimate understanding both of what it means to "be human" and what it is that makes an individual person what he or she is. Furthermore, that information will be expressed in chemical terms. One will be able to say that mental retardation is caused, in some cases, by a base sequence of ATTAACGGGCTA . . . rather than ATTAATGGGCTA. . . . Expressed in those terms, changes in an individual's mental capacity may be modifiable by chemical methods, namely the replacement of one nitrogen base by another.

Changes in a person's genome to correct a medical defect are likely to be relatively popular among the general public. Who would not want to have this kind of repair made on a young child with mental retardation if the procedure could make the child "normal"? But having the capability to make genetic "corrections" also means the ability to make genetic "modifications." It means allowing a parent to choose the sex of a child, the color of his or her eyes or hair, and any number of other characteristics. The ethical issues surrounding these decisions are much more complex.

References

Bishop, J., and M. Waldholz. *Genome: The Story of the Most Astonishing Scientific Adventure of Our Time—The Attempt to Map All the Genes in the Human Body.* New York: Simon and Schuster, 1990.

Collins, F., and D. Galas. "A New Five-year Plan for the U.S. Human Genome Project." *Science,* 1 October 1993, 43–46.

Friend, Tim. "New Effort to Decode the Human Genome on the Fast Track." *USA Today,* 9 June 1998, 6D.

Hall, Carl T. "Gene Library: How DNA Data Fuels Drug Industry." *San Francisco Chronicle,* 16 July 1996, C1+.

"Human Genome Project Information." http://www.ornl.gov/hqmis/about.html. 21 May 1999.

Kevles, Daniel J., and Leroy Hood, eds. *Code of Codes: Scientific and Social Issues in the Human Genome Project.* Cambridge: Harvard University Press, 1993.

Wade, Nicholas. "It's a Three-legged Race to Decipher the Human Genome." *New York Times,* 23 June 1998, F3.

Wilkie, T. *Perilous Knowledge: The Human Genome Project and Its Implications.* Berkeley: University of California Press, 1993.

Hydrochlorofluorocarbons

The hydrochlorofluorocarbons (HCFCs) are a group of chemical compounds with molecules containing carbon, chlorine and/or fluorine, and hydrogen atoms only. A typical HCFC is HCFC-22, chlorodifluoromethane ($CHClF_2$). The numerical suffix "-22" is an industry designation for a specific member of the HCFC family. HCFCs differ from chlorofluorocarbons (CFCs) in that their molecules contain one or more atoms of hydrogen.

The HCFCs have been known for many years, but interest in their practical applications did not develop until the 1980s. At that time, continuing evidence for the role of CFCs in the depletion of the ozone layer had been collected. Chemical companies began to search for substitutes for an enormously successful and profitable group of products—the CFCs—that were soon to be banned. They hoped that HCFCs might fill that need.

Indeed, HCFCs are now widely viewed as at least a temporary replacement for the CFCs. They, like the CFCs, also escape into the Earth's atmosphere, break down, and release a chlorine atom that can react with ozone molecules. However, they do so much less efficiently than do the CFCs. The ozone depletion potential (ODP) for HCFC-22, for example, is 0.35, indicating that a single gram of HCFC-22 is only 35 percent as efficient as a gram of CFC-11 (the standard) in destroying ozone.

On a long-term basis, however, even an ODP of 0.35 is too large. As a result, provisions of the Copenhagen Amendment of 1992 to the Montreal Protocol on Substances that Deplete the Ozone Layer capped the production of HCFCs in 1996, called for their reduction by 35 percent in 2004, and required a complete phase-out of their production by 2030.

Some concerns about the potential health hazards of HCFCs have also been expressed. In 1997, a group of researchers at the Catholic University of Louvain in Belgium reported on an "epidemic" of liver disorders apparently associated with exposure to HCFC-123 and HCFC-124. A total of nine workers at a smelting plant were diagnosed with a variety of liver disorders apparently caused by exposure to HCFCs. The significance of this report is still not clear since scientists had previously been relatively confident about the safety of these compounds.

See also: Chlorofluorocarbons; Ozone Depletion; Ozone Depletion Potential

References

"CAMP-HCHC; Hydrochlorofluorocarbon Measurements." http://www.cmdl.noaa.gov/noah/flask/hcfc.html. 23 June 1998.

Cole-Misch, Sally. "Hydrochlorofluorocarbons." In *The Gale Encyclopedia of Science,* edited by Bridget Travers. Detroit: Gale Research, 1996, 1855–1857.

Pickering, Kevin T., and Lewis A. Owen. *An Introduction to Global Environmental Issues.* London: Routledge, 1994, Chapter 3.

Wallington, Timothy J., et al. "The Environmental Impact of CFC Replacement—HFCs and HCFCs." *Environmental Science & Technology,* July 1994, 320A–326A.

Zurer, Pamela. "Liver Damage Tied to CFC Substitute." *Chemical and Engineering News,* 25 August 1997, 8.

Indoor Air Pollution

Indoor air pollution consists of pollutants that are found inside homes, office buildings, and other structures. These pollutants come from both natural sources and human activities. Their effects indoors can be dramatically different from those outdoors, however, because they are trapped within a structure and their concentrations can reach much higher levels than they do out of doors. Indoor air pollution can be a serious problem because most people tend to think of their homes and workplaces as environmentally "safe" places. In some circumstances, however, an individual may actually be more at risk at home or at work than on a busy street corner.

Radon

The most common form of indoor air pollution from natural sources is radon gas. Homes and other structures are sometimes built on soil that contains high concentrations of radioactive elements, such as uranium and thorium. When these elements undergo radioactive decay, they produce a combination of other radioactive elements, including radon. Radon is a gas that tends to escape from the soil and seep upward through the floors of buildings above it. As radon itself decays within the building, it releases radiation. This radiation is believed to be responsible for anywhere from 5,000 to 20,000 cases of lung cancer a year in the United States, making it the second most important cause (after smoking) of that disease.

Tobacco Smoke

The threat posed by radon in a structure depends to a large extent on the presence of a second indoor air pollutant: tobacco smoke. Tobacco smoke consists of about two dozen known carcinogens, most in relatively small concentrations. When tobacco smoke is trapped inside a structure, however, those concentrations tend to increase and may pose a threat not only to a smoker but also to nonsmokers in the same structure. Some experts believe that the majority of deaths from lung cancer because of exposure to radon can be attributed to synergistic effects between that element and tobacco smoke.

Chemical Pollutants

Some of the most serious forms of air pollution occur as the result of ordinary, everyday human activities, such as those involved in cooking and heating. When natural gas is burned in a gas stove or a gas furnace, for example, very small amounts of carbon monoxide gas are produced. Carbon monoxide is a toxic gas that, at low concentrations, may produce headaches, nausea, and disorientation. At high concentrations, the gas can cause unconsciousness and even death. A properly functioning stove, furnace, or other appliance releases such low concentrations of carbon monoxide as to pose no threat to human health. However, such appliances do malfunction from time to time, and when they do, they can represent a serious threat to human health.

Much the same can be said for other everyday activities in the home. People use a large variety of chemicals in cleaning and washing, protecting against pests, painting, and personal health and cosmetic care. Many of these chemicals contain potentially dangerous chemicals. When one sprays the kitchen for ants, for example, the fumes that remain can cause discomfort and, in susceptible individuals, more serious health problems.

In such cases, as in all instances of indoor air pollution, an important factor in the effects on human health of pollutants is the construction of the building itself. The materials used in construction, how tightly they fit, the presence or absence of doors and windows that open, the type of insulation used, and the presence or absence of an air conditioner all affect the levels of pollutants that remain trapped in a structure.

Somewhat ironically, the "new and improved" materials now used in much construction have greatly increased the problem of indoor air pollution. Many new building materials and insulation materials contain synthetic chemicals that are known carcinogens or that have other potential health effects. Typical of these materials and one of the most commonly used is formaldehyde. Formaldehyde is a simple, inexpensive organic compound used as a raw material in the manufacture of plywood, pressboard, imitation wood paneling, carpeting, insulation, clothing, and many other products. In its combined form, it poses little or no threat to human health. However, in its uncombined form, it is a relatively toxic compound.

While perhaps the most common pollutant in building and furnishing materials, formaldehyde is by no means the sole chemical culprit. A significant number of other organic compounds that vaporize easily and are hazardous to human health—the so-called volatile organic compounds (VOCs)—also contribute to indoor air pollution. These chemicals pose a health hazard primarily in new structures, where they continue to escape from materials into a room long after construction has been completed. It is no longer uncommon to

hear of people who have had to move out of their new homes or offices because of such problems. In some cases, whole offices or buildings have had to be evacuated briefly because employees have become ill after working in them. This phenomenon has even been given a name, "new building syndrome" or "sick building syndrome." Again, new building syndrome is an effect observed differentially in different individuals with different levels of susceptibility. Those most sensitive to certain indoor air pollutants may not be able to return to, work in, or live in a structure for weeks, months, or ever.

See also: Formaldehyde; Radon

References

American Lung Association, Environmental Protection Agency, Consumer Product Safety Commission, American Medical Association. *Indoor Air Pollution: An Introduction for Health Professionals.* Washington, DC: U.S. Government Printing Office, Publication No. 1994–523–217/81322, 1994.

Baro, Madeline. "Workers Gasp for Fresh Air in Sealed Office Buildings." *Oregonian,* 17 April 1997, A16.

"Indoor Air Quality: Basics for Schools" (EPA Document #402-F-96-004). Washington, DC: U.S. Environmental Protection Agency, 1996.

"The Inside Story—A Guide to Indoor Air Quality" (EPA Document #402-K-93-007). Washington, DC: U.S. Environmental Protection Agency, 1995.

Leslie, G. B., and F. W. Lunau, eds. *Indoor Air Pollution.* New York: Cambridge University Press, 1994.

Moffat, Donald W. *Handbook of Indoor Air Quality Management.* Englewood Cliffs, NJ: Prentice Hall, 1997.

"Protect Your Family and Yourself from Carbon Monoxide Poisoning" (EPA Document #402-F-96-005). Washington, DC: U.S. Environmental Protection Agency, 1996.

"Secondhand Smoke" (EPA Document #402-F-93-004). Washington, DC: U.S. Environmental Protection Agency, 1993.

"Sick Building Syndrome (SBS) (EPA Document #6607J; Indoor Air Facts No. 4

[revised]). Washington, DC: U.S. Environmental Protection Agency, 1991.

"Targeting Indoor Air Pollution: EPA's Approach and Progress" (EPA Document #400-R-92-012). Washington, DC: U.S. Environmental Protection Agency, 1993.

The U.S. Environmental Protection Agency maintains a web site on indoor air pollution at http://www.epa.gov/iaq. See especially "Where to Go for Additional Information on Indoor Air Quality." http://www.epa.gov/iaq/moreinfo.html.

Intact Dilation Evacuation

Intact dilation evacuation (IDE) is an abortion procedure that came to public attention in early 1996. The procedure is one used relatively late in a pregnancy, often in the third trimester of gestation (after the twenty-sixth week of pregnancy). The procedure involves the removal of a fetus from the uterus feet first. An incision is made in the fetus's skull to allow removal of its contents. The collapsed skull then permits removal of the fetus to be completed. At the time IDE is performed, the fetus may or may not still be alive and viable (capable of life outside the mother's body).

IDE is performed for a variety of reasons. In some cases, the mother's health or life may be at risk. In other cases, tests show the fetus to have a serious genetic disorder that might result, for example, in severe physical deformities and/or mental disorders. Finally, some IDE procedures are performed as elective surgeries by women who have waited until late in their pregnancies to seek an abortion.

No statistics are available on the number of IDE procedures done in any one year. Some authorities place the number at no more than about 500 annually, while others claim the rate is much higher than that.

Medical and popular opinion on IDE is strongly divided. Some physicians claim that the procedure is the only one available for certain problems that arise late in pregnancy. For example, some genetic disorders may not be manifested or detectable until the fetus is well developed. At that point, it may be too late to use abortion procedures that are effective earlier in a pregnancy. Some people also argue that IDE, like any other abortion procedure, involves a decision that should be left to a woman, her physician, and, perhaps, her mate.

Critics argue that IDE is a brutal procedure that is always—or almost always—medically unnecessary. In the majority of cases, they say, the fetus is able to survive outside the body and will not be malformed. The procedure is primarily a matter of convenience for women who have made a decision to abort their fetus too late in their pregnancies.

The debate over IDE became a political issue in early 1996. The U.S. Congress passed a bill that would have made the procedure illegal except in cases where a woman's life was at risk. President Bill Clinton at first favored the bill but eventually changed his mind and vetoed it. He had insisted that the bill be modified to permit the procedure when a woman's health (not just her life) was at risk. Congress was not able to override the president's veto. The same sequence of events was repeated in 1997 when a similar bill was passed by Congress and vetoed by the president.

See also: Abortion; Early-Term Surgical Abortions

References

Hunt, Terence. "Clinton Vetoes Bill on Abortion." *Oregonian*, 11 April 1996, A1+.

"Language in 'Partial-birth Abortion' Ban Would Criminalize Most Procedures, ACLU Explains." http://www.aclu.org/news/n052297a.html. 28 May, 1998.

Partial-Birth Abortion Information. http://www.prolifeinfo.org/pba.html. 21 May 1999.

Intelligence and Race

See IQ

Internet and Privacy

See Privacy and the Internet

IQ

IQ is an abbreviation for *intelligence quotient*. The term was first proposed in 1920 at a time when psychologists were searching for ways to measure a person's "intelligence." The intelligence quotient was defined as the ratio of the score a person received on an intelligence test compared to the typical score obtained by other persons of the same age group, multiplied by 100. That is, suppose that a person received a higher score on an intelligence test than his or her peers. Then that person's IQ score would be more than 100. A person who received a lower score than his or her peers would have an IQ score of less than 100.

The Origins of Intelligence Testing

IQ and intelligence testing have become one of the most controversial issues in all of human psychology and education. That issue had its origins at a relatively simple level in the late 1890s when French psychologist Alfred Binet began to look for a way to help students with learning difficulties. Binet invented a test that contained intellectual exercises that he deemed suitable for various age levels. Any student who was not able to perform the exercises set as a standard for his or her age level was regarded as having a learning deficit. Binet used this information to help students improve their learning skills.

Binet's original tests made no effort to measure any fixed trait that we might think of as "intelligence." They simply attempted to determine the level of achievement students had reached at various ages. In fact, Binet warned against thinking of his tests as measures of any permanent capacity to learn.

Not long after Binet died in 1911, however, the intelligence testing movement took a quite different direction. Inspired to some extent by the international eugenics movement, some scholars began to think of intelligence as a natural, inborn, heritable characteristic determined in some way by a person's genes. According to this theory, a person who did not seem to be "intelligent" had limited genetic capability for being intelligent.

A leading U.S. psychologist, Lewis Terman, adopted a version of this attitude. Terman was responsible for modifying the Binet intelligence tests and introducing them into the U.S. educational system. At one point, Terman wrote that low intelligence

is very common among Spanish-Indian and Mexican families of the South-West and also among negroes. Their dullness seems to be racial, or at least inherent in the family stocks from which they come. . . . Children of this group should be segregated in special classes. . . . They cannot master abstractions, but they can often be made efficient workers. . . . There is no possibility at present of convincing society that they should not be allowed to reproduce, although from a eugenic point of view they constitute a grave problem because of their unusually prolific breeding. (Lewontin 1977, 12)

IQ and Race

Terman had expressed a view that was to reappear over and over in history. In Great Britain, psychologist Sir Cyril Burt held views similar to those of Terman. He believed that Christians were inherently more intelligent than Jews, the English more intelligent than the Irish, members of the upper class more intelligent than members of the lower class, and so on.

Burt supported his theories with some of the most comprehensive and powerful studies on genetics ever conducted. One of his most famous studies focused on genetic twins who were separated at birth. This study supposedly allowed Burt to work out the differences between character traits that were due to inheritance (which would, therefore, be similar in both members of a twin pair) and those due to upbringing or learning (which would be different in the members of a pair). Burt argued that his results showed conclusively that intelligence was, to a very considerable extent, inherited. That is, he said, people are essentially either born smart or dumb.

During his lifetime, Burt was accorded the highest level of respect by his colleagues and was knighted for his work. After his death, questions began to arise about his research. In 1978, U.S. psychologist D. D. Dorfman published a study reanalyzing Burt's data. Dorfman's study raised questions about the results Burt reported and his interpretation of them. The general consensus among psychologists today is that Burt probably falsified much of his data and, in all probability, actually invented much of the data he reported. He apparently was so convinced of his own theoretical concepts that he was willing to "fudge" his research to make it "come out right."

The primary goal behind Burt's research was to discover the relationship between intelligence and other characteristics, such as ethnicity, race, nationality, and economic status. Other researchers have been interested in the same question. One of the most famous articles on the subject ever written appeared in the *Harvard Educational Review* in 1969. The *Review* devoted most of one issue—123 pages—to an article by psychologist Arthur Jensen. In this article, Jensen purported to show that genetic makeup is responsible for a very high fraction of intelligence. In addition, Jensen argued that some groups of people were *inherently* more intelligent than others. Specifically, he claimed that Blacks were genetically less intelligent than Whites.

Anti-Jensenism and Neo-Jensenism

Jensen's article, later expanded to book form in *Genetics and Education*, released a flood of criticism. Much of that criticism related to technical aspects of Jensen's work. Other psychologists questioned the statistical techniques used by Jensen to reach his conclusions. Others doubted that his understanding and use of genetics principles were complete or accurate. For many psychologists, Jensen's work provided an opportunity to discuss and debate a scholarly topic about which qualified professionals might well disagree.

But Jensen was also attacked on a totally different level. He was criticized as a racist who was using psychology (or pseudopsychology) to promote his own social beliefs about people of different race, ethnicity, and economic status. These objections often took the form of closely reasoned arguments in scholarly journals and the general press. But they also took the form of more violent, more personal attacks in the written word and in person.

For a short period of time, the debate over Jensen's views faded from public view. Then, in 1994, another book was published with largely the same thesis as Jensen's. The book was *The Bell Curve*, by Richard Herrnstein and Charles Murray. Herrnstein and Murray presented arguments broadly similar to those of Jensen, namely that intelligence is genetically determined to a very significant degree and that some groups are genetically more intelligent than others. In particular, Herrnstein and Murray pointed out that Blacks have trailed Whites in IQ tests by about fifteen points for as long as those tests have been administered. They suggested that one reason for this pattern is that Blacks are genetically inferior to Whites by about that amount.

IQ Testing: So Who Cares?

Perhaps the most important point about this whole debate is that it does not focus simply on an academic issue that gives scholars something to write about. IQ testing has a very important function in the educational system in the United States (and in those of other nations). One's future academic course may very well be determined by the score one obtains on an early IQ test. In fact, as far back as the 1920s, students were "tracked" into one kind of academic program or another—business, trades, or college—depending on the scores obtained on an IQ test.

Indeed, all of the authors who have written about IQ and race (and ethnicity and economic status) point out that IQ tests can be and are used predictively. That is, people with high IQs tend to go to better schools, do better academically, and have greater economic success in later life.

It is of some interest that the title of Jensen's original article in the *Harvard Educational Review* was "How Much Can We Boost IQ and Scholastic Achievement?" The focus of this article was the huge amounts of money then being poured into educational programs by the federal government, much for "disadvantaged" children. Jensen's argument was that such appropriations were a waste of money. If the people on whom this money was spent—Blacks and the poor—were *genetically* inferior, there was not a lot that educational programs could do to help them.

Among the many objections that critics have raised about the Terman/Jensen/Burt/Herrnstein/Murray argument is the uncertain meaning of the term *intelligence* itself. Ask 100 psychologists to define the term, and one is likely to get 100 different answers. For Jensen, for example, intelligence is defined as whatever it is that is measured by intelligence tests.

But psychologists have long known and pointed out that intelligence tests are very much race and class biased. For many years, they were written by, administered by, and interpreted by middle- and upper-class White men and women. The questions asked and the "right" answers accepted in such tests inevitably and usually unconsciously reflected the thinking of the test makers and interpreters.

For example, one might ask which of the following words does not belong with the other words:

James   John   Judy   William

A test maker might very well have intended the "right" answer to be *Judy*, since

that is the only female name in the list. But by another criterion, the "right" answer might be *William*, since that is the only name that does not start with a *J*.

Test makers have been trying for many years now to remove race bias and biases of other kinds from intelligence testing. But these efforts have probably not been totally effective, and the hidden assumptions of even the most conscientious test makers may still be introducing elements of bias into their products.

See also:
Biological Determinism; Eugenics; Nature versus Nurture

References
The Ann Arbor Science for the People Editorial Collective. "Race and IQ." Minneapolis: Burgess Publishing Company, 1977, 21–55.
"APA Task Force Examines the Knowns and Unknowns of Intelligence." http://www.apa.org/releases/intell.html. 6 July 1998.
Crenson, Matt. "Scientists Downplay Genes' Role in Setting Intelligence." *Oregonian*, 31 July 1997, A7.
Herrnstein, Richard, and Charles Murray. *The Bell Curve.* New York: The Free Press, 1994.
Jensen, Arthur. "How Much Can We Boost IQ and Scholastic Achievement?" *Harvard Educational Review,* Spring 1969, 1–123.
Kamin, Leon. *The Science of Politics and I.Q.* New York: Halsted Press, 1974.
Lewontin, Richard C. "Biological Determinism as a Social Weapon," in the *Ann Arbor Science for the People Editorial Collective.* Minneapolis: Burgess Publishing Company, 1977.
Murray, Charles. "'The Bell Curve' and Its Critics." http://www.cycad.com/cgi-bin/Upstream/People/Murray/bc-crit.html. 7 July 1998.
Naureckas, Jim. "Racism Resurgent: How Media Let *The Bell Curve*'s Pseudo-science Define the Agenda on Race." http://www.fair.org/extra/9501/bell.html. 6 July 1998.
Quigley, Margaret. "The Roots of the I.Q. Debate." http://www.publiceye.org/pra/magazine/eugenics.html. 5 July 1998.

"Racism Resurgent." http://www.fair.org/extra/9501/bell.html. 6 July 1998.

Reilly, Philip R. "A Look Back at Eugenics." *The Gene Letter*, November 1996; also at http://www.geneletter.org/1196/eugenics.htm. 6 July 1998.

"Two Views of *The Bell Curve*" [book reviews]. *Contemporary Psychology*, May 1995; also at http://www.apa.org/journals/bell.html. 07/07/98.

Irradiation of Food

Irradiation of food is a method for destroying microorganisms that can cause food to spoil and that cause disease. From time to time, especially serious instances of food-borne disease occur. In August 1997, for example, Hudson Foods of Columbus, Nebraska, recalled 25 million pounds of beef that were suspected of being contaminated with the bacterium *Escherichia coli O175:H7*. The bacterium can cause nausea, vomiting, abdominal pain, and, in some cases, even more serious intestinal disorders.

Food-borne disease of all kinds is a serious problem in the United States and throughout the world. The U.S. Centers for Disease Control and Prevention estimate that at least 6 million Americans become ill each year from food-borne organisms and up to 4,000 of these individuals die from their illness. The need for protecting food from such risks is obvious.

History of Food Irradiation

Food irradiation is not a new technology. The practice was begun in Europe in the 1920s. It has taken much longer to catch on in the United States. Part of the delay is due to a decision by the U.S. Congress in 1958 that food irradiation was equivalent to adding a chemical to food. Under this definition, the U.S. Food and Drug Administration (FDA) is required to make a separate decision each time irradiation is proposed for use with a specific food.

The first such decision came in 1963 when the FDA approved irradiation for the treatment of wheat and flour to kill insects. Since that time, the agency has approved irradiation for eight other applications, including its use on white potatoes, spices, pork, fresh fruit, and poultry.

On an international scale, food irradiation is a relatively common method for protecting food. It has been approved for general use in forty nations, although it is commonly practiced in only twenty-eight.

Method

Food irradiation is accomplished by means of gamma rays emitted by radioactive materials. Gamma rays are a high-energy form of electromagnetic radiation, somewhat similar to, but more powerful than, X-rays. Gamma rays pass through the cells of a microorganism and destroy deoxyribonucleic acid (DNA), enzymes, and other essential chemicals needed for the cell's survival. The organism dies as a result of the irradiation.

The usual source of gamma rays used in food irradiation is cobalt-60 or cesium-137. The radioactive isotope is placed above a conveyor belt that carries the food to be treated. As the food passes under the isotope, it is exposed to gamma rays, and disease-causing organisms are destroyed. The food itself does not become radioactive and can be handled and packaged immediately upon leaving the conveyor belt.

Pros and Cons

The irradiation of food is a matter of concern to many people. One source of concern is the use of radioactive materials. Some people worry that irradiated food itself may become radioactive and, therefore, be dangerous to eat. Most scientists believe, however, that the level of radiation used in food irradiation is so low that there is virtually no possibility of the foods themselves becoming radioactive.

There are concerns also that irradiation may change the taste, texture, nutritional value, and/or shelf life of food. Such fears appear to be relatively baseless. Almost any technique used to process food does have some effect on taste, texture, and/or nutritional value. Indeed, nutrients must often be returned to processed foods in the form of food additives just for this reason. But irradiation appears to have less effect on these food properties than do most other methods of food processing, such as freezing, dehydrating, or boiling. With regard to taste and texture, tests have shown that consumers are unable to distinguish between irradiated and nonirradiated foods.

Finally, irradiation actually extends the shelf life of foods. As an example, non-treated pork can be stored in a refrigerator for about forty days, while irradiated pork can be stored for up to ninety days under the same condition.

Many health authorities now conclude that the overriding issue for food irradiation is the potential danger of food-borne illness versus the potential risk of radiation from the food irradiation process. The preponderance of opinion currently seems to be that microorganisms present a greater threat to human health than does irradiation. Among the organizations that have concluded that food irradiation is a safe and effective means of protecting foods are the World Health Organization, the American Medical Association, the American Council on Science and Health, the American Dietetic Association, and the National Aeronautics and Space Administration.

References

Blumenthal, Dale. "Food Irradiation: Toxic to Bacteria, Safe for Humans." *FDA Consumer*, November 1990.

Braus, Patricia. "Food Irradiation." In *The Gale Encyclopedia of Science*, edited by Bridget Travers. Detroit: Gale Research, 1996, 1515–1517.

"Facts About Food Irradiation." http://www.iaea.or.at/worldatom/inforesource/other/food/index.html. 27 May 1998.

"Food Irradiation Bibliographic Database." http://www.nal.usda.gov/fnic/foodirad/intro.html. 27 May 1998.

"Food Irradiation 2 Wholesomeness Collection: U.S. Government Reports from the 1950s and 1960s." http://www.nal.usda.gov/fnic/.Foodirad-v2/food2.html. 27 May 1998.

Loaharanus, Paisan. "Cost-benefit Aspects of Food Irradiation." *Food Technology*, January 1994, 104–108.

Manning, Anita. "Food Safety Regulators Warm to Radiation Idea." *USA Today*, 4 November 1997, 4D.

O'Neil, Patrick. "Digesting Irradiation." *Sunday Oregonian*, 19 October 1997, E1+.

"Position of the American Dietetic Association: Food irradiation." http://www.eatright.org/airradi.html. 27 May 1998.

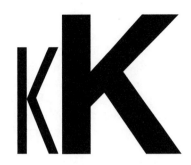

Kennewick Man

Kennewick Man is the name given to a nearly complete skeleton found on the shores of the Columbia River in Kennewick, Washington, in July 1996. The skeleton is regarded as the oldest complete skeleton ever found in the Pacific Northwest. It was dated by radiocarbon techniques and was found to be about 9,200 years old. Scientists estimate that the skeleton is one of a male who stood about five feet nine inches tall and died at the age of about forty-five to fifty-five. The discovery has led to a long and acrimonious debate between scientists and Native Americans as to what should eventually be done with the skeleton.

Native American Claims

At first, there seemed to be little question as to what was to become of the skeleton. The U.S. Corps of Engineers, to whom it was turned over, announced plans to return the remains to a group of five tribes, the Yakama, Umatilla, Nez Perce, Colville, and Wanapum. The Corps' decision was based on the Native American Graves and Repatriation Act, which requires that remains of Native Americans be returned to the tribe to which they belong.

On 3 October 1996, the Kennewick Man seemed headed for his final resting place. The Umatilla tribe was awarded custody of the skeleton and announced plans to inter the bones in an undisclosed location.

Scientific Objections

Hopes for a quick resolution of the Kennewick case were soon dashed, however.

From the outset, a number of scientists had expressed interest in conducting scientific tests on the skeletal remains. They felt that the remains might provide valuable information as to the early history of humans in North America. The first migrations to this continent from Asia are sometimes placed at about 14,000 years ago. The bones of Kennewick Man might hold clues as to the details of how that migration occurred, some scientists argued.

As a consequence, a group of eight anthropologists filed suit on 18 October 1996 to prevent transfer of the skeletal remains to the Umatilla tribe. They asked permission to examine the bones.

Over the next year, a wide array of individuals and institutions became involved in the debate over the fate of the Kennewick Man. Representatives from major media outlets visited Kennewick to find out more and then write about the controversy. The U.S. House of Representatives and Senate both became embroiled in the debate, eventually expressing their views on the subject in amendments to legislation. And the case finally ended up in the courts when first one side, then the other, sued to gain possession of the bones.

The Issue Posed by Kennewick Man

The discovery of the Kennewick Man raises some interesting and fundamental questions about the limits of scientific research. Native American leaders object strongly to any kinds of scientific tests' being conducted on the skeleton. They claimed that studies of the skeleton would make Native

Americans feel like laboratory animals being subjected to experimentation. Jerry Meninick, a member of the Yakama Indian Nation, suggested to reporters: "Let the anthropologists study their own bones" (*Tri-City Herald* 1996).

Scientists acknowledged that the skeleton might be a sacred object to Native Americans. But, they pointed out, there was still not enough information to indicate that the skeleton was actually a Native American or the tribe to which the man belonged. It would be inappropriate without more information, then, to turn the bones over to any specific tribe. Besides, the scientific tests envisioned by anthropologists were nondestructive and noninvasive. After the tests had been completed, a tribe would still be able to bury the skeleton with all the honors it deserved.

In July 1998, the chief archaeologist of the National Park Service announced that scientific tests were necessary to identify the ethnicity and tribal connection of the Kennewick Man. He said that such tests could take up to a year. After the tests had been completed, a final decision as to the skeleton's disposition would be made.

This decision satisfied almost no one. It denied tribal demands for possession of the skeleton, and, in the view of other archaeologists, it provided for too long a delay in completing the necessary identification tests.

The final decision in this case will provide additional input on the question as to what limits there are, if any, on the kinds of scientific research that can be conducted.

References

The most complete record of the Kennewick Man story can be found in the local newspaper closest to the find, the *Tri-City Herald* in Kennewick, Pasco, and Richland, Washington. The paper's Kennewick Man Virtual Interpretive Center can be found at http://www.tri-cityherald.com/bones/news/. Also see Richard L. Hill. "More Tests Asked to Decide Future of Kennewick Man." *Oregonian*, 17 July 1998, C9.

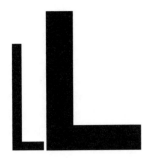

Land Use

See Barrier Islands; Clearcutting;
Conservation and Preservation; Grazing
Legislation; Headwaters Forest; Logging
Roads; Multiple Use/Sustained Yield;
Sagebrush Rebellion; Stream
Channelization; Water Rights; Wetlands;
Wild Horses and Burros; Wildlife
Management

Late-Term Abortions

See Intact Dilation Evacuation

Legalization of Drugs

Virtually every human society about which
historians have information has used drugs
in one form or another. Throughout history,
people have smoked and chewed tobacco
products, drunk alcoholic beverages, or
taken drugs in some other form. Some Native American tribes, for example, eat psy-
chedelic mushrooms because of the mental
state they produce. Drug users in these
tribes claim to have mystical, religious,
"out-of-body" experiences as a result of
eating these mushrooms.

In other cultures, people have used
drugs purely for recreational purposes, for
the mental, physical, or emotional pleasure
they bring. Cocaine is the drug of choice
among some people because of the sense of
euphoria it provides.

The Legal Status of Drugs in the United States

Drug use in the United States has long been
highly restricted. Only alcohol and nicotine

(in tobacco) among the major drugs have
been legally available. Other drugs with
mind-altering potential (such as lysergic
acid diethylamine [LSD], cocaine, and
heroin, for example) have never been
legally available to the general public for
nonmedical purposes.

Beginning in the 1970s, a few people
started to question this kind of drug policy.
They argued that drug laws ought to be revised or abolished. Some critics have called
for the repeal of laws against all currently
illegal drugs. Others have focused on the
elimination of marijuana laws only.

By the early 1990s, calls for the legalization of marijuana and other drugs had become more widespread. Such diverse public
figures as conservative columnist William F.
Buckley, Jr., economist Milton Friedman,
Congressman George W. Crockett, astronomer Carl Sagan, former New York City
police chief Patrick V. Murphy, federal judge
Robert Sweet, former secretary of state
George Schulz, and Baltimore mayor Kurt
Schmoke had all spoken out in favor of legalizing drug use. In addition, support for this
position among the general public had been
slowly growing for more than two decades.
Various public opinion polls showed that
anywhere from 10 percent to one-third of all
Americans had come to support decriminalization of marijuana and/or other drugs.

For many people, the idea of legalizing
drugs seems absurd. Most of us have
grown up knowing about the horrors of
drug addiction, they point out. How can
any reasonable person recommend decriminalization of drugs?

Arguments in Favor of Decriminalization

At the outset, most proponents of legalization state very clearly that they do not favor drug use. Their argument is that the "war against drugs" is immoral, expensive, and ineffective. They raise some of the following specific arguments.

First, some critics argue that drug use is a normal human behavior, observed in one form or another in all human cultures. In his book, *Intoxication: Life in Pursuit of Artificial Paradise*, University of California at Los Angeles psychopharmacologist Ronald K. Siegel argued that the craving for mind-altering substances is inherent in the human race. He has concluded that "winning the war on drugs by eradicating non-medical drug use is neither possible nor desirable. We need intoxicants—not in the sense that an addict needs a fix, but because the need is as much a part of the human condition as sex, hunger, and thirst" (Siegel 1989, 288).

Second, those who favor decriminalization argue that governments spend far too much money on drug enforcement programs, with too little return. In the early 1990s, the federal government alone was spending more than $10 billion a year to control illegal drug use.

Third, as long as drugs are illegal, some people say, they will continue to be a major source of income for organized crime. By some estimates, organized crime syndicates now make about $50 billion annually through the sale of drugs.

Fourth, efforts to control drug use simply do not work. After decades of trying to cut off drug supplies, educate drug users, provide detoxification programs, and increase penalties for drug use, the problem is no closer to solution today than it was forty years ago, critics say.

Finally, some people say that the question of drug use is really an issue of personal privacy. People should have the right to do what they want with their own bodies, as long as they cause no harm to others. In a 1972 essay in *Newsweek* magazine, Professor Friedman argued that the government has no right to tell people not to drink or to commit suicide or to use illegal drugs.

Arguments against the Decriminalization of Drugs

The arguments for legalization of drugs have clearly not convinced the majority of people. Those who support the continuation of existing drugs laws have their own arguments. Some of those arguments include the following:

First, legalization of drugs will probably lead to an increase in drug use. With drugs legally and readily available, many current nonusers might be tempted to start using some drugs. Some experts have predicted that legalization would result in a doubling of the number of people who use drugs for recreational purposes.

Second, proponents of decriminalization do not give adequate attention to the possible harmful health effects of drugs. Many experts believe that even relatively mild drugs like marijuana can cause serious and irreversible brain damage. Those who use marijuana may also move on to more dangerous drugs, such as cocaine and heroin, according to this argument.

Third, opponents of legalization do not agree that the nation is losing the war on drugs. James Q. Wilson, chairman of President Nixon's 1972 National Advisory Council for Drug Abuse Prevention, has claimed that there is no evidence of a "nationwide [drug] plague that threatens to engulf us all" (Wilson 1990, 25).

Fourth, advocates of the status quo argue that drug education and treatment programs would become meaningless if drugs were made legal. Why would people go through all the trouble of staying in drug rehabilitation programs if there were no legal reason for them to try getting rid of the drug habit? And what would drug education programs tell young people about the dangers of using *legal* substances?

Finally, the critical issue for many opponents of legalization is a moral argument. How can we possibly give official approval to a practice (drug use) that is so harmful to individuals, their friends and neighbors, and the community at large, they ask.

References

"Against Legalizing Drugs." http://www.commonlink.com/users/carl-olsen/DPF/gray–002.html. 25 June 1998.

Baer, Donald. "A Judge Who Took the Stand: It's Time to Legalize Drugs." *U.S. News and World Report,* 9 April 1990, 26–27.

Dennis, Richard J. "The Economics of Legalizing Drugs." *The Atlantic,* November 1990, 126–132.

"Drugs." http://www.gargaro.com/drugs.html. 21 May 1999.

"The Federal Drug Store." *National Review.* 5 February 1990, 34–41. Also see responses in Letters to the Editor, 1 April 1990, 42–44+; 14 May 1990, 58–60; 28 May 1990, 49–52; and 25 June 1990, 44.

Friedman, Milton. "Prohibition and Drugs." *Newsweek,* 1 May 1972, 104.

"Legalizing Drugs: Just Say Yes." *The National Review,* 10 July 1995, 44–51; also at http://www.pdxnorml.org/review.html. 25 June 1998.

Nadelmann, Ethan A. "Drug Prohibition in the United States: Costs, Consequences, and Alternatives." *Science,* 1 September 1989, 939–947, with responses in Letters to the Editor, 1 December 1989, 1102–1105.

Siegel, Ronald K. *Intoxication: Life in the Pursuit of Artificial Paradise.* New York: E. P. Dutton, 1989.

"When War Doesn't Work." http://www.cwrl.utexas.edu/~dliss/student . . . jects/final_project/rebecca2/solution.htm. 25 June 1998.

Wicek, Walter. "Biting the Bullet: The Case for Legalizing Drugs." *The Christian Century,* 8–15 August 1990, 736–739.

Wilson, James Q. "Against the Legalization of Drugs." *Commentary,* February 1990, 21–28.

Lie Detector

See Polygraph

The Limits to Growth

Rising concerns about population issues in the 1960s occurred at a time when the power of electronic computing was becoming apparent. It should be of little surprise, then, that some scholars began to explore the possibility of using computers to project the Earth's future under various population growth scenarios. Probably the most famous of these studies was one conducted under the auspices of the Club of Rome in the early 1970s.

Predicting the Consequences of Population Growth

The Club of Rome was an exclusive group of about 100 scholars and corporate leaders from twenty-five nations interested in important international issues. The club commissioned a study of future population trends to be conducted by computer experts Jay Forrester and Dennis and Donella Meadows. Forrester and the Meadows developed a computer program that included a number of basic variables, such as birth and death rates, reserves of natural resources, and environmental factors. They then asked the computer what consequences might be expected under a variety of future population growth scenarios. Among these scenarios were continued exponential population growth with various changes in technological capacity and reduced population growth with the same technological changes.

Using the data it was given, the computer program predicted disastrous future events under any circumstances of exponential population growth. It warned of loss of natural resources, vastly increased pollution, and, ultimately, dramatic declines in population as a result of food shortages and climbing death rates. Only if population growth were severely limited and a number of governmental actions were taken to control the distribution of wealth and natural resources was the computer program able to predict a "livable" future for humans on Earth.

The results of this study were published in 1972 in a book entitled *The Limits to Growth.* The book was an immediate popular success and ultimately sold more than 3 million copies in twenty-three languages.

Certain population experts turned to the book as proof of their warnings about the dangers of runaway population growth.

Questions about *The Limits to Growth*

Criticisms of the Forrester-Meadows research began to appear almost immediately. One charge raised was that computer modeling was not yet well enough developed to make valid projections of the type attempted by Forrester and the Meadows. More important, critics pointed out that the researchers had neglected to consider the possibility that technological innovations would be found that would deal with population-related problems and that these innovations would grow at least as fast as population itself.

Indeed, when this hypothesis is introduced into the Forrester-Meadows computer program, a very different future results. In that future, population and pollution both rise modestly and then level off by about 2100. In the meanwhile, the quality of life rises slowly in 2000, and then begins to improve rapidly after that time.

References

Boyd, Robert. "World Dynamics: A Note." *Science,* 11 August 1972, 516–519.

Meadows, Donella H., et al. *Groping in the Dark: The First Decade of Global Modeling.* New York: John Wiley, 1982.

———. *The Limits to Growth.* New York: Universe Books, 1972.

Mesarovic, Mihajlo, and Eduard Pestel. *Mankind at the Turning Point,* New York: E. P. Dutton, 1974.

Oltmans, Willem L. *On Growth.* New York: G. P. Putnam, 1974.

Starr, Chauncey, and Richard Rudman. "Parameters of Technological Growth." *Science,* 26 October 1973, 358–364.

Logging Roads

Logging roads are roads built into forested areas by lumber companies in order to allow them to reach trees scheduled to be harvested. It is obvious that roads of this kind have to be built if companies are going to reach the live and fallen trees that are to be removed from an area.

Logging roads represent a very considerable financial expense. More than 370,000 miles of such roads have been built in national forests alone. That network of roads is more extensive than the U.S. interstate highway system. Estimated cost for the construction of a typical logging road is about $30,000 per mile. Like other road and highway systems, logging roads require continual repair and replacement. By the late 1990s, the U.S. Forest Service was estimating that more than 230,000 miles of the logging road system in national forests needed repair or replacement. The issue that has long been raised is who it is that should pay for the building of those roads.

Government Subsidies for Road Building

Historically, the U.S. government has assumed a major responsibility for financing the construction and maintenance of logging roads. Timber companies have been allowed to deduct the cost of building logging roads from the fee they pay for removing trees from public lands. By some estimates, timber companies save as much as $40 million each year as a result of the federal government's subsidy for the construction and maintenance of logging roads by companies themselves.

Timber companies argue that the subsidy is a fair one. Logging roads are used

Logging roads serve some essential functions for the timber industry, but they can also create some serious environmental problems for an area. (Corbis/Layne Kennedy)

extensively by the general public, they point out, for hiking, bicycling, and other recreational activities. People would not be able to gain access to remote areas of the national forests were it not for these roads, they say.

In addition, local governmental agencies benefit from road construction, which allows timber companies to harvest and sell trees, providing state and local government with valuable tax revenues. If logging roads were not built, these revenues would decrease significantly. Boise County (Idaho) officials have pointed out, for example, that 29 percent of their budget comes from taxes on logging in the state.

Objections to Governmental Support for Logging Roads

Environmentalists have long objected to the practice of having the federal government provide subsidies for the construction and maintenance of logging roads. In the first place, they say, companies should build their own roads if they need them. The federal government should not have to pay for an activity that, for a timber company, should be part of the cost of doing business.

Also, logging roads have been found to be a major cause of many environmental problems in forests. They tend to cut across the topography of a hilly area, disrupting the natural flow of water down the hill. This change in water flow, in turn, causes erosion of the land, washing soil into lakes and rivers. Streams become clogged, and fish and other aquatic life die off as a result. The habitat of land animals is also disturbed.

Recently, environmentalists have been joined in their fight against logging roads by fiscal conservatives. In a time when fiscal resources are limited, the government should no longer be providing "corporate welfare" for timber companies, conservatives say.

In the late 1990s, therefore, an interesting coalition developed between two groups that have often been opposed to each other: environmentalists and fiscal conservatives. These two groups joined forces to try eliminating federal subsidies for logging road construction and maintenance. In July 1998, the House of Representatives passed an amendment to the Department of the Interior's funding bill ending the government's subsidy program for logging roads.

References

Brinckman, Jonathan. "Crumbling Forest Roads Lead to Trouble." *Oregonian*, 20 February 1998, E4.

Cushman, John H. "House G.O.P. Leaders Agree to End Timber Road Subsidy." *New York Times*, 4 June 1998, A16.

"Logging Roads: Senate Should Take Its Chain Saw to a Foolish Subsidy." *Detroit Free Press*, 16 September 1997.

Schouten, Fredreka. "Logging Roads: 'Corporate Welfare' for Timber Industry Coming to an End?" Gannet News Service, 7 March 1997.

"Streams, Landslides, Logging Roads and Rain." http://www.efn.org/~jpreed/landsl.html. 28 July 1998.

Marbled Murrelet

See Headwaters Forest

Masked Bobwhite Quail

The masked bobwhite quail is classified scientifically as *Colinus virginianus ridgwayi*. The story of its discovery, near extinction, and ultimate survival provides some important object lessons in debates over endangered species.

History

The masked bobwhite quail was first seen in Arizona in the 1860s but was not officially named until two decades later. Early surveys of the bird showed that its habitat covered southern Arizona and northern Sonora in Mexico. Few studies of the quail were completed, but it was determined that the bird preferred a grassy environment common in much of the western United States before widespread cattle grazing began.

When cattle were brought into the quail's habitat, its future was bleak. Biologists who studied the quail decided that they could not survive on land that cattle had stripped clean of vegetation. So it was no surprise to discover that the bird had become essentially extinct through most of its Arizona range by the early 1900s. It had little or no chance of survival against the numbers of cattle—estimated to be between 900,000 and 1,500,000 head in 1890—that roamed Arizona.

A few interested biologists held out hope for the quail's survival when a small breeding colony of the birds was found in Sonora in 1964. Efforts to create a rescue program in the area were unsuccessful, however, and the remaining birds were given to the U.S. Fish and Wildlife Service (FWS) in 1966.

By that time, the FWS had begun to plan for a restoration program for the masked bobwhite quail on the Buenos Aires National Wildlife Refuge in Arizona. The FWS hoped to restore a grasslands habitat on a 45,000-hectare (111,000 acre) portion of the refuge. Restoration efforts included the removal of old fences and corrals left by cattlemen and the burning of vegetation that had grown up in place of the original grassland ecosystem.

Ongoing Controversies

By the late 1990s, the Buenos Aires project had become the focus of an intense controversy among environmentalists and cattle

The threat posed to the masked bobwhite quail highlights the question as to whether humans have an obligation to save every species possible and, if not, which ones are to receive that attention. (Corbis/Peter Johnson)

ranchers and state legislators. One topic of that debate was the long-standing controversy between the first two groups. Cattlemen were angry that a wide stretch of highly desirable grazing land had been taken out of production. The land could be used much more profitably, they said, if used for cattle ranching than for protecting a relatively insignificant little bird.

A second topic of discussion was the cost effectiveness of the restoration program. Over the lifetime of the Buenos Aires project, more than 30,000 hand-grown quail had been released to the wild. Many of those birds had not survived, however, but had been taken by coyotes, rattlesnakes, and other predators. Critics complained that the millions of dollars spent on the restoration program had been wasted. In a time of economic cutbacks, they said, the money could better be spent on any number of social programs.

But those in charge of restoration continued to feel strongly about the need and ultimate success of the restoration effort. As refuge manager Wayne Shifflett was quoted as saying, "We're nourishing the earth with ashes, which will create new native grass and better cover for the quail. But we're also slowly putting a fragmented ecosystem back together here, the only Sonoran savanna grasslands left in the United States. . . . Cattle will never come back to this refuge" (Sahagun 1997).

References

Hawks, Troy T. "Grazing and the Endangered Masked Bobwhite Quail in Sonora, Mexico." http://www.gamebird-alliance. org/maskedpaper.html. 8 August 1998.
"Masked Bobwhite Quail." http:// www.gamebird-alliance.org/ maskedbob.html. 8 August 1998.
Sahagun, Louis. "Quail Release Triggers Arizona Range War." *Los Angeles Times*, 2 April 1997, A5.

Medical Issues

See Abortion; Alternative Therapies; Anencephalic Babies; Assisted Reproductive Technologies; Early-Term Surgical Abortions; Euthanasia; Faith Healing; Fetal Tissue Research; Health Hazards of Electromagnetic Fields; HIV and AIDS; Human Experimentation; Human Gene Therapy; Intact Dilation Evacuation; Medical Uses of Marijuana; "Morning-After" Pill; Organ Transplantation; Right to Die; Sterilization, Human; Vitamins and Minerals

Medical Uses of Marijuana

Many Americans are very much concerned about drug abuse. They see drugs as the cause of a large part of the nation's crime problem, as a contributing factor in workplace inefficiency, and as a threat to the health and well-being of future generations. For over two decades, the federal government has conducted an aggressive campaign to reduce both the demand and the supply of drugs to American citizens.

Ironically, some of the drugs of greatest concern have the potential for beneficial as well as harmful uses. The amphetamines, as an example, have some important medical uses. They have been used to treat a form of epilepsy known as petit mal, to treat children who are hyperkinetic (overactive), and to suppress appetite (diet pills). But amphetamines are also used as recreational drugs, to produce a sense of euphoria in users.

Marijuana as a Drug of Abuse

In fact, drugs are classified by the federal government as to their potential for both beneficial and harmful uses. Those classified as *controlled substances* are subdivided into three categories: Schedule 1, Schedule 2, and Schedule 3 drugs. Schedule 1 drugs are those with no medical uses, such as heroin, lysergic acid diethylamine (LSD), and mescaline. Schedule 2 drugs are those with some medical uses, such as morphine and the amphetamines. Schedule 3 drugs are prescription drugs that are often abused. Examples of Schedule 3 drugs are valium and phenobarbital.

Marijuana is only one of many drugs that can both offer medical benefits and pose the risk of misuse by drug abusers. (Corbis/Todd Gipstein)

There is probably no illegal drug about which there has been as much dispute as marijuana. Marijuana is currently classified by the government as a Schedule 1 drug. It is regarded by the government as a drug with no medical benefits and a high potential for abuse.

Many authorities have long argued that marijuana is a "gateway" drug, one that is easy to obtain and easy to use. Once a person becomes a user of marijuana, he or she tends to experiment with and become addicted to more dangerous drugs, according to the "gateway" theory. The claim is that people move on from marijuana to cocaine and heroin, as an example.

Other people reject this argument. They say that marijuana is a pleasant, harmless recreational drug, no more dangerous than nicotine in cigarettes or alcohol, both of which are legal. A number of individuals who hold this view argue that marijuana should be legalized as a drug.

Medical Uses of Marijuana

Within recent years, a new dimension has been added to this argument. Some people are claiming that marijuana has distinct medical benefits. The drug should be made available by prescription to those who can benefit from it, they say, just as are other beneficial drugs. In 1996, voters in Arizona and California passed referenda making the medical use of marijuana legal.

Benefits from marijuana therapy supposedly lie in four general areas. The drug is said to reduce pain and muscle spasms associated with epilepsy and multiple sclerosis, reduce nausea associated with chemotherapy, stimulate appetite in patients with human immunodeficiency virus (HIV) infection, and reduce pressure on the eyeball associated with glaucoma. As with almost any form of medication, side effects may accompany these beneficial effects. For example, marijuana may decrease blood flow to the optic nerve, reducing its ability to function properly and contributing to the loss of sight associated with glaucoma.

Scientists now understand the process by which marijuana produces these effects. An active ingredient in the drug, tetrahydrocannabinol (THC), mimics the behavior of naturally occurring chemicals in the human brain. THC is accommodated in receptor cells in much the same way as are the natural chemicals, inducing hunger, changes in motor activity, and other effects.

Legal Status of Marijuana as a Drug

Marijuana has been illegal in the United States since 1937, and the federal government has long maintained an active campaign to reduce use of the drug. There has been only one exception to that campaign. In 1976, the U.S. Food and Drug Administration (FDA) initiated a "compassionate-use" program of supplying marijuana to people with unusual medical conditions. In 1992, the agency discontinued the program because of a flood of applications from HIV patients and because of fears that it was sending "the wrong message" about drug use. Of the thirteen people enrolled in the program at the time, five have since died.

Passage of the 1996 Arizona and California referenda produced an aggressive response by President Bill Clinton's administration. Gen. Barry McCaffrey, director of the Office of National Drug Control Policy; Secretary of Health and Human Services Donna E. Shalala; Attorney General Janet Reno; and President Clinton himself all held news conferences at which they warned about the dire consequences of the Arizona and California votes.

Shalala spoke about the serious health consequences of marijuana use on the body, Reno threatened legal action against doctors who prescribed marijuana, and McCaffrey claimed that "there is no proof that smoked marijuana is the most effective available treatment for anything" (Cowley 1997, A12). Within weeks, even the Internal Revenue Service became involved in the controversy. The agency announced that it would not allow the cost of marijuana treatments to be deducted as a legitimate medical expense even in Arizona and California.

McCaffrey went on to suggest that medical researchers begin to examine potential health benefits from compounds found in marijuana, such as THC. If such benefits were found to exist, he said, then "we must immediately make them available to the American medical community" (Cowley 1997, A12). General McCaffrey even announced a $1 million grant to the National Institutes of Health to begin such studies.

Support for Marijuana as a Medical Treatment

For some medical researchers, McCaffrey's arguments were difficult to understand. They pointed out that more than seventy-five peer-reviewed studies on the health effects of marijuana were already available. The National Academy of Sciences (NAS) had reviewed a number of those studies and acknowledged as early as 1982 that

marijuana does have medical value in some circumstances. These researchers claimed that McCaffrey's real goal in recommending further research was to delay implementation of the new Arizona and California laws for the term of the new studies, a period of at least eighteen months.

Support for the promarijuana position came in early 1997 from one of the two most prestigious medical journals in the United States, the *New England Journal of Medicine*. Editor Jerome P. Kassirer wrote that "a federal policy that prohibits physicians from alleviating suffering by prescribing marijuana for seriously ill patients is misguided, heavy-handed, and inhumane" (Kassirer 1997, 366). He pointed out that the Drug Enforcement Administration (DEA) had already reviewed scientific evidence on the medical effects of marijuana as early as 1988. One of the agency's own administrative law judges had written at the time that "it would be unreasonable, arbitrary, and capricious for DEA to continue to stand between those sufferers and the benefits of this substance in light of the evidence in this record"(Kassirer 1997, 366). In spite of that recommendation, the DEA refused to move marijuana to a Schedule 2 classification and continued to oppose its medical use.

References

Annas, George J. "Reefer Madness—The Federal Response to California's Medical-marijuana Law." *New England Journal of Medicine,* 7 August 1997; also at http://www.nejm.org/content/1997/0337/0006/0435.asp. 21 May 1999.

"Cannabis Dependence: Marijuana." *The Harvard Medical School Mental Health Letter,* November 1987; also at http://www.mentalhealth.com/magl/fr51.html. 21 May 1999.

Conrad, Chris. *Hemp for Health: The Medicinal and Nutritional Uses of Cannabis Sativa.* Los Angeles: Inner Traditions International, 1997.

Cowley, Geoffrey. "Marijuana: Is It Good Medicine?" *Oregonian,* 19 February 1997, A12.

"Good Medicine?" [transcript of a PBS *Newshour* segment of 30 December 1996].

http://www.pbs.org/newshour/bb/
health/december96/mccaffrey_12–30.
html. 21 June 1998.

Grinspoon, Lester, and James B. Bakalar.
Marijuana, the Forbidden Medicine. New
Haven: Yale University Press, 1997.

Kassirer, Jerome P. "Federal Foolishness and
Marijuana." *New England Journal of
Medicine,* 30 January 1997, 366.

Martin, Glen. "Judge Orders Shutdown of
S.F. Pot Club." *San Francisco Chronicle,* 16
April 1998, A1+

"Medical Use of Marijuana: Assessment of
the Science Base." National Academy of
Sciences. [Study inaugurated in
December 1997] http://www2.nas.
edu/medical-mj/. 21 June 1998.

Paddock, Richard C. "Is Smoking Pot Good
Medicine?" *Los Angeles Times,* 26 February
1995; also at http://turnpike.net/
~jnr/goodmed.htm, 06/21/98.

"Say It Straight: The Medical Myths of
Marijuana." http://www.usdoj.gov/
dea/pubs/sayit/myths.htm. 21 June
1998.

"Statement by General Barry R.
McCaffrey . . ." 2 December 1996. http://
www.whitehousedrugpolicy.gov/news/
testimony/dope.html. 21 June 1998.

Szwarc, Rina. "Views Are Mixed on
Marijuana Use by Some Patients." *San
Diego Daily Transcript,* 19 September 1995;
also at http://www.sddt.com/files/
library/95headlines/DN95_09_19/
STORY95_09_19_04.html. 21 June 1998.

"Uncle Sam Is the Pot Supplier for Sick
Florida Stockbroker." *San Francisco
Chronicle,* 18 November 1996, A8.

Wren, Christopher S. "Doctors Criticize
Move against State Measures." *New York
Times,* 31 December 1996, D18.

Zeese, Kevin B., and Tom O'Connell, M.D.
"A Superfluous Study." *Oregonian,* 4
February 1997, B9.

Zimmer, Lynn, and John P. Morgan.
*Marijuana Myths Marijuana Facts: A Review
of the Scientific Evidence.* New York:
Lindesmith Center, 1997.

Mediterranean Fruit Fly

The Mediterranean fruit fly is a member
of the biological family known as the true
flies. Its systematic name is *Ceratitis capi-*
tata. It is commonly referred to as the
Medfly.

The Medfly is considered by some authorities to be the most serious agricultural
pest in the world. It attacks more than 250
kinds of fruit, vegetable, and nut. Its arrival
in the United States in 1975, then, produced
a near-panic response by agricultural experts. The U.S. Department of Agriculture
predicted that a full-scale invasion by the
pest could cost U.S. farmers more than $1.5
billion in lost crops. Officials in California,
where the fly first appeared, determined to
take immediate and aggressive action to
prevent the spread of the pest.

Controlling the Medfly

The first approach to controlling the Medfly was with aerial spraying of areas in
which the pest had been found. The spray
used was malathion, a well-known pesticide. Malathion is a member of the
organophosphate pesticides that also include the more toxic parathion. Because
these chemicals attack the nervous system
in both insects and humans, it is recommended that they be used only with great
caution in areas where humans are present.

A number of professional scientists and
residents of the sprayed areas questioned
the use of malathion for the control of the
Medfly. They suggested that the pesticide
was far too toxic to be used over inhabited
areas. They pointed out that other methods
of control were safer to humans and probably as effective as the use of malathion. For
example, sterilized male flies could be released to the environment to mate with female flies. This approach would prevent
flies from reproducing, thus reducing or
eliminating their threat to agriculture.

In fact, state agencies later switched
from aerial spraying to the use of sterilized
males as a preferred method of control. But
spraying remained the treatment method of
choice early on in the Medfly incursion because it provide a quick, easy, and efficient
method of killing off flies. Officials either
decided that the threat to humans was low
enough to ignore or that the potential bene-

One of the issues raised by the problem of Mediterranean fruit fly controls is how far a governmental body has the right to go to protect one group (such as farmers) while imposing on the rights of the general public. (Corbis/Ted Streshinsky)

fit to farmers outweighed any potential danger of humans exposed to the spray.

Current State of Affairs

By early 1998, the combination of spraying, trapping, and release of sterilized males appeared to have ended the threat of the Medfly to California agriculture. To some extent, the inability of the fly to adapt to the California climate may have been a factor in its demise in the state. In any case, the state department of agriculture lifted its quarantine on the fly in April 1998.

Ironically, the state of Florida experienced its first outbreak of Medfly infestation at almost the same time. By June 1998, the state agricultural department had begun aerial spraying of malathion in Manatee County, one of the first locations in which Medflies had been discovered. According to observers from California, it was an example of history repeating itself. Florida agriculturists took advantage immediately of a tried and true weapon—malathion—while outcries from other scientists and the general public questioned whether the general public was being exposed to an unacceptable level of toxic pesticide.

References

Hollingsworth, Jan. "Experts Disagree on Air War." *Tampa Tribune*, 6 June 1998; also at http://www.tampatrib.com/news/fly110 30.htm. 13 July 1998.

Judson, David. "Congressional Panel Hears Debate over Medfly Spraying." Gannett News Service, 5 May 1994.

Reed, Susan, and Dan Knapp. "Controversy: Californians Take Issue with the Pesticide Malathion, Which Comes in on Medfly Wings and a Sprayer." *People*, 2 April 1990, 67.

"USDA Lifts Last California Medfly Quarantine." Reuters, 27 April 1998.

Methyl Bromide

Methyl bromide is a chemical compound with the molecular formula CH_3Br. It is one of the few halogenated hydrocarbons to occur naturally. A halogenated hydrocarbon is a chemical compound containing carbon, hydrogen, and at least one halogen (fluorine, chlorine, bromine, or iodine). Methyl bromide is produced as the result of biological activities in the ocean. It is then released to the atmosphere, where it has a lifetime of about 1.5 years.

Methyl bromide is also produced synthetically for commercial use, the most important of which is fumigation. Nearly 90 percent of this methyl bromide is used in agriculture, for the fumigation of soil. Much smaller amounts of the compound are used in the fumigation of commodities and structures. The compound is a highly effective pesticide for the control of insects, nematodes, rodents, pathogens, and weeds. In 1997, about 27 million kilograms (60 million pounds) of the product were being used annually.

Environmental Effects

Methyl bromide, like other halogen-containing compounds, has the capability of attacking and destroying ozone (O_3) in the stratosphere. The compound escapes efficiently from the soil in which it has been used, diffuses into the stratosphere, and breaks down to release a bromine atom. The bromine atom then attacks and destroys ozone molecules:

$$CH_3Br \rightarrow CH_3 + Br$$
$$Br + O_3 \rightarrow BrO + O_2$$

Because of this action, one might have expected methyl bromide to have been included among substances whose production and use were banned by the Montreal Protocol on Substances that Deplete the Ozone Layer of 1987 or the London Amendments to that protocol of 1990. However, such was not the case, primarily because no satisfactory substitute for the compound exists. The Copenhagen Amendments of 1992 to the protocol, however, capped production of the compound by 1995 and called for further studies and recommendations for a complete phaseout of production. Eventually, a date of 2005 was set for that action. The deadline for developing nations was extended to 2015.

Debate over Limitations on Methyl Bromide Use

Farmers in the United States faced a special dilemma with regard to proposed bans on the production of methyl bromide. Amendments to the U.S. Clean Air Act adopted in 1990 had set an even earlier date of 1 January 2001 for an end to the manufacture and importation of methyl bromide. Farmers could continue to use the fumigant after that date, but no more of the product would be available to them. U.S. farmers argued that depriving them of the only chemical effective for this particular use—soil fumigation—would place them at a serious disadvantage vis-à-vis farmers in other nations. Even more serious, they pointed out, was the extra ten years (to 2015) that had been granted to farmers in developing nations for the production and use of methyl bromide.

Environmentalists argued for a strict adherence to the Clean Air Act deadline. They pointed out that farm workers are already being exposed to unacceptable levels of the chemical and that the United States should not create a pesticide-use policy based on international regulations.

See also: Ozone Depletion

References

Adler, T. "Methyl Bromide Doesn't Stick Around." *Science News,* 28 October 1995, 278.

"California Study Admits Methyl Bromide Safety Standard Inadequate." http://www.ewg.org/pub/home/reports/DPRweb/dprweb.html. 14 September 1997.

"Methyl Bromide." Pesticide Action Network. http://www.igc.apc.org/panna/campaigns/MBPage.html. 21 May 1999.

Ralof, Janet. "U.N. to Oversee Methyl Bromide Phaseout." *Science News,* 16 December 1995, 405.

The U.S. Environmental Protection Agency maintains a website on methyl bromide at http://www.epa.gov/ozone/mbr/mbrqa.html.

Montreal Protocol

The Montreal Protocol on Substances that Deplete the Ozone Layer is a treaty originally signed by twenty-four nations on 16 September 1987. Those twenty-four nations are responsible for the production of 99 percent of the chlorofluorocarbons (CFCs) produced in the world and for 90 percent of the consumption of CFCs. The primary provisions of the protocol were to define certain substances as ozone-depleting chemicals, to establish a program for the phase-out of the production of those chemicals, and to create an ongoing administrative body to handle provisions of the treaty.

In its original form, the protocol called for the reduction in the production of five chemicals, CFC-11, CFC-12, CFC-113, CFC-114, and CFC-115, to 20 percent of their 1986 levels by 1995 and to 50 percent in 2000.

The protocol was amended twice, at London in 1990 and at Copenhagen in 1992. The London amendments accelerated the phase-out of the original ozone-depleting chemicals, calling for a more rapid rate of reduction by 1996. In addition, it added a host of new ozone-depleting chemicals to be banned. These included CFC-13, CFC-111, CFC-112, CFC-211, CFC-212, CFC-213, CFC-214, CFC-215, CFC-216, and CFC-217; halons 1211, 1301, and 2402; carbon tetrachloride (CCl_4); and methylchloroform (CH_3CCl_3). Phaseout provisions differed for each category of ozone-depleting chemical, but production of all except methylchloroform was to be completely banned by 2000. The ban on methylchloroform was to occur by 2005.

The Copenhagen amendments advanced phase-out timetables even more. Production of all CFCs, carbon tetrachloride, and methylchloroform was to be completed by 1996, and of halons, by 1994. In addition, a phase-out schedule for the production of hydrochlorofluorocarbons (HCFCs) was created, calling for a complete ban on the production of these chemicals by 2030. Finally, limitations on the production of another ozone-depleting chemical, methyl bromide (CH_3Br), were first established.

See also: Chlorofluorocarbons; Hydrofluorochlorocarbons; Methyl Bromide; Ozone Depletion

References

Benedick, Richard E. *Ozone Diplomacy: New Directions in Safeguarding the Planet.* Cambridge: Harvard University Press, 1991.

Gribbin, John. *The Hole in the Sky: Man's Threat to the Ozone Layer.* New York: Bantam Books, 1988.

The full text of the Montreal Protocol and later amendments can be found at http://www.unep.org/unep/secretar/ozone/montreal.htm. 23 June 1998. A summary of the protocol can be found in David E. Newton, *The Ozone Dilemma.* Santa Barbara, CA: ABC-CLIO, 1995, 102–123.

"Morning-After" Pill

Women who choose to have abortions do so usually because they have decided that they are unable or unwilling to care for a child. In the majority of cases, such women prefer to have their abortions as soon as possible after leaning that they are pregnant. The ideal technique for such women might be a "morning-after" pill, a pill that could be taken the morning after unprotected sexual relations. A "morning-after" pill would allow a woman to avoid pregnancy even if she had forgotten or been distracted from using her usual method of birth control.

In fact, a "morning-after" pill has been known and used in Europe for a number of years. The pill is not really a new drug, but any one of six contraceptives generally available to women in most parts of the world. The difference is that the contraceptives must be taken in much larger doses after sexual intercourse in order for them to function as "morning-after" pills. In such cases, the contraceptive prevents implantation of a fertilized egg. Used in this context, the drug is also known as an *emergency contraceptive.* More than 4 million women in Great Britain alone have already taken emergency contraceptives with no serious medical problems having been reported.

Approval of Emergency Contraceptives in the United States

In early 1997, the U.S. Food and Drug Administration (FDA) announced approval for the labeling of six contraceptives as "safe and effective 'morning-after' pills." The decision was an important one because manufacturers of the six drugs had previously declined to advertise or sell their products as emergency contraceptives. They feared possible legal problems resulting from the lack of governmental approval for such labeling of their products.

As a result of the FDA's decision, physicians were able to begin prescribing the six contraceptive pills in increased doses as a means of preventing pregnancies. The drugs are effective up to seventy-two hours following intercourse and have an effectiveness of about 75 percent. According to one estimate, 2.3 million pregnancies—of which 1 million would otherwise have ended in abortions—would be prevented annually with the availability of the new "morning-after" drugs.

The FDA decision further complicates the debate between those who approve and those who oppose access to abortion. Many individuals who oppose abortion believe that life begins at the moment an egg is fertilized. For those individuals, the "morning-after" pill is still a means of abortion that should, therefore, be opposed. The labeling of such pills as "emergency contraceptives" is, critics say, merely a semantic game to hide the real intent and function of the pills.

For those who oppose abortion, however, the existence of "morning-after" pills presents some new and difficult tactical problems. It is much more difficult to protest and act against women who take pills in the privacy of their homes than it is to protest and act against clinics in which surgical abortions take place.

See also: Abortion; Early-term Surgical Abortions; Intact Dilation Evacuation; RU486.

References
"Birth control: Emergency Contraception." http://www.plannedparenthood.org/birth-control/ec.htm. 24 June 1998.
"Emergency Birth Control." http://www.tv.cbc.ca/healthshow/pastitem/emergbc.html. 24 June 1998.
"Emergency Contraceptive Pills." http://opr.princeton.edu/ec/ecp.html. 24 June 1998.
"Morning After Pill." http://www.carleton.ca/health/text/services/med/morn_aft.html. 24 June 1998.
"'Morning-after Pill' Gets FDA Go-ahead." *San Francisco Chronicle,* 25 February 1997, A1+.
"What's the Morning After Pill?" http://www.electra.com/drans005.html. 24 June 1998.
For more information on the pill, see http://opr.princeton.edu/ec/. 28 May 1998.

MTBE

Methyl-tertiary-butyl ether (MTBE) is a chemical compound widely used as a gasoline additive. Its use has increased dramatically over the last decade as a result of legislation designed to reduce air pollutants. In the late 1990s, however, questions were raised about the compound's possible adverse health and environmental effects.

Structure and Properties

At room temperature, MTBE is a clear liquid with a characteristic turpentinelike odor and a boiling point very close to that of gasoline itself. The latter property is an important advantage to the gasoline industry, since no special precautions are required for the handling of a gasoline-MTBE mixture. Other forms of gasoline, such as those containing ethanol (ethyl alcohol), do require special handling because of the significant difference in boiling point and rate of evaporation between gasoline and ethanol.

MTBE, unlike gasoline itself, is soluble in water. This property creates one disadvantage in using MTBE as a gasoline additive. If the additive leaks into soil, it dissolves in groundwater and can be transported long distances into natural water supplies and reservoirs used as water sources for human communities.

The chemical formula for MTBE is CH_3-O-$C(CH_3)_3$. That is, a molecule of MTBE consists of a methyl group ($-CH_3$) and a tertiary butyl group ($-C[CH_3]_3$) joined to each other through a single oxygen atom. The R-O-R' structure, with two hydrocarbon groups joined through an oxygen atom, is typical of the chemical family known as *ethers*. Another gasoline additive, for example, is ethyl-tertiary-butyl ether (ETBE). In ETBE, an ethyl group ($-CH_2CH_3$) and a tertiary butyl group ($-C[CH_3]_3$) are joined through an oxygen atom.

Uses

MTBE was first used as a gasoline additive in the early 1970s. At the time, tetraethyl lead was being phased out as a gasoline additive because of growing concerns about that compound's health and environmental effects. Until that time, tetraethyl lead ("lead") had been widely used in gasoline as an antiknock additive that improved gasoline octane ratings and engine performance. In a search for replacements for tetraethyl lead, the gasoline industry turned to MTBE (among other products).

Very quickly, a second advantage of MTBE became apparent. The presence of an oxygen atom in the MTBE molecule raises the temperature at which gasoline burns and improves the efficiency of combustion. Undesirable combustion products, such as carbon monoxide and ozone, are also reduced when MTBE is added to gasoline.

In 1990, the U.S. Congress accelerated demands for MTBE as a gasoline additive when it passed a number of amendments to the Clean Air Act. One of these amendments required the use of an oxygenate in gasoline sold in parts of the country where air pollution was most severe. An oxygenate is a compound that is able to supply oxygen. A powerful force behind the adoption of this amendment was the nation's corn farmers, who hoped to provide the ethanol that was then expected to be the most likely oxygenate to be adopted. Instead, the gasoline industry expanded its use of MTBE, adding anywhere from 11 to 15 percent of the compound to gasoline sold in polluted areas. As a result, the consumption of MTBE rose between 1990 and 1996 from about 80,000 barrels a day to nearly 250,000 barrels per day.

Criticisms of MTBE

As the consumption of MTBE increased, so did complaints about its possible health and environmental effects. For example, a number of individuals began to report respiratory and other health effects that they attributed to the presence of MTBE in gasoline. Existing groups, such as the American Lung Association, began to take stands against the use of MTBE as a gasoline additive. In addition, a nationwide string of "Oxy Busters" organized to demand a ban on the compound. Oxy Buster founder Barry Grossman, a salesman in New Jersey, has attributed the success of his group to frustration among private citizens who have experienced new cases of asthma and other respiratory disorders, headaches that never seem to end, and other health problems that can be traced (they say) to the introduction of MTBE in gasoline.

Other health experts point out that experimental animals exposed to MTBE have developed lymphomas (tumors of the lymph tissue), leukemia, and cancers of the liver, kidneys, and testicles. Although no epidemiological studies on MTBE in humans have been conducted, animal studies suggest that the compound may be a human carcinogen as well. At the least, they point out, more complete studies should have been conducted before MTBE was adopted as a gasoline additive.

Perhaps the most troublesome turn of events in recent years has been the discovery of MTBE in public water supplies in some regions of the country. In July 1996, for example, the city of Santa Monica, California, was forced to shut down part of its public water system because the concentration of MTBE had reached more than 600 ppb (parts per billion) in some of its wells. The U.S. Environmental Protection Agency (EPA) has established a maximum recommended level of 35 ppb of the compound.

As the possible harmful effects of MTBE were being debated, a new report on the effectiveness of the compound as a gasoline addition was being released. In August 1995, a study sponsored by the automobile and oil industry found a vast improvement in air quality with gasoline currently used in California. However, the report also found that the same improvement could be achieved with the new gasoline formulation whether or not it contained MTBE. In other words, it was not clear that adding MTBE actually made any contribution to achieving the clean air goal for which it was intended.

Proponents of MTBE Respond

Proponents of MTBE as a gasoline additive acknowledge that more complete studies of the compound might have been—and still are—desirable. But they point out that most levels of exposure to the additive are still far below any existing standards. In California, for example, more than 1,800 drinking water sources were tested for MTBE in 1992. Only twenty-six of those sources contained MTBE at the most sensitive level attainable (5 ppb). Eleven of those sources are sites at which motor boat activity is permitted, suggesting that fuel loss might be a factor in measured MTBE concentrations. Only two municipal water supplies—Santa Monica and Marysville—were found to have MTBE levels greater than EPA standards.

The gasoline industry has also argued that gasoline itself poses certain health and environmental hazards. Its vapors can cause skin and eye irritation, may be car-cinogenic in long-term exposure, and can even cause death if improperly handled. Overall, the industry has said, gasoline mixed with MTBE is probably no more of a health and environmental risk than is gasoline without the additive.

Finally, many observers believe that the benefits of using MTBE probably far exceed any risks it poses. They believe that reductions in carbon monoxide, ozone, and other air pollutants known to cause respiratory disorders, cancer, and other health problems fully justify any minimal threats posed by the compound. It is this view that has prompted a number of health and environmental groups, such as the Sierra Club and the Natural Resources Defense Council (NRDC), to join with automobile and oil companies to oppose any ban on the use of MTBE. As Janet Hathaway, a representative of NRDC has said, "It's one of those unpleasant choices" (Carlsen 1997, A8) her group has had to make.

References

Carlsen, William. "Gas Additive's Needless Risk." *San Francisco Chronicle*, 15 September 1997, A1+; continued on 16 September 1997, A1+.

"Chevron Gasoline Questions and Answers: Methyl Tertiary Butyl Ether (MTBE)." http://anonymous.chevron.com/prodserv/gas_qanda/mtbe.html. 7 January 1998.

"MTBE in Drinking Water." http://www.virginpure.com/news/mtbe.html. 7 January 1998.

"Santa Monica Water Supply Threatened by MTBE." http://www.uswaternews.com/archives/arequality/6smonica.html. 21 May 1999.

"Why a MTBE Bibliography?" http://wwwsd.cr.usgs.gov/nawqa/vocns/mtbe/bib/. 7 January 1998.

Oxy Busters home page on the Internet is at http://www.oxybusters.com/oxybustr.htm.

Multiple Use/Sustained Yield

Multiple use/sustained yield is a concept of land management set forth in a 1960 act carrying that name. The Multiple Use/Sustained Yield Act represented an attempt by the U.S. Congress to define the ways in which public lands, particularly public forests, are to be developed for public use.

History of Forest Use Policy

U.S. forest use policy has undergone some dramatic changes over the past century. Prior to the 1870s, no such policy even existed. Timber companies were essentially free to follow such lumbering practices as they chose on land they owned and, in some cases, on publicly owned land.

As the supply of forests decreased rapidly in the mid-nineteenth century, a few farsighted individuals began to call for some kind of control on the cutting of timber by lumbering companies. In 1876, the U.S. Department of Agriculture created a forestry division whose task it was to develop and maintain policies for the harvesting of timber on public lands. Over the next century, the policies developed by that division, later renamed the U.S. Forest Service, changed in various ways depending on individuals involved in decision making, on national needs, and on other factors.

For example, the first chief of the U.S. Forest Service was Gifford Pinchot, a firm believer in the concept of forest conservation. By forest conservation, Pinchot meant using forest reserves in such a way that they would be able to supply the nation's needs for new lumber without completely depleting our supply of trees. He saw the national forests as nationally owned and operated "tree farms" that private companies could log but which those companies were also obligated to care for.

That policy was probably never carried out in exactly the form that Pinchot imagined. By the time World War II had ended, the demands for lumber placed pressures on the U.S. Forest Service that were difficult to resist. Before long, trees were being harvested at a rate that threatened the total destruction of some forest regions. Timber sales increased by 800 percent between 1941 and 1966 in order to meet the nation's demand for wood products.

The Multiple Use/Sustained Yield Act of 1960

In 1960, the U.S. Congress acted to cut back on the excesses of lumbering companies and to protect national forests from unwarranted depletion. In the Multiple Use/Sustained Yield Act, Congress declared that national forests were to be managed to produce two equally important objectives: *multiple use* and *sustained yield.*

By multiple use, Congress said that it meant managing forests in order to make "the most judicious use of the land for some or all of these resources or related services ... and harmonious and coordinated management of the various resources, each with other, without impairment of the productivity of the land, with consideration being given to the relative values of the various resources, and not necessarily the combination of uses that will give the greatest dollar return or the greatest unit output" ("The Evolution of U.S. Forest Policy").

That is, national forest managers were required to give consideration to all of the ways that forests might be used: as storehouses of water resources; as habitat for fish, birds, and other forms of wildlife; as places of recreation for the public; as protection against soil erosion; and as a source of wood products. All of these forest benefits had to be compared with each other, and some balance had to be developed so that all were available in a forested region.

The law did not require that every forest area provide every benefit. For example, some forests might be given over primarily to lumbering companies for the removal of timber. But parts of that forest might be set aside for fishing and hunting, as wildlife preserves, or left totally undeveloped.

The second part of the Multiple Use/Sustained Yield Act called for "the achievement and maintenance in perpetuity of a high-level annual or regular periodic output of the various renewable resources of the national forests without impairment of the productivity of the land." In other words, forest managers were required to develop policies that made sure that a forest area would *always* be able to provide some trees for harvesting, *always* contain the biodiversity of wildlife traditionally found there, *always* provide the kind of recreational opportunities there. A lumber company could not take away so many trees that various species of fish or birds could no longer live there or that soil erosion clogged rivers and lakes. Nor could recreational facilities be developed that made it impossible to continue growing and harvesting trees in the future.

Policy to Reality

The goal of the Multiple Use/Sustained Yield Act must almost certainly be regarded as an admirable one. It outlines a policy by which everyone concerned with forest use gets consideration when forest use decisions are made. But making the everyday decisions needed to put multiple use/sustained yield into action can be a nightmare. One can certainly determine the value per acre of forestland used for timber production. But how does one put a dollar value on the use of forests for protection of endangered species, for promoting recreation, or for preserving soil and water resources?

In addition, the U.S. Forest Service has at times been criticized as having an industry bias, more concerned with timber harvesting than other comparable forest uses. That charge was often raised during the administration of President Ronald Reagan, whose secretary of the interior, James Watt, believed strongly in the commercial development of forest areas. Reagan's own views about forest preservation are reflected in a quotation attributed to him that "once you've seen one giant redwood, you've seen them all." The implication of this statement is that huge public forests devoted to the preservation of redwood trees may not be necessary. As long as there are a few specimens of the trees around to look at, the rest could as easily be sacrificed for lumber.

Over a period of three decades, since the adoption of the Multiple Use/Sustained Yield Act, Congress and presidential administrations have attempted to move the U.S. Forest Service away from an industry bias, or a perceived industry bias, and return to a more evenly balanced policy for national forest use. Part of the problem has been that one factor in this equation—the timber industry—is a large and very powerful political force. By contrast, those who would speak for other forests uses—recreation, wildlife, soil, and water—tend to be less well organized and less politically influential. Without much doubt, the conflict between those who would use national forests primarily as a source of commercial products versus those who see a more complex mix of uses is likely to continue well into the future.

See also: Conservation and Preservation; Headwaters Forest; Logging Roads; Northwest Forest Plan

References

Clawson, Marion. *Forests for Whom and for What?* Baltimore: Johns Hopkins University Press, 1975.

Dana, Samuel T., and Sally K. Fairfax. *Forest and Range Policy: Its Development in the United States.* 2d edition. New York: McGraw-Hill, 1980.

"The Evolution of U.S. Forest Policy." http://bioag.buy.edu/Botany/Rushforth/www/conbio/forests2.htm. 21 May 1999.

Hewett, Charles E., and Thomas E. Hamilton, eds. *Forests in Demand: Conflicts and Solutions.* Boston: Auburn Publishing House, 1982.

Miller, G. Tyler, Jr. *Living in the Environment.* Belmont, CA: Wadsworth Publishing, 1985, Chapter 10.

"Multiple Use of Forestlands." http://www.safnet.org/policy/psst/psst27.html. 25 July 1998.

Petulla, Joseph M. *American Environmental History.* San Francisco: Boyd and Fraser Publishing, 1977.

"What Does Sustained Yield Mean?" http://www.iww.org/iu120/local/Yield.html. 10 April 1998.

Nature versus Nurture

The expression "nature versus nurture" was first coined by the English psychologist Francis Galton in 1874. It refers to a very old debate about the relative importance of the traits and characteristics with which a person is born (his or her *genetic potential*) versus the knowledge and skills that one is able to acquire later in life, primarily through education.

This debate has gone on for centuries, although in a somewhat ambiguous format before the nineteenth century. For example, the philosophical belief in the "divine right of kings" is based on one position in this debate. It suggests that some people are "born to be rulers" because they have "royal blood" or because they inherit from their parents some other qualities that make them natural leaders. By contrast, other people are doomed by their ancestry to be workers, slaves, or other members of the lower class in a society.

By contrast, other thinkers have argued that human personalities are infinitely malleable. Individuals all come into the world with essentially the same potential for learning. The French philosopher Rousseau, for example, referred to the human mind as a *tabula rasa,* a "blank slate" on which anything could be written. It was on this view of humanity that the Age of Enlightenment was built, an age when all humans were thought to be capable of reaching a state of perfection if they were only willing to strive for that goal.

These views applied to groups of people as well as to individuals. Throughout history, some clans, societies, nations, and other groups have often been thought of as inherently superior to others. The whole concept of slavery, for example, is based at least in part on the notion that one race, sex, or nationality is inherently inferior to another and the latter is justified, therefore, in demanding that the former be totally subservient to its "natural masters."

Genetics and Nature versus Nurture

The rise of Darwinism and modern genetics brought a clearer focus to this age-old debate. The discoveries of Darwin, Mendel, and other pioneers of genetics made it more clear as to precisely how physical and biological characteristics were passed down from one generation to the next. Many early geneticists saw a certain inevitability in the hereditary tradition with which an individual and a group of individuals were saddled. They argued that the class divisions one finds in society are not the result of hard work on the part of some or sloth on the part of others but the natural result of inborn advantages and disadvantages. It was out of this philosophy that the theory of eugenics was born. Eugenics was the science that strove to improve the overall quality of a society by weeding out those with inferior genetic backgrounds and promoting the reproduction of those with superior backgrounds.

Modern science has provided fodder for those who argue the nurture side of the nature versus nurture debate. For example, data from a very long history of learning studies have shown that experimental animals can be trained to perform an impressive array of sophisticated tasks. One might conclude from such studies that humans, too, have a very significant capacity to learn new information and skills.

Many people have long held a middle view of this debate. They accept the fact that humans are born with certain genetic capabilities but believe that those capabilities are developed or not depending on the educational opportunities available. That is, two people might be born with the same capacity to learn Latin. But if one is never exposed to the language and the other is, the results will be very different.

Intelligence

The controversy over nature versus nurture has, therefore, often come down to an attempt to measure the relative contributions of each factor to one's total personality. The trait that has been studied more than any other in this context is intelligence, as measured by the intelligence quotient or IQ. Scholars have asked, for example, what part of an IQ score of 180 can be attributed to a person's genetic background and what part to his or her educational experiences.

Obtaining an answer to this kind of question is very difficult, given that human development is involved and humans cannot be manipulated as can experimental animals. One approach is to study identical twins (twins with the same genetic makeup) who have been separated at birth. The principle is that the two members of a twin pair have the same genetic background but different educational experiences. Any differences between the two members of the pair can be attributed, therefore, to learning and not inheritance.

Although studies of this kind would appear to be relatively straightforward, they have seldom been accepted as providing the "final answer" to the nature versus nurture debate. They are usually criticized as having been improperly conducted, improperly analyzed, or, in some cases, actually falsified.

Practical Consequences of the Debate

The issue of nature versus nurture is hardly a purely academic matter. The debate is by no means a scholarly issue about which well-mannered researchers exchange reasoned scientific papers that summarize research, present data, and draw tentative conclusions. After all, the issue involves some of the most fundamental issues in all of human society. Is it really true that some individuals are fundamentally and basically "better" or "worse" than others? Do some inherit a greater level of intelligence that makes them "natural masters" over those whose genes are more limited?

If these questions are answered in the affirmative, then a number of social institutions and traditions may, perhaps, be justified. What is the point of spending enormous amounts of money on educating the less advantaged, for example, if their genetic background sets a limit beyond which they cannot hope to achieve. If Whites are, in fact, genetically superior to Blacks (or Asians to Whites), then is a social structure in which one group provides the rule makers and the other the rule followers not a "natural" and "correct" one?

It is hardly surprising, then, that the nature versus nurture debate has spilled out of the fields of genetics and psychology into the area of politics and social science and that it has provided the basis of fiery exchanges in the popular press. One of the questions that arises in this debate is whether it is even permissible to consider the possibility that the "nature" side of the argument is correct or, at least, largely correct. Should one be allowed to carry out research of the type that might lead to "yes" answers to the questions in the preceding paragraph? Or are the social consequences so serious that the issue is one that should not even be examined scientifically?

Like the controversy over abortion, then, the debate over nature versus nurture has two important aspects. One is the scientific aspect: What scientific information can we collect with regard to each side of the issue? The other aspect is political: How does the profound political significance of the debate affect the way the scientific aspect is framed, carried out, interpreted, and presented to the general public?

See also: Biological Determinism; IQ; Sociobiology

References

Csongradi, Carolyn. "A New Look at an Old Debate." http://www.gene.com/ae/21st/SER/BE/what.html. 8 July 1998.

"Environmental-heredity Controversy." In W. F. Bynum, E. J. Browne, and Roy Porter, *Dictionary of the History of Science*. Princeton, NJ: Princeton University Press, 1981, 125–126.

Fukuyama, Francis. "Is It All in the Genes?" *Commentary*, September 1997, 30–35.

Herbert, Wray. "Politics of Biology." *U.S. News and World Report*, 21 April 1997; also at http://www.usnews.com/usnews/issue/970421/21natu.htm. 12 July 1998.

Hundert, Edward M. *Lessons from an Optical Illusion*. Cambridge: Harvard University Press, 1995.

Lewontin, Richard C. "Biological Determinism as a Social Weapon." In The Ann Arbor Science for the People Editorial Collective, *Biology as a Social Weapon*, 6–20. Minneapolis: Burgess Publishing Company, 1977.

Needle Exchange Programs

Human immunodeficiency virus (HIV) infection is spread when two individuals share bodily fluids, such as blood or semen, with each other. If one of those individuals is infected with HIV, the virus may be transmitted to the uninfected person by means of the fluid. The two most common means of sharing bodily fluids are sexual acts and sharing of needles among drug abusers.

Significant financial resources have been invested in programs to encourage safer sex techniques to reduce the spread of HIV through sexual contacts. These efforts have, many authorities believe, led to a significant reduction in the number of HIV cases spread by sexual activity. The spread of HIV through needle sharing, however, presents somewhat different issues.

Needle Sharing as a Means of HIV Spread

Individuals who use heroin or cocaine illegally often do so by injection. They use ordinary hypodermic needles to inject drugs directly into the bloodstream. Such needles are inexpensive to purchase, usually about twenty cents each, but they can be purchased only with a prescription. They are, therefore, somewhat difficult for many drug abusers to purchase on a regular basis. This fact is especially relevant for the homeless, who may be both poor and addicted.

As a result, it is not uncommon for drug abusers to share needles among themselves. A single needle may be shared many times over among such individuals. If one of the persons sharing a needle is infected with HIV, the virus is likely to be passed in blood left in the needle to others using the same needle. Some authorities estimate that more than half of all new cases of HIV infection today can be attributed to needle sharing.

Providing Free Needles to Drug Abusers

One method for dealing with this problem is by providing drug abusers with clean needles. The intent of this program is to ensure, insofar as possible, that two individuals will never share the same needle. To the extent that that goal is achieved, the spread of HIV through needle sharing could be dramatically reduced.

Needle exchange programs have now been implemented in more than 100 cities across the United States and in a few European countries. In Europe, these programs are often sponsored, paid for, and run by governmental agencies. In the United States, such programs are generally illegal. Needles used for the injection of cocaine and heroin are regarded as "drug paraphernalia" and are, therefore, illegal without a prescription. In many states and cities, however, law enforcement officials have turned a blind eye to needle exchange programs run by private citizens and private agencies. For a variety of reasons, they have chosen not to arrest or prosecute those who operate such programs.

Many people oppose needle exchange programs, even when scientific evidence supports their value, simply because the concept seems ethically wrong. (Corbis/Annie Griffiths Belt)

The goal of needle exchange programs is to obtain one used (dirty) needle for each clean needle handed out. That goal is difficult to achieve. A homeless person who uses seven needles a week may not be able to safely carry all seven needles with him or her until the weekly needle exchange program occurs. In some instances, fewer than half as many dirty needles are returned as there are clean needles handed out.

Pros and Cons of Needle Exchange Programs

Nonetheless, most public health officials feel that needle exchange programs have been very successful in reducing the spread of HIV among drug abusers. A number of scientific studies of needle exchange programs have been conducted, and almost without exception, they have showed that the spread of HIV is reduced with such programs. Of equal importance, these studies tend to show that needle exchange pro-

grams have no tendency to increase the illegal use of drugs among those participating in the program.

As a consequence of these findings, needle exchange programs have now been endorsed by a wide variety of health and political organizations, including the American Medical Association, the American Bar Association, the U.S. Conference of Mayors, the Centers for Disease Control and Prevention, the National Institutes of Health, the Institute of Medicine, and the General Accounting Office.

Support for needle exchange programs is, however, far from unanimous. Many individuals and organizations feel that such programs tend to encourage drug use. By handing out free needles, they say, authorities are giving tacit approval to drug abuse. The most common criticism offered is probably the comment that needle exchange programs "send the wrong message" to the public. When the United States is spending billions of dollars to fight drug abuse, they say, it should not simultaneously be giving away the tools that make drug abuse possible.

Critics of needle exchange programs tend to discount scientific studies confirming their usefulness. They claim that such studies are flawed or that they simply fly in the face of common sense. For example, Robert Maginnis of the conservative Family Research Council rejects the results of research and relies on his own intuition regarding this issue. "I have no doubt," he has said, "that these programs increase drug use" (Painter 1997, D1).

Drug abusers themselves take a somewhat different position than that of Maginnis. Their response to critics of needle exchange programs is that needle users are already addicts and, in most cases, have been so for years. Needle exchange programs will not encourage them to use drugs, although it may save their lives.

Needle exchange programs tend to create a difficult dilemma for politicians. Any decision they make is likely to offend one important group or another: public health leaders and AIDS activists on the one hand and conservative political groups and antidrug organizations on the other. For example, President Bill Clinton announced in April 1998 that he had become convinced that needle exchange programs had been proven to be effective. He acknowledged that such programs were already saving thirty-three people per day from contracting HIV. Having made this acknowledgement, Clinton then declared that he would not recommend the use of any federal funds to support such programs.

A similar battle was going on simultaneously in the U.S. Congress. On one side, Representative Nancy Pelosi (D-Calif.) asked "How can we turn our back on science?" in her support for funding of needle exchange programs. On the other side, Representative Tom DeLay (R-Texas) responded that a congressional ban was necessary to counter "a deadhead president that supports a program that gives free needles to drug addicts." The House of Representatives eventually agreed with DeLay's position and voted 287–140 to permanently ban federal funds for any needle exchange program.

See also: HIV and AIDS

References

"The Clinton Administration's Internal Reviews of Research on Needle Exchange Programs." http://www.lindesmith.org/library/clinton/review.html. 20 June 1998.

"Does Needle Exchange Work?" http://www.cafis.ucsf.edu/NEPrev.html. 21 May 1999.

Freedberg, Louis. "Needle Swap Surprise from White House." *San Francisco Chronicle*, 21 April 1998, A1+.

"House Votes to Ban Needle Distribution Funds." http://www.pathfinder.com/CQ/bills/H19980114.html. 20 June 1998.

"Needle Exchange Programs Have Nearly Doubled in Number since 1993, New Report Finds." http://chanane.ucsf.edu/capsweb/nepnews.html. 20 June 1998.

"Needle Exchange Programs Reduce HIV Transmission Among People Who Inject Illegal Drugs." http://www.nas.edu/onpi/pr/sept95/need_exch.html. 20 June 1998.

"Needle Exchange Questions and Answers." http://www.sfaf.org/prevention/needlexchangeqanda.html. 21 May 1999.

Painter, Kim. "Needle Exchanges in the Eye of AIDS Debate." *USA Today*, 17 September 1997, 1D+.

"The Public Health Impact of Needle Exchange Programs in the United States and Abroad: Summary, Conclusions, and Recommendations." http://www.caps.ucsf.edu/capsweb/publications/needlereport.html. 20 June 1998.

"U.S. Conference of Mayors Reports on Needle Exchange Policy Implications and Pilot Programs." http://www.ndsn.org/FEB95/MAYORS.html. 20 June 1998.

New Building Syndrome

See Indoor Air Pollution

Noise Pollution

The concept of *noise* is a difficult one to define. It is sometimes described as "unwanted sound." But one person's "noise" can be another person's "music." The devotee of opera and the fan of hard metal music would probably disagree as to which of these is "noise" and which is "music."

Loudness

A somewhat more objective way to talk about noise is to describe its loudness. Regardless of the subjective quality of sound, it is possible to measure objectively how loud that sound is.

The loudness of a sound is measured in decibels. The decibel scale is a geometric scale rather than an arithmetic scale. As an example, the decibel rating for ordinary conversation is about 60. A sound with a decibel rating of 70 is 10 times as loud as ordinary conversation. A sound with a decibel rating of 80 is 10 x 10 (or 100) times as loud as ordinary conversation. A sound with a decibel rating of 90 is 10 x 10 x 10 (or 1,000) times as loud as ordinary conversation. And so on.

The decibel scale is easier to understand when various points on it are compared to everyday sounds. For example, a decibel rating of 10 is sometimes compared to the sound of rustling leaves. Softly played background music has a decibel rating of 40. A dishwasher operates at about 80 decibels, and the sound of a jet airplane taking off may range anywhere from 100 decibels (at a distance of 300 meters) to 150 decibels (at a distance of 25 meters).

Sources of Noise Pollution

The sounds to which we are exposed in everyday life sometimes become annoying. They may even pose a threat to human health and/or the environment. The presence of such sounds is said to constitute *noise pollution.* Many people seem to regard noise pollution as an unavoidable part of modern society. Many of the gadgets and machines we use every day are noisy. In addition, the noise level of modern music and

ordinary conversation appears to have risen over time. Today, noise is part of the home, where washing machines, blenders, radios and televisions, leaf blowers and lawn mowers, and power tools are constantly in use. Noise is also generated by all forms of transportation, ranging from cars and trucks on freeways to jet airplanes on runways and overhead. It is also part of our recreational activities, whether from snowmobiles, jet skis, or trail bikes.

In fact, some people question whether the concept of "peace and quiet" any longer has meaning in modern societies. Even an escape to a remote national park no longer guarantees an escape from noise. In Grand Canyon National Park, for example, there are now more than 90,000 commercial flights each year for tourists, including both small aircraft and helicopters. It is unlikely that a visitor to the park would be able to experience the peaceful seclusion that one might expect from such a setting.

Health Effects

Many people object to noise pollution because it is an annoyance. They want to be able to sit on their front porches, walk in the forest, or fish on a lake without being immersed in loud sounds. But beyond that, scientists have now determined that exposure to loud noises can have clear and identifiable physiological effects. Long-term exposure to noise levels of 80 decibels or more, for example, can lead to loss of hearing. Some experts estimate that 10 percent of all Americans experience some hearing loss as a result of simply living in our noisy society. The American Speech, Language and Hearing Association claims that hearing loss problems among Americans have increased by 14 percent since 1971.

People exposed to louder sounds are susceptible to more serious problems. Individuals who have played in, worked with, or listened to loud rock bands, for example, tend to suffer hearing problems later in life at a rate greater than that of those without those experiences.

Exposure to loud noises may also lead to other problems that are still not well understood. Scientists have long known that loud sounds can cause constriction of blood vessels, tensing of muscles, increase in heart beat and blood pressure, stomach spasms, and other physiological responses. How those responses affect one's long-term health is still not clear, although some studies suggest they may lead to higher rates of cardiovascular and other disorders.

Dealing with Noise Pollution

What is to be done about noise pollution? Some obvious steps are possible. For example, one of the sources of noise pollution that has received the most attention is the leaf blower. Citizens in a number of communities have become outraged at being aroused on Sunday morning (or any other time) by the roar of these machines. More than 300 U.S. cities have now banned their use.

Barrier walls along major highways may also be helpful. A number of states are now including four-meter-high (twelve feet) concrete barriers along any new highway being built. These barriers tend to decrease noise levels by about half (10 decibels), but have somewhat limited effectiveness for those living or working above ground level. In addition, they tend to cost an average of about $2 million per mile, significantly increasing the cost of any new highway.

To some extent, individuals and the federal government have admitted defeat on the issue of noise pollution. For example, the U.S. Environmental Protection Agency once had an Office of Noise Abatement and Control. That office was eliminated under the administration of President Ronald Reagan. The federal government no longer has a comparable agency concerned with the issue of noise pollution. It appears to have accepted the premise that increasing levels of noise pollution are an inevitable consequence of a higher standard of living.

References

Cohen, Sheldon. "Sound Effects on Behavior." *Psychology Today,* October 1981, 38–49.

Duckworth, Carolyn. "Noise Pollution." *Gale Encyclopedia of Science.* Detroit: Gale Research, 1996, 2524–2526.

Lehigh, Scott. "Noise Pollution: 'Say What?'" *Oregonian,* 22 November 1996, A26.

McCabe, Michael. "Anti-noise Protesters Turn up the Volume." *San Francisco Chronicle,* 19 February 1998, A1+.

Morse, M. "Have We Surrendered in the War against Noise?" *Utne Reader,* September/October 1990, 22–23.

Shapiro, Sidney. "Rejoining the Battle against Noise Pollution." *Issues in Science and Technology,* Spring 1993, 73–78.

Suter, Alice M. "Noise Wars" *Technology Review,* November/December 1989, 42–49.

Turk, Jonathan, and Amos Turk. *Environmental Science.* 4th ed. Philadelphia: Saunders College Publishing, 1988, 534–539.

"The Unhealthy Sound of Victory." *Discover,* May 1988, 10+.

Northwest Forest Plan

The Northwest Forest Plan is a blueprint proposed by President Bill Clinton in April 1993 for the management of forestlands in the Pacific Northwest owned by the federal government. The plan was developed in an attempt to resolve long-standing disagreements as to how forest resources should be used. In the short time that it has been in existence, the plan has met with relatively modest acceptance and success.

History of the Logging-Environment Dispute

Logging has long been a very important activity in the United States. The North American continent contains some of the richest temperate forests in the world. The timber obtained from these forests has been used, at different times in our history, for the operation of blast furnaces, to run railroad engines, and for construction.

For more than two centuries, timber companies treated the nation's forests as if they were an endless source of wood. Vast portions of the nation's wooded areas were cut down and converted to farmland, rangeland, or other uses. In the early 1900s, the first efforts were made to set aside parts of the nation's forests as protected areas. These areas became the core of the nation's national forest and national parks systems.

Various philosophies were espoused at the time as to how the timber resources on these "protected" lands were to be used. Some experts argued that the lands should be thought of as carefully managed tree farms, where the major emphasis ought to be on ensuring a supply of trees that would last essentially forever. Other experts argued that at least some portions of these lands should be left in their pristine conditions and that the lands should be developed for other purposes: as wildlife sanctuaries, recreation areas, and research sites, for example.

The conflict between these two viewpoints has never satisfactorily been settled. The 1990s saw, in fact, an even more vigorous debate as to how national forestlands should be used. One motivation for that debate was the concern by environmentalists that important forestlands, such as old-growth forests, were being destroyed at an alarming rate. In addition, environmentalists felt that something had to be done to rein in what they saw as a strong industry bias in the agencies that control lumbering in national forests, the U.S. Forest Service and the Bureau of Land Management.

Legislative History

The U.S. Congress has made clear its position in this debate for nearly four decades. In 1960, it passed the Multiple Use/Sustained Yield Act that required forest managers to take into consideration all possible uses of forests and to give some weight to each of these uses in deciding how forests were to be used. That is, Congress said that lumbering might be the most important way to use some parts of some forests at some times, but other uses of forestland were also equally or even more important in other situations.

Congress was apparently dissatisfied with the results of the Multiple Use/Sustained Yield Act because, in 1976, it

adopted an even stronger statement about forest use, the National Forest Management Act (NFMA). The NFMA imposed restrictions on the most destructive logging activities, such as clearcutting and logging on fragile lands, and repeated its intent that forest managers should give attention to a broader use of forest lands.

A decade later, questions were still being raised as to the U.S. Forest Service's commitment to the concept of "multiple use." Timber sales from federal lands reached the highest point ever, 12.7 billion board feet, in 1987, and agency plans called for a doubling of production in the Pacific Northwest over the following decade. In addition, nearly 150,000 kilometers (100,000 miles) of new logging roads were being proposed for the next fifty years, a decision that would open up vast new areas of forestland for harvesting.

The Clinton Initiative

As these new lumbering initiatives were being planned, environmentalists became more and more concerned not only about the forests themselves but about the diversity of animals and plants typically present in the forest ecosystem. Confrontations between lumbering interests and those trying to save forest lands or endangered species, such as the northern spotted owl and the marbled murrelet, became more serious. Reasoned discussions became less common and violent confrontations more common.

As part of his campaign for the presidency, Bill Clinton promised that he would take action as president to resolve this escalating conflict. That action was the Forest Plan for the Pacific Northwest, released on 1 July 1993. The plan later became known almost universally as the Northwest Forest Plan (NWFP). The plan had been developed as the result of a conference held in Portland, Oregon, on 2 April 1993 involving many of the groups involved in the dispute over forest management. Some of the major elements in the NWFP are as follows:

Reserve Areas: Certain forest areas are to be designated as Reserve Areas in which very limited activities are to be allowed. Certain old-growth forests, for example, are to be included in such Reserve Areas. Minimal logging, such as timber salvage and thinning, is to be allowed in Reserve Areas.

Adaptive Management Areas: Ten Adaptive Management Areas are to be created to provide "intensive ecological experimentation and social innovation to develop and demonstrate new ways to integrate ecological and economic objectives." The mechanism by which these areas are to be developed is not specified.

Harvest Levels: The NWFP sets an annual limit of 1.2 billion board feet of lumber to be harvested in forests inhabited by the spotted owl. This section of the plan has met with considerable opposition. To begin with, the area is already under injunction by a federal court to prevent further lumbering. How the courts will respond to the NWFP provisions is not clear. Also, the annual limit specified by NWFP is considerably less than the level reached in earlier years, and timber companies complain that they and those who depend on lumbering for a livelihood will suffer a severe economic hardship with the new limits.

Economic Development: In response to industry concerns, the NWFP includes a Northwest Economic Adjustment Initiative to assist communities and individuals in making a transition from lumber-based to other types of occupations. A fund in the amount of $1.2 billion spread out over five years will be used to help workers and their families, communities, business and industry, and projects to develop local ecosystems.

NWFP also contains directives as to how various federal agencies are to work with each other to make this plan work successfully.

Later Developments

Given the long history of disagreement between logging companies and environmentalists, it would not be surprising to find that implementation of the NWFP might encounter problems. And such has been the case.

Perhaps the most serious roadblock to NWFP encountered so far was an act passed by the U.S. Congress in 1995, the so-called Salvage Rider bill attached to a bill designed to reduce federal spending. The Salvage Rider bill allows lumber companies to remove "salvage" trees from national forests, including forestlands that have already been declared Reserve Areas. "Salvage" trees are dead trees, trees that have been blown over, or sick trees. The Salvage Rider bill also exempted lumber companies from abiding by the provisions of most environmental acts, such as the National Environmental Protection Act and the Endangered Species Act.

The timber industry has quickly taken advantage of this exception to the NWFP and moved into many "protected" areas to remove salvage trees. Almost without exception, however, environmentalists have complained that companies are taking healthy as well as "salvage" trees and are causing extensive damage to the land and wildlife where they are working. Whether Clinton's hopes for a resolution to the industry-environmental conflict in the Pacific Northwest will see reality will probably not be known for many years.

See also: Headwaters Forest; Logging Roads; Multiple Use/Sustained Yield; Rain Forests

References

Bishop, Ellen Morris. "Environmental Groups Rip Northwest Forest Plan." *Columbian,* 5 February 1998, B4.

Gorte, Ross W. "The Clinton Administration's Forest Plan for the Pacific Northwest." Congressional Research Service; Report for Congress. http://www.cnie.org/nle/for–3.html. 29 June 1998.

Jewett, Joan Laatz. "Fate of Umpqua River Basin Tests President's Northwest Forest Plan." *Oregonian,* 8 April 1996, A1+.

Mize, Jeff. "Environmental Group Blasts Money-losing Timber Sales." *Columbian,* 8 January 1998, B1.

"Northwest Forest Plan-Related Documents." http://www.or.blm.gov/ForestPlan/NWFPTitl.htm. 28 July 1998.

"Our Forests and Our Futures." *Wild Oregon,* February 1998, 2.

Sonner, Scott. "Lawsuit over Spotted Owl May Test Cuts in Logging." *Rocky Mountain News,* 2 March 1997, 52A.

Nuclear Disarmament

See Nuclear Weapons

Nuclear Fusion

See Cold Fusion; Nuclear Power Plants; Nuclear Weapons

Nuclear Power Plants

A nuclear power plant is a facility at which nuclear power is converted to electrical power. The nuclear power is produced at the center of the plant in a nuclear reactor.

Components of a Nuclear Power Plant

A nuclear reactor consists of three basic parts: the core, a moderator, and control rods. The *reactor core* contains the fuel rods used in the power plant. They contain uranium-235 or plutonium-239. Uranium-235 and plutonium-239 are isotopes, or forms of the elements uranium and plutonium respectively. These isotopes are used because they have the rare ability to undergo nuclear fission. In a nuclear fission reaction, a neutron collides with the nucleus of a uranium-235 or plutonium-239 atom and causes it to break apart. The products of this fission reaction include two smaller atomic nuclei, one or more neutrons, and large amounts of energy. The reaction can be represented very generally by the following equation:

$$\,^{1}_{0}n + \,^{235}_{92}U \rightarrow \,^{145}_{56}Ba + \,^{90}_{36}Kr + x \,^{1}_{0}n + energy$$

Two conditions are essential for the operation of a nuclear reactor. First, the neutrons used to initiate a fission reaction must be traveling at speeds within a somewhat limited range. If they travel too fast, fission will not occur. The reactor core must contain, therefore, a *moderator,* a substance that will slow down the neutrons that pass through it. Two of the most common moderators used in a nuclear power plant are water and graphite.

The number of neutrons present at any one time in the reactor core is also critical. If neutrons are produced too slowly, nuclear fission reactions will cease to occur. If they are produced too rapidly, fission reactors occur too rapidly. Energy is produced so quickly that temperatures within the reactor core climb very rapidly and the whole system experiences a *meltdown.* In a meltdown, all of the major components of the reactor core turn to liquid or gas. The only major accident of this kind occurred at the V. I. Lenin Nuclear Power Facility at Chernobyl, in Ukraine, in 1986.

Disasters of this type can be prevented by the use of *control rods,* long cylindrical tubes containing an element with a tendency to absorb neutrons. Boron and cadmium are two common elements used in control rods. Control rods are suspended in a reactor core by means of systems that allow them to be raised and lowered. The deeper the rods extend into the core, the more neutrons are absorbed and the more slowly nuclear fission proceeds. The more control rods are removed from the core, the more neutrons remain within the core, and the faster fission occurs.

Ideally, nuclear fission should occur in a reactor core in such a way that for every neutron used up in a fission reaction, about one new neutron is produced. Under this condition, a self-sustaining *chain reaction* occurs. That is, each fission event generates the means—a new neutron—for the initiation of another fission reaction.

Heat Transfer

The heat energy generated in a reactor core is ultimately used to boil water. Steam from the boiling water is used to turn turbines that, in turn, drive electrical generators. These components of a nuclear power plant are essentially similar to those found in a conventional fossil fuel power plant. The

Problems at Chernobyl, Three Mile Island, and a few other settings have raised the question as to whether nuclear power will ever play a role in the production of energy in the United States. (U.S. Department of Energy)

difference in the two systems is that fossil fuel plants derive their heat from the combustion of coal or oil rather than from nuclear fission reactions.

Nuclear power plants generally make use of one additional step not found in a fossil fuel power plant: a heat exchange system. A heat exchange system consists of a fluid that picks up heat energy from the reactor core and transfers it to a second unit. In the second unit, the heat carried away by the fluid is used to boil water, producing steam to operate turbines and generators.

Promises and Problems

The 1950s and 1960s were a period of great optimism about the role of nuclear power

plants in the United States. It was an era characterized by great concerns about environmental issues, such as air and water pollution. Fossil fuel power plants were an important source of some of the most troubling of those environmental problems. Many people saw nuclear power plants as a solution to these problems, because such plants produce none of the harmful emissions released by fossil fuel power plants. Many energy experts forecast a time in the then not-too-distant future when a significant portion of the energy needs of the United States would be met by nuclear power.

Interestingly enough, that scenario has become a reality in some parts of the world. France, for example, generates about half of all its electrical power in nuclear power plants. Such has not been the case in the United States, however. No new nuclear power plants have been built in this country since the mid-1980s. By contrast, a number of older plants have been closed and torn down.

Two factors have led to the abandonment of hopes for a nuclear-powered energy future. One factor has been a growing concern about the safety of such plants. Americans will probably always associate in their minds nuclear power plants and nuclear weapons. Some people worry that a nuclear power plant might explode like an atomic (fission) bomb, even though such an event is physically impossible.

Better-informed people recognize and are troubled by other kinds of accidents, however, such as the meltdown at the Chernobyl power plant. Some experts believe that the accident at the Three Mile Island Nuclear Power Plant in Pennsylvania in 1979 sounded the death knell for nuclear power generation in the United States, at least for the foreseeable future.

In any case, the regulatory burdens involved in the construction and operation of a nuclear power plant in the United States today have become so great that the facilities are no longer economically viable. As a result, few energy companies any longer give much thought to the construction of nuclear power plants as a means of generating their electricity. It is not clear how soon, if ever, such plants will once again become part of this nation's energy future.

Nuclear Wastes

A second issue blocking nuclear power plant development has been the problem of nuclear wastes. The operation of a nuclear power plant results in the formation of "ashes," somewhat similar to those produced in a coal-fired power plant. The "ashes" from a nuclear power plant consist of elements removed from the reactor core that are no longer capable of sustaining nuclear fission reactions. These elements are highly radioactive. They release forms of radiation that can cause cancer, other illnesses, and death in humans. Further, they have very long half-lives, meaning that they will continue to be radioactive for hundreds or thousands of years.

The question is where and how these radioactive wastes are to be stored. Remarkably, that problem has not yet been satisfactorily solved. Thousands of tons of radioactive waste have been produced since the discovery of nuclear power in the 1940s. Since that time, those wastes have usually been stored in "temporary" facilities, many of which have now become unsafe or inadequate for continued use. It was not until late 1998, in fact, that the first permanent storage site for certain types of radioactive waste—the Waste Isolation Pilot Plant (WIPP) in New Mexico—opened. WIPP handles only a very small fraction of all radioactive wastes that have been produced, however. Until a more satisfactory solution to the storage of nuclear wastes has been found, it seems unlikely that nuclear power plants will ever be built again in the United States.

See also: Chernobyl; Nuclear Wastes; Three Mile Island

References

Eriksson, Henrik. "Control the Nuclear Power Plant." http://www.geocities.com/CollegePark/Union/2830/nuclear.html. 21 May 1999.

"Frontline: Nuclear Reaction; Why Do Americans Fear Nuclear Power?" http://www.pbs.org/wgbh/pages/frontline/shows/reaction/. 23 June 1998.

Kaku, Michio, and Jennifer Trainer. *Nuclear Power: Both Sides*. New York: W. W. Norton, 1982.

Kraul, Chris. "Twilight of the Nukes?" *Los Angeles Times*, 13 October 1996, B1+.

Miller, G. Tyler, Jr. *Environmental Science*. 3d ed. Belmont, CA: Wadsworth Publishing, 1985, 420–432.

Nero, Anthony V., Jr. *A Guidebook to Nuclear Reactors*. Berkeley: University of California Press, 1979.

Newton, David E. "Nuclear Power Plants" *Gale Encyclopedia of Science*, edited by

Bridget Travers. Detroit: Gale Research, 1996, 2551–2554.

Turk, Jonathan, and Amos Turk. *Environmental Science.* 4th ed. Philadelphia: Saunders College Publishing, 1990, Chapter 14.

"U.S. Commercial Nuclear Power Reactors." http://www.nrc.gov/AEOD/pib/states. html. 23 June 1998.

Home page for the Nuclear Energy Institute is at http://www.nei.org/. The institute can also be contacted at 1776 I Street, N.W., Suite 400; Washington, D.C., 20006-3708; 202-739-8000.

Nuclear Warfare

See Nuclear Weapons

Nuclear Wastes

Nuclear wastes are radioactive materials left over from or produced during various nuclear processes. These processes include the manufacture of nuclear weapons, the generation of electricity in nuclear power plants, and the use of nuclear materials in medical research, diagnosis, and treatment. As an example, nuclear power plants convert heat produced as a result of nuclear fission reactions into electrical energy. The elements used to carry out those fission reactions become depleted after a period of time—usually a few years—and must be removed from the nuclear power plant. But they are still radioactive and pose a serious threat to human health and the environment.

Historical Methods of Nuclear Waste Treatment

Most power companies and government agencies appear to have give relatively little attention to the issue of nuclear waste disposal early in the history of nuclear power and nuclear weapons development. Most of those who used the technology appeared to feel that a solution to the storage of nuclear wastes would eventually be found and implemented without too much difficulty.

What is to be done with the nation's nuclear wastes when even the most geographically remote regions of the nation are regarded by some people as being unsafe for their permanent burial? (U.S. Department of Energy)

Such as been anything but the case. Instead, thousands of tons of radioactive wastes have been accumulating for more than half a century in "temporary" storage facilities across the United States.

In many cases, "nuclear waste disposal" has meant sealing radioactive materials in metal canisters or concrete casing and stacking them in piles near the facilities where they were generated. The U.S. government is still trying to decide, for example, how to get rid of radioactive wastes produced during the development of the first atomic bombs in the early 1940s. Some of those wastes are currently being stored in 46,000 huge metal cylinders in Paducah, Kentucky; Oak Ridge, Tennessee; and Piketon, Ohio. The cylinders weigh up to fourteen metric tons each and have been slowly rusting away since the 1940s. They contain highly radioactive uranium and other elements as well as toxic hydrogen fluoride gas. The U.S. Department of Energy announced plans in 1998 to offer these cylinders for sale to potential recyclers of these materials. Many authorities, however, doubted that anyone would be interested in that offer.

Planned Disposal Sites

Two of the most ambitious plans for nuclear waste disposal are those for the storage of high-level wastes at Yucca Mountain, Nevada, and for low-level wastes in southeastern New Mexico. The Yucca Mountain repository plan grew out of a congressional decision in 1982—the Nuclear Waste Policy Act—requiring the Department of Energy (DOE) to develop a plan for storing spent (used) fuel rods removed from nuclear power plants. As of 1997, more than 32,000 metric tons of such wastes had accumulated and were being stored close to the reactors from which they were removed or at other "temporary" sites.

After an extensive survey, DOE selected Yucca Mountain as the most feasible location for a long-term disposal site. The site is located on Nellis Air Force Base, the Nevada Nuclear Test Site, and land managed by the Bureau of Land Management. Plans called for burial of high-level nuclear wastes in 200 kilometers (130 miles) of tunnels buried 300 meters (1,000 feet) underground in the mountain. Wastes would be hauled into the mountain tunnels by railroad cars and sealed off in side tunnels running off a main tunnel. The site was designed to hold up to 77,000 metric tons of wastes, enough to handle the nation's high-level waste products for fifty years or more. The cost of the design and construction of the project was to be paid for by a 0.1 cent tax on each kilowatt hour of electricity generated by nuclear power in the United States.

Problems at Yucca Mountain

As to be expected, the Yucca Mountain proposal had its opponents. Residents of Nevada and politicians in the state were violently opposed to the use of their state as a "nuclear graveyard." Environmentalists worried about the escape of radioactive wastes into the ground and, then, into groundwater. They pointed out that the wastes would continue to pose a health and environmental threat for thousands of years into the future. No one knew, they said, how natural forces might cause the movement of wastes. Some support for this concern was announced in early 1998 when geological reports confirmed that the land around Yucca Mountain was stretching at ten times the rate originally estimated. The earth's surface in the region was found to be pulling apart at a rate about ten times faster than originally expected.

In any case, progress at Yucca Mountains has been very slow indeed. Originally scheduled to open in 1998, the facility was less than 5 percent completed on 1 January of that year. Opening date had by then been moved to 2010. The possibility of further delays beyond that point, however, could not be discounted.

In the meanwhile, power companies had lost patience with the federal government. They had collected and paid to the government more than $4 billion for development of a nuclear waste disposal site. And, by 1998, construction on that site had barely begun. Companies were still having to spend tens of millions of dollars to store nuclear wastes on their own facilities. In desperation, a group of twenty-five utilities and eighteen state commissions had filed suit as early as 1996 against DOE for its lack of action. In July of that year, a federal appeals court ruled that DOE was legally responsible for nuclear waste, "ready or not," on 31 January 1998. If Yucca Mountain were *not* ready (as it turned out not to be), DOE was potentially responsible for tens of billions of dollars in refunds and storage costs to utilities and states.

By mid-1999, the final status of Yucca Mountain had still not been determined. Secretary of Energy Bill Richardson had submitted a Viability Assessment Plan to the U.S. Congress in December 1998. The plan outlined DOE's efforts to determine whether or not Yucca Mountain was a satisfactory site for the disposal of nuclear wastes. Secretary Richardson told a House committee on 12 March 1999 that the department's research had found no "show-stoppers" and that research was "on target to decide in 2001 whether Yucca Mountain is suitable to be the location of a repository and to submit a license application" for opening of the site ("Nuclear Waste Storage" 1999).

Storage of Low-Level Wastes in New Mexico

Meanwhile, a somewhat similar scenario was being played out with regard to low-level radioactive wastes. Low-level radioactive wastes include gloves, tools, and other materials used in working with plutonium and other so-called transuranic elements (elements heavier than uranium). DOE began planning as early as the 1970s for a waste disposal site for such wastes in southeastern New Mexico. The site—the Waste Isolation Pilot Plant (WIPP)—was to consist of a huge burial vault in salt beds 650 meters (2,150 feet) below the Earth's surface. Construction of the site was begun in the late 1980s and completed in 1988. Total cost of the project was $1.8 billion.

WIPP could not open, however, until DOE wrote an environmental impact statement acceptable to the U.S. Environmental Protection Agency (EPA). Ten years after construction on the site had been completed, that approval had still not been granted. One of the major concerns expressed by the EPA was that radioactive materials might escape into the salt dome and, from there, into underground water systems. After all, critics said, the burial process is designed to seal off wastes for thousands of years. But who knows what geological changes might occur over that time? Besides, they said, DOE had already developed an even more secure burial site at Yucca Mountain. Why could low-level wastes not also be stored there?

DOE officials argued that neither of these arguments was valid. First, the New Mexico salt dome was laid down during the Permian Period 225 million years ago, and it has remained stable ever since. Further, the Yucca Mountain facility was not designed to have the capacity needed to store low-level as well as high-level wastes.

Resolution of the New Mexico site debate appears to have been achieved in May 1998 when the EPA announced its approval of the DOE site plan. Most citizens in the local area surrounding the site were ecstatic and began making plans to greet the arrival of the first shipment of radioactive wastes. That shipment arrived on 26 March 1999. It consisted of radioactive waste materials generated by clothing, tools, rags, debris, residues, and other disposable items contaminated with radioactive elements, primarily plutonium.

The wastes had come from the Los Alamos National Laboratory, also in New Mexico. They had been stored at the laboratory since the 1940s and were the first of seventeen such shipments scheduled for delivery to WIPP.

Over the planned life of the site, an additional 37,000 shipments would arrive over a thirty-year period from "temporary" holding sites in California, Colorado, Idaho, Illinois, Nevada, New Mexico, Ohio, Tennessee, South Carolina, and Washington.

References

Barlett, Donald L., and James B. Steele. *Forevermore: Nuclear Waste in America.* New York: W. W. Norton, 1985.

Brooke, James. "Underground Haven or a Nuclear Hazard? *New York Times,* 6 February 1997, A12.

Long, James, and Jim Barnett. "WareHousing Nuclear Waste." *Oregonian,* 12 March 1997, A12.

"No Quiet Burial for Nation's Nuclear Waste Debate." *San Francisco Chronicle,* 21 January 1997, A5.

"Nuclear Waste Storage." Statement of Secretary Bill Richardson, U.S. Department of Energy, before the Subcommittee on Energy and Power on the Committee on Commerce, U.S. House of Representatives, 12 March 1999.

Sahagun, Louis. "N.M. City Embraces Radioactive Waste Plant." *Oregonian,* 9 June 1998, A12.

Turk, Jonathan, and Amos Turk. *Environmental Science.* 4th ed. Philadelphia: Saunders College Publishing, 1990, 399–403.

Warrick, Joby. "N.M. Nuclear Waste Dump Gains License to Operate." *Oregonian,* 14 May 1998, A9.

Nuclear Weapons

Nuclear weapons can be classified into either of two major categories: fission weapons (atomic bombs) and fusion weapons (hydrogen bombs). Fission weapons tend to be simpler and less powerful devices than fusion weapons. They were also the first kind of nuclear weapons to be made. They were developed in the early 1940s in the United States in a massive military research program codenamed *Manhattan Project.*

One of the debates about the development of nuclear weapons is how necessary it is to continue developing larger and more effective weapons in order to avoid ever having to use those weapons. (U.S. Department of Energy)

Fission Weapons

Fission weapons exploit the explosive energy released as the consequence of an uncontrolled nuclear fission reaction. Nuclear fission occurs when a neutron collides with the nucleus of certain large atoms, such as uranium-235 or plutonium-239. The numerical designation "-235" or "-239" represents a specific isotope, or variation, of the named element. Other isotopes of the same elements, such as uranium-238 or plutonium-240, are generally not capable of undergoing fission.

Fission reactions always result in the formation of three kinds of products: energy, more neutrons, and nuclei of elements lighter than uranium or plutonium. Such a reaction can be represented by the following equation:

$$\,^{1}_{0}n + \,^{235}_{92}U \rightarrow \,^{142}_{56}Ba + \,^{90}_{36}Kr + x\,^{1}_{0}n + energy$$

The two lighter nuclei formed (barium, Ba, and krypton, Kr, in this example) are known as fission products.

The unique value of fission as a source of energy depends on the production of neutrons in the process. Note that neutrons are needed to *initiate* the reaction and are then *produced* as a result of it. This phenomenon makes possible a *chain reaction* in uranium-235, plutonium-239, or any other fissionable material. Neutrons produced during the first stage of fission are available to initiate a second stage of the reaction. Similarly, neutrons released in the second stage can initiate a third stage, and so on. In actual practice, this process can occur millions of times in a fraction of a second. The amount of energy released during this whole sequence of reactions is enormous, unmatched by almost any other process known to humans.

A multitude of practical problems must be solved in order to convert the principle of nuclear fission to an actual weapon. One of the first challenges for scientists was finding out how large a piece of uranium-235 or plutonium-239 would be needed for a successful chain reaction to occur. If the piece were too large, a spontaneous ignition of the material might occur as a result of free neutrons in the air. In such a case, a premature explosion would occur. If the piece were too small, too many neutrons would be lost to the surrounding environment, preventing the continuation of a chain reaction. The exact size needed for a successful chain reaction—the *critical size*—for any given isotope was at first calculated and then confirmed experimentally.

That size provides the ultimate limit for any fission weapon. For example, an atomic bomb may contain two pieces of uranium-235, each half critical size. The two pieces can then be forced together after release of the bomb and prior to its hitting the ground. Or a bomb could contain three pieces of uranium-235, all less than critical size. Before long, however, the physical problems of arranging pieces of uranium-235 of less than critical size becomes more troubling than it is worth . . . particularly in view of the availability of fusion weapons as an option for more powerful explosive devices.

Fusion Reactions

Nuclear fusion is the process by which two or more small atomic particles are joined to each other ("fused") to make one larger particle. That process can be represented very simplistically by the following equation:

$$4 \text{ H} \rightarrow \text{He} + \text{energy}$$

This equation indicates, somewhat incompletely, that four hydrogen nuclei can be combined into a single nucleus of helium. The fusion process, like fission, results in the formation of very large amounts of energy. The process is, indeed, thought to be the primary mechanism by which stars such as our own Sun produce their energy.

One conceptually simple problem must be overcome in order for humans to make use of fusion power. Combining two (or more) hydrogen nuclei means forcing together two particles with the same (positive) electrical charge. Very large amounts of energy are needed to overcome this electrical repulsion. In fact, the amounts of energy required are found naturally in only

one environment: the very hot interior of stars. At the temperatures found inside stars—a few ten millions of degrees—fusion does occur. For this reason, such reactions are often called *thermonuclear* ("heat-nuclear") reactions.

For better or worse, temperatures of this magnitude can be found in one other environment, the center of a nuclear fission bomb. In the smallest fraction of a second after such a weapon has been exploded, the temperature reaches 10–20 million degrees. Sufficient energy is released to initiate fusion reactions.

A fusion weapon, then, is not much more than a fission weapon surrounded by hydrogen. When the fission weapon is ignited, very high temperatures are produced, and fusion in the surrounding hydrogen begins. That fusion releases, in turn, even larger amounts of energy.

Fusion weapons, unlike fission weapons, are not limited in size by the amount of fuel they contain. In theory, one could surround a fission weapon with one kilogram of hydrogen, ten kilograms of hydrogen, a hundred kilograms or hydrogen, or any other quantity of hydrogen that one might imagine. The limiting factor with a fusion weapon is the capacity of the device by which it is to be delivered to an enemy: the rocket or aircraft on which it is to be transported.

The size of nuclear weapons is usually expressed in kilotons or megatons. These terms do not designate the weight of the weapon itself, but its destructive capacity. For example, a 1-kiloton weapon is one that has the destructive equivalent of 1 kiloton (1,000 tons) of trinitrotoluene (TNT), one of the most powerful chemical explosives known. The first fission bombs dropped on Hiroshima and Nagasaki, Japan, at the end of World War II were rated at about 15 kilotons each. The first fusion weapon tested by the United States at Bikini Atoll in March 1954, by contrast, had an explosive power of 15 megatons (15 millions of tons), a thousand times greater than the Hiroshima and Nagasaki fission weapons.

Issues

The invention and development of nuclear weapons have been surrounded by intense controversy for more than half a century. There are no questions, of course, about their effectiveness as instruments of war. Their destructive capacity far exceeds anything that humans had previously discovered. Yet, that very property raised questions as to their use. Were they, in fact, too terrible to use even in warfare? Certainly, the results of their first use by the United States against Japan gave many people pause as to what role such weapons might have in future conflicts. One of the most fundamental issues faced by nations since World War II, then, is what role, if any, nuclear weapons should have in the military systems of the world's nations.

A second, but closely related, issue has involved the testing of nuclear weapons. At first, such tests were conducted in the atmosphere, under conditions that would not be so different from those in actual warfare. The problem with atmospheric testing, however, was the release of radioactive fission products to the environment. These fission products are released during an atmospheric test in the form of very fine powder that may circulate in the atmosphere for weeks, months, or even years before settling to the Earth. Many of those fission products remain radioactive for dozens or hundreds of years, posing threats to human health and the environment.

As an example, radioactive strontium-90 is a common fission products of nuclear explosions. When released to the Earth, it is ingested by cows, becomes part of their milk, and then, in turn, is ingested by humans. The problem with strontium is that it behaves chemically like calcium. When ingested by children, for example, the radioactive strontium-90 is used to build bones and teeth, just as is calcium. These new structures, however, are radioactive because of the strontium-90 they contain. The horror resulting from atmospheric testing, then, was that young children were incorporating radioactive materials into the very substances from which their growing bodies were made.

Because of these concerns, the world's nations struggled for more than fifty years to find a way of monitoring the use and testing of nuclear weapons. Progress was extraordinarily slow as nations worked to protect their own military interests while trying to protect all nations from the horror of a world nuclear conflict. The process was made more difficult by what became an even greater distrust of other nations' motives and promises than is even normally the case in international negotiations.

Over time, small steps forward were made toward nuclear disarmament. The following list summarizes the most important of the nearly two dozen treaties dealing with nuclear disarmament that have been signed.

The Limited Test Ban Treaty of 1963 was agreed to by three nations: the United States, the United Kingdom, and the Soviet Union. The treaty prohibited the testing of nuclear weapons in the atmosphere, outer space, or under water. By 1998, 116 nations had signed the treaty. The two major non-signing nations are France and China.

The Nuclear Non-Proliferation Treaty of 1968 was a temporary agreement designed to prevent the spread of nuclear weapons. It was also designed to prevent nations from diverting nuclear materials from peaceful applications to weapons development. Three major nuclear powers, the United States, United Kingdom, and Soviet Union, along with 133 nations without nuclear weapons eventually signed the treaty. In May 1995, the treaty was made permanent by action of the United Nations.

The Antiballistic Missile Treaty of 1972 was a bilateral agreement between the United States and the Soviet Union. It limited the regions in which antiballistic missiles (ABMs) can be deployed and prohibits the development of many critical features of ABM technology.

The Strategic Arms Limitation Treaty (SALT) I of 1972 was a bilateral agreement between the United States and the Soviet

Union, to extend over a five-year period. Its major component was a freeze on the number of strategic ballistic missiles to be developed. SALT II of 1979 was a bilateral agreement between the United States and the Soviet Union providing further limitations and ceilings on the production of strategic nuclear weapons.

The Strategic Arms Reduction Treaty I (START I) of 1991 was a bilateral treaty between the United States and Soviet Union further reducing the number of nuclear weapons to be developed by both nations. The Strategic Arms Reduction Treaty II of 1993 was a bilateral agreement between the United States and Russia. It was signed by Presidents Bush and Yeltsin in January 1993 and further reduced the number of nuclear weapons to be developed by both nations.

The Comprehensive Test Ban Treaty (CTBT) of 1996 was finally approved in 1996 by the United Nations after having been discussed by that body for more than four decades. The treaty was signed by 149 nations.

Signing of these treaties by member nations has often been only half the battle in moving toward nuclear disarmament. Those treaties must then be ratified also by national legislatures in each signatory nation. Getting those legislatures to agree to treaties is sometimes a difficult task. For example, it took the U.S. Senate more than three years to ratify the START II Treaty. By 1998, the Senate had not even begun to consider ratification of the CTBT. In fact, Senator Jesse Helms (R-N.C.), chairman of the Foreign Relations Committee, announced that consideration of the treaty was "very low" on the committee's list of priorities. In fact, he was then working on legislation designed to nullify or weaken earlier weapons treaties.

By mid-1999, action on the CTBT had still not been taken. The Senate had spent a significant portion of its 1998 session dealing with issues surrounding the impeachment of President Bill Clinton. A number of substantive issues, the CTBT among them, had to be put on hold until the impeachment issue was resolved.

Weapons Developments in India and Pakistan

The fragility of agreements over nuclear weapons development was illustrated in 1998 when both India and Pakistan conducted underground tests of such weapons. Neither nation had signed the CTBT, and both announced that they needed to develop nuclear weapons to protect themselves from each other and from China, their neighboring nonsignatory and nuclear power. The new weapons crisis arose on 11–13 May when India detonated a series of five underground nuclear tests. India's Prime Minister Atal Bihari Vajpayee announced that the tests were needed to protect India from aggressive actions by Pakistan.

Two weeks later, Pakistan carried out its own nuclear tests. Its justification for these tests was concerns about the aggressive stance of India. Since the two countries had been at war three times between 1947 and 1998, the rising uneasiness between them was a source of serious concern throughout the world. The editors of the *Bulletin of the Atomic Scientists,* for example, moved their Doomsday Clock forward to read 11:51 P.M. The clock has been shown on the front page of the journal since its inception in 1947 and shows how close the editors feel the world has come to nuclear warfare, or Doomsday.

Nuclear Weapons in the United States

The signing and ratification of nuclear weapons treaties has also not marked the end of nuclear weapons testing and development in the United States. As the world's sole remaining first-class nuclear power, the United States is torn between a desire to reduce or eliminate its weapons and an understanding that its control of those weapons may be the most powerful single force in ensuring peace in the world.

As the 1990s drew to a close, an ongoing debate continued among politicians, scientists, and military leaders as to how the nation's nuclear weapons supply could be tested and maintained. Major research programs were being developed to find ways of testing nuclear weapons without actually exploding them. In addition, the nation made a commitment to resume weapons production by the year 2003 in order to guarantee that some minimum number of effective weapons would be available to the military for the foreseeable future.

References

"Arms Control Treaties." http://www.clw. org/contro.html. 21 May 1999.

"Background Information [on START II Ratification]." http://infomanage. com/nonproliferation/treaties/startii. html. 28 May 1998.

Davidson, Keay. "U.S. May Conduct Nuke Tests in Nevada." *San Francisco Examiner,* 1 December 1996, A1+.

Gibson, James N. *Nuclear Weapons in the United States: An Illustrated History.* Atglen, PA: Schiffer Publications, 1996.

Gordon, Michael R. "U.S. Proposes Deeper Cuts in Atom Arms to Russians." *New York Times,* 9 March 1997, A9.

Nichols, Bill. "Sanctions on India Likely." *USA Today,* 13 May 1998, A1+.

Sagan, Scott Douglas, and Kenneth Neal Waltz. *The Spread of Nuclear Weapons: A Debate.* New York: W. W. Norton, 1995.

Schroeer, Dietrich. *Science, Technology and the Nuclear Arms Race.* New York: John Wiley and Sons, 1984.

Turner, Stanfield. *Caging the Nuclear Genie: An American Challenge for Global Security.* Boulder, CO:Westview Press, 1997.

"U.N. Passes Nuclear Test Ban Treaty." *San Francisco Chronicle,* 11 September 1996, A1+.

Vartabedian, Ralph. "U.S. Plan to Resume Bomb-making Is Hitting Snags." *San Francisco Chronicle,* 26 February 1997, A5.

Wald, Matthew L. "Lab's Task: Assuring Bombs' Quality without Pulling Nuclear Trigger." *New York Times,* 3 June 1997, A16.

Wright, Robin. "India Tests Strength of Nuclear Pacts." *Sunday Oregonian,* 17 May 1998, A5.

The text of the Comprehensive Nuclear Test-Ban Treaty can be found on the Internet at http://www.acda.gov/treaties/ramaker. htm. 28 May 1998.

Nuclear Winter

Nuclear winter is a term invented to describe environmental effects that might occur as the result of a war in which nuclear weapons are used on a large scale. The potential for such effects was explored on a theoretical basis by a group of respected scientists in the mid- and late 1980s.

Mechanism of Nuclear Winter

The analysis was based on well-known environmental effects of nuclear weapons explosions of given magnitude. In general, such explosions result in the very rapid incineration of all combustible materials within a given range (depending on the size of the weapon) of the blast. For example, all such materials within a fifty-kilometer radius are incinerated within seconds by a one-megaton weapon.

A major product of this form of combustion is soot, consisting of very fine particles with a diameter of less than one micron (one micrometer). Clouds containing large quantities of soot have properties somewhat different from those of rain and ice clouds. In particular, they tend to reflect a larger fraction of sunlight back into space than do water clouds. At the same time, they tend to be about as transparent to thermal (heat) radiation given off by the Earth's surface.

The combination of these two factors produces an effect just the opposite of the natural greenhouse effect that occurs in the atmosphere. That is, less solar energy would reach the Earth's surface (because of the greater reflectivity of the soot clouds), while the loss of heat from the Earth's surface and lower atmosphere would be just as great as it is now. The overall effect of these two factors, then, would be an "anti–greenhouse effect," in which the lower atmosphere would become cooler and the upper atmosphere warmer.

Estimates of Nuclear Winter Effects

Nuclear winter researchers estimated that a large-scale nuclear conflict might result in the formation of clouds containing anywhere from 10 million to 100 million metric tons worldwide. Clouds of that magnitude, they said, would blot out the sun for weeks and might cause a drop in surface temperatures of up to 25 degrees C. The nature of the cooling by soot clouds is such, they warned, that the effect could well be self-sustaining and continue for months or even years. In the meanwhile, crops deprived of sunlight would begin to die out, and mass starvation would occur around the planet. One researcher set the number of human casualties of hunger and disease in the first year following a nuclear conflict at 3 billion. He then warned that "the future of civilization beyond that point would seem grim, with little infrastructure remaining to support a long-term recovery" (Turco 1997, 461).

References

Crutzen, P. J., and J. Birks. "Twilight at Noon: The Atmosphere after a Nuclear War." *Ambio* 11 (1982): 114.

Harwell, M., and T. Hutchinson. *Environmental Consequences of Nuclear War.* Vol. 2: *Ecological and Agricultural Effects.* New York: Wiley, 1985.

Pittock, A., et al. *Environmental Consequences of Nuclear War.* Vol. 1: *Physical and Atmospheric Effects.* New York: Wiley, 1985.

Sagan, Carl, and Richard P. Turco. *A Path Where No Man Thought: Nuclear Winter and the End of the Arms Race.* London: Century Press, 1990.

Turco, R., et al. "The Climatic Effects of Nuclear War." *Scientific America,* 251 (1984): 33.

Turco, Richard P. *Earth Under Siege.* Oxford: Oxford University Press, 1997.

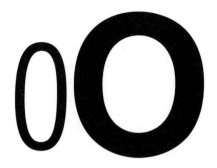

The Oceans

The oceans are perhaps the most poorly understood, least appreciated resource on our planet. It has been said that humans know more about the face of the Moon than they do about the ocean floor.

Yet, the oceans are an enormously valuable resource. They cover nearly three-quarters of the Earth's surface to an average depth of about 4 kilometers (2.5 miles). They are one of the richest sources of food on Earth and a virtually unexplored treasure trove of mineral resources. The oceans also have important, but poorly understood effects on the Earth's climate and weather patterns.

The Law of the Sea Treaty of 1982

It has been only recently that politicians have begun to think seriously about problems related to the oceans. One factor in producing this interest has been the dramatic decline in food fish that has occurred over the past twenty years. As an example, about 500 million metric tons of the East Atlantic bluefish were harvested in 1972. By 1998, that number had dropped to nearly zero. Highly sophisticated technology developed for massive fishing operations has contributed to this decline. Many fishing ships are now capable of trawling lines 130 kilometers (80 miles) long and then immediately processing the fish caught in on-board facilities.

Almost unimaginable stores of mineral resources can be found on the ocean floors also. By some estimates, billions of metric tons of nodules containing manganese, copper, cobalt, and nickel rest on the ocean bed for the taking by anyone who develops a method for doing so.

Perhaps the most direct cause for interest in the oceans' future was the issue of ownership. Traditionally, there has been no widely agreed upon definition as to the extent to which any one nation has authority over the oceans around it. Some nations claimed ownership to a distance of 320 kilometers (200 miles) from their coastlines, while others asked for dominion over no more than a nineteen-kilometer (twelve miles) stretch from the coast.

In an effort to deal with these issues, the United Nations held a Law of the Sea Conference in 1982. That conference produced a document known as the Law of the Sea Treaty that set out agreements about ownership and use of ocean resources. Ratification of that treaty came only very slowly, however, as individual nations raised questions about one or another of its component. For example, the United States originally refused to sign the treaty because of concerns about the regulation of seabed mining. Even after treaty changes were made to deal with these concerns, the U.S. Senate refused to ratify the treaty. As of the end of 1998, 124 nations had approved the treaty, but the United States was still not one of them.

Ocean Pollution

One of the fundamental problems in dealing with ocean-related problems is the extent of our ignorance about the oceans. The issue of ocean pollution is an example.

Many environmentalists have become very concerned about ocean pollution. In some cases, that pollution can be pinpointed to specific causes as, for example, when oil leaks from an offshore oil drilling platform. Those cases are relatively easy to identify and, in some cases, easy to remedy. Much more difficult is the problem of nonpoint sources of pollution.

The term *nonpoint sources of pollution* refers to a large variety of locations that release pollutants that eventually flow into the ocean. For example, nearly all rivers and streams carry some pollutants in them as they flow to the seas. The amount may be very small and often has little or no effect on the water quality of the river or stream. But those pollutants accumulate once they reach the ocean. Unlike the situation with rivers and streams, ocean water has no place to which to flow. It may evaporate into the atmosphere, but the pollutants released to it remain in the oceans.

An example of the kind of problem troubling ocean scientists is the one that can be found in the Gulf of Mexico. A region covering thousands of square kilometers west of the Mississippi River's outlet in Louisiana has been declared a "dead zone." The term refers to the fact that fish, shellfish, and other marine life are unable to survive in this region. When first measured in 1989, the dead zone covered an area of about 9,000 square kilometers (3,500 square miles). After the 1993 rainy season, the zone doubled in size. Since that time, it has receded somewhat but still covers about 16,000 kilometers (6,000 square miles).

The immediate cause of fish deaths in this zone is algal blooms. An algal bloom is a relatively rapid and extensive growth of algae in seawater. As algae grow, they use up oxygen dissolved in the water, leaving little or none for other marine organisms to breathe. Without oxygen, fish and other organisms literally suffocate.

The fundamental question facing scientists and politicians is what it is that causes the algal blooms. Most scientists tend to believe that the answer lies farther up the Mississippi River, in the broad fields of corn, wheat, soybeans, and other crops that spread across the midwestern states. More than half of the farmland in the United States occurs in the Mississippi River basin which, in turn, covers about 40 percent of the lower forty-eight states.

The explanation that scientists have developed begins with fertilizers used by U.S. farmers on these lands. Some of that fertilizer washes off the land and into rivers and streams. It is eventually carried into the Mississippi and then into the Gulf of Mexico. There, the nitrogen found in fertilizer serves as food for algae. Other sources of nitrogen in the river and the gulf are wastes from feedlots and grazing lands, outflow from sewage treatment plants, and chemical products of fossil fuel burning.

This scenario seems very reasonable to many scientists. The problem is that actual measurements have so far been unable to show a direct connection between farming techniques and algal blooms. Connections between algal blooms and decline in fish and other marine life are also not as clear as scientists would hope.

Reducing Ocean Pollution

In the absence of more definite data, policy makers have been uncertain as to what steps to take. They are reluctant to tell farmers to cut back on the amount of fertilizer they use on their lands. Farmers have responded, when such suggestions have been made, that a stronger case needs to be made for fertilizers as a cause of ocean pollution. Environmentalists argue that making the fertilizer/algal bloom connection would have a secondary benefit. That connection could be used to show farmers how much fertilizer they are wasting and how they could save money by learning how to use their fertilizer more efficiently.

The Gulf of Mexico "dead zone" is not unique. Indeed, scientists have discovered about fifty similar zones along ocean coasts in North America, Europe, Japan, the Middle East, and a few other parts of the world.

An additional problem in trying to deal with such zones is the long lag time between reducing pollution and seeing a rebirth of marine life. Chesapeake Bay, for example, is one of the first coastal regions in which algal blooms were studied in detail. Scientists found that even after nitrogen runoff had been reduced, levels of the mineral remained high in rivers, streams, and the bay for years. Any hope of solving this problem at its source, therefore, has to be tempered by the realization that improvements in the ocean may not appear for many years.

References

Barnum, Alex. "Sounding a Siren for the Sea." *San Francisco Chronicle,* 8 June 1998, A1+.

Borgese, E. M. "The Law of the Sea." *Scientific American,* March 1983, 42–49.

Daley, William M. "A Move Toward Healing the Ocean." *San Francisco Chronicle,* 20 June 1998, A19.

Dalton, John H. "Ratify the Law of the Sea Treaty." *Oregonian,* 20 July 1998, B7.

Kristof, Nicholas D. "Tension in Asia over Wealth under Sea." *New York Times,* 19 May 1996, A6.

Malakoff, David. "Death by Suffocation in the Gulf of Mexico." *Science,* 10 July 1998, 190–192.

McQuaid, John. "Reaching the Limit." *San Francisco Examiner,* 7 July 1996, A10.

Olestra

Olestra is the trade name for a fat substitute manufactured by the Proctor & Gamble Company (P&G). It was approved for use by the U.S. Food and Drug Administration (FDA) in 1996. It is now used in certain snack foods, such as potato chips, corn chips, and crackers under the name Olean®.

Fats in the Human Diet

The development of a substitute for natural fats has been a high priority among food manufacturers for many years. Fats are an essential part of the human diet. They have a number of important functions in the body, such as providing calories (energy) for a number of body functions; dissolving certain "fat-soluble" vitamins, such as vitamins A, D, E, and K; serving as an important structural material in bodies, providing both insulation and protection for body organs; and providing essential substances needed in normal growth and development. Fats are also important in supplying desirable qualities to natural and processed foods. They make foods more tasty and easier to eat and swallow, they give foods consistency, and they increase the stability of certain foods.

But an excess of fat in the diet can lead to a number of health problems. One of those problems is obesity. Any fats included in a person's diet that are not digested and metabolized are stored as body fat. A diet consistently high in fats leads to weight problems that may, in turn, lead to other health problems, such as diabetes, heart disease, and some forms of cancer.

Official government guidelines suggest that Americans not consume more than 30 percent of their daily diet in fats. Yet research shows that many Americans exceed that limit by significant amounts. Food manufacturers have, therefore, long recognized the market for a product that has all the qualities of fat with few or none of its disadvantages.

Biochemistry of Olestra

Fats are chemical molecules that consist of two parts: glycerol and fatty acids. Glycerol belongs to the organic family known as alcohols. It contains three hydroxyl (-OH) groups attached to a carbon backbone. In a fat, a fatty acid is attached at each of the three hydroxyl locations. The general structure of the fat is somewhat like an extended letter E:

```
|— — — — — — — — — — —
|— — — — — — — — — — —
|— — — — — — — — — — —
```

where each of the long horizontal extensions is a fatty acid.

Even though Olestra was approved for public sale only after years of development and testing, many individuals feel the product is still not safe enough for people to consume. (Courtesy of Procter & Gamble)

Olestra has a chemical structure somewhat like that of a natural fat. The difference is that the glycerol molecule is replaced by a sucrose molecule. A sucrose molecule has seven hydroxyl groups rather than three. In olestra, fatty acids have been attached at the location of all seven hydroxyl groups:

One property this structure gives to olestra is indigestibility. That is, an olestra molecule passes through the human digestive system before it can be completely broken down. Since it is not digested, it cannot be metabolized, and it provides no calories to the body. On the other hand, olestra is enough like natural fat to provide the physical tastes and sensations associated with fats, such as smoothness, flavor, and consistency.

Regulatory Issues

Fat substitutes other than olestra have been invented and certified by the FDA for use in certain foods. Two such products are known as Simplesse and Avicel. These fat-substitutes have one major disadvantage,

however. They break down under heat and cannot be used, therefore, in foods that must be fried or heated to high temperatures for other reasons. Olestra, by contrast, is stable at high heats and can be used in such foods.

P&G filed a petition with the FDA in 1987 asking for approval to use olestra in certain snack foods, such as potato chips, cheese puffs, and snack crackers. It provided the agency, as is always required, with extensive research evidence that the new product was both safe and effective. Almost a decade later, the FDA granted approval to P&G to use olestra in this restricted range of foods.

Almost immediately, however, critics demanded a reconsideration of the FDA's decision. They said the agency had neglected three important points about the food additive. First, olestra may cause mild gastrointestinal (GI) problems, such as abdominal cramping and loose stools, in some consumers. One advocacy group, the Center for Science in the Public Interest, said that more than 8,000 people had reported GI problems in the first two years that olestra products had been on the market.

Second, some scientists also worried about potential health problems from olestra-containing products. The compound is not able to dissolve fat-soluble vitamins the way natural fats can. Although these fats are actually added to olestra-containing products, the experts felt that the ingestion of olestra might lead to vitamin-deficiency problems.

Third, some researchers were concerned about the potential long-term health effects of using olestra instead of natural fats in the diet. The amounts of olestra needed to carry out valid animal studies are so large that such studies were impossible. That may suggest, according to some critics, that too little is still known about the problems some individuals may experience after eating olestra-containing products for ten, twenty, or thirty years.

In 1998, the FDA decided to appoint a panel to reconsider the data used in approving olestra. The panel decided that the data were adequate to support the agency's original decision, and the product continues to be available in the marketplace.

References

"Facts About Olestra." American Dietetic Association. http://www.eatright.org/nfs.18.html. 5 July 1998.

"FDA Panel Member Voices Olestra Concerns." *USA Today,* 2 February 1998, D1.

Hellmich, Nanci. "Olestra Has Passed Safety Test, FDA Says." *USA Today,* 18 June 1998, D1.

Horovitz, Bruce. "Fake Fat's Big Test: Olestra." *USA Today,* 19 June 1997, B1+.

"Issue in Focus: Olestra." http://www.acsh.org/publications/olestra/index.html. 7 July 1998.

"New Actions Opposing Olestra." http://www.cspinet.org/new/oles_6_9.htm. 7 July 1998.

"Olestra and Other Fat Substitutes." http://www.fda.gov/opacom/backgrounders/olestra.html. 5 July 1998.

Silverstein, Ken. "How the Chips Fell." *Mother Jones,* May/June 1997, 13–14.

The olestra home page is at http://www.olestra.com/.

Organ Transplantation

Organ transplantation refers to the procedure by which a heart, kidney, liver, pancreas, lung, or other organ is transferred from the body of one organism to that of another organism. In theory, organ transplants could take place between any two organisms or kinds of organisms. For example, one could theoretically transplant the heart of one dog into a second dog or into a monkey. In fact, transplants of this kind are done only for experimental purposes, to find out more about the transplantation process itself. Almost the only practical application of organ transplantation occurs in order to provide a healthy replacement to a human who has a damaged organ. That replacement organ may come from another human or from an animal other than a human, such as a chimpanzee. Transplantations between unlike species are known as *xenotransplants.*

History of Transplantation

Medical workers have explored the possibility of transplants for centuries. As early as the sixteenth century, for example, techniques were developed to replace noses lost during battle or as the result of infection by syphilis. New noses were constructed from skin and tissue taken from other parts of the patients' bodies.

Research on the transplantation of vital organs, such as heart, kidney, and liver, into humans did not really begin, however, until the early twentieth century. The major problem in early research was a mechanical one: learning how to reconnect an organ in the recipient's body after it had been removed from the donor. One of the earliest successful experimental organ transplants occurred in 1902 in Germany, where researchers transplanted a kidney from one dog to another.

The first human-to-human kidney transplant was attempted in 1933, but the procedure failed. The kidney never functioned in the recipient's body. The first successful transplant of this kind occurred in 1954. Joseph Murray, at Brigham and Women's Hospital in Boston, transferred the kidney from one twin to another. The transplant was successful because of the close biochemical relationship of donor and recipient.

Over the next decade, the first successful transplants of other organs also occurred. The first successful liver transplant was performed in 1967 by Thomas Starzl at the University of Colorado Health Sciences Center. And in 1967, Christiaan Barnard of South Africa performed the first successful hearth transplant.

Problems of Rejection

Learning how to attach a transplanted organ in a patient's body was only the first problem facing surgeons in learning how to do organ transplants. Once they had developed the techniques to perform this procedure, they faced a second challenge: rejection by the recipient's body. Organ rejection occurs because, as far as the recipient's body is concerned, a new organ is a "foreign object." The human body has at its disposal a host of techniques for fighting back against foreign bodies, such as those that cause infection. In an overwhelming number of cases, those techniques are highly effective and desirable. They save lives from diseases that would otherwise be fatal.

Organ transplantation is one of the few situations in which scientists would prefer that the body *not* act so efficiently against foreign bodies. When it does so with a donated organ, it can prevent that organ from functioning successfully. Surgeons may be able to get the organ into its proper position in a body and "hooked up" properly. But if the recipient's body rejects the organ, the transplantation is a failure.

Until 1983, the success of an organ transplantation depended strongly on how closely the donor and recipient were related biologically. The closer this relationship, the less likely the recipient's body was to reject the donated organ. Then, in that year, a new drug—cyclosporine—was placed on the market and changed the future of transplantation. Cyclosporine essentially shuts down a patient's immune system that would otherwise reject a donated organ. Transplantations became possible between any two individuals, and the number of such procedures increased dramatically.

Organ Transplantation: Donor Issues

The transplantation of an organ from one human or other animal into a second human has always been surrounded with a number of difficult ethical and social issues. One of the first and most difficult questions is how to obtain the needed organs. The obvious answer is to allow a person to designate that his or her organs may be removed after death and given to another person who needs them.

This concept is a simple one, but it involves some complex decisions. First, a person has to express very clearly that organ

donation is acceptable to him or her. If such decisions are left to relatives of the deceased, they might find it difficult to permit this procedure soon after their relative dies. The adoption of the Uniform Anatomical Gift Act in 1968 has made this issue somewhat simpler. Any responsible person can sign a Donor Card allowing a person to donate his or her organs after death.

The second problem with organ donation is that the process has to occur very quickly. If an organ is not removed and then implanted into the recipient very quickly, the organ will die and not function in the new body. For example, a kidney can be preserved outside a body for about seventy-two hours and a liver for about twenty-four hours. But a heart will survive no more than about four to six hours. It is critical, therefore, that the donor and recipient both be present at the same time for a successful transplantation to occur. Just handling the physical details of preparing for a transplantation, then, can be very difficult.

In the early years of transplantation, the need for speed in beginning the procedure created some problems. Physicians were sometimes not able to agree as to when a prospective donor had actually died. A number of criteria were used to define "death," including loss of respiration, loss of heartbeat and/or pulse, and loss of brain activity. This issue was finally resolved in 1978 when the Uniform Brain Death Act was passed. This act decreed that a person was to be declared dead when all brain activity ended.

Questions still remain, however, as to whether medical workers declare a person dead, even unconsciously, before death has actually occurred. The desire to remove an organ in order to save someone else's life may at times be so great, some critics say, that medical workers are less than totally cautious about waiting until a donor has actually died.

Organ Transplantation: Recipient Issues

Probably the most difficult question regarding prospective recipients is who should get an organ that becomes available. Since organ transplantation became a procedure that many qualified surgeons could do, there have never been nearly enough donors to meet the needs of everyone who needs an organ transplant. At the beginning of 1998, there were 37,944 Americans awaiting a kidney transplant, 9,406 awaiting a new liver, 356 waiting for a pancreas, 3,895 waiting for a heart, 2,607 waiting for a lung, and 1,996 waiting for some other organ or combination of organs. But less than half of that number will ever receive the organ they need to live. In 1996, for example, just over 4,000 people died while they were waiting for an organ transplantation that might have saved their lives.

The question, then, is which of the more than 56,000 Americans waiting for an organ transplantation will get one?

Until recently, this decision was made on a case-by-case basis in areas where prospective recipients lived. For example, Dr. A. in Chicago might have a list of six patients waiting for a heart transplant. When Dr. A. found out that a heart was available, he or she would make the necessary decision as to which of the patients would get the heart.

In 1997, this procedure changed. The U.S. Department of Health and Human Services (HHS) decided that organs that became available would go to the sickest patients first. The process for making this decision was assigned to the United Network for Organ Sharing (UNOS), which was created in 1984 to act as a central clearinghouse for organ transplantation in the United States. The 1997 HHS decision shifted responsibility for deciding who gets transplants from individual doctors, transplantation committees, and medical facilities to the centralized computer system at UNOS.

Proponents of the plan argued that it was a fairer way to make life-and-death decisions about organ allocations. The personal biases of individual and institutions would be avoided, and organs would be sent to those that were determined objectively to be the most needy. Opponents of the plan argued that the new system would make a shambles of organ transplantation in the United States. Individual doctors and transplantation committees can make more informed decisions than can a computer, some say. In addition, some hospitals now doing transplantations might choose to close down if they do not have more control over the procedures they can perform.

In March of 1998, HHS announced its intentions to implement its new regulations on transplant policies. The public had until May 1999 to comment on those regulations, after which they would be approved in their original or modifed form.

Other Issues

Transplantation involves a number of other issues as well. One is economic. Transplantation is a very expensive medical procedure. The average heart transplant, for example, costs more than $250,000 in the United States. Although most medical workers would like to provide all patients with equal access to transplantation, the reality is that people with more financial resources are more likely to receive an organ than are those without those resources. Thus far, no governmental body has been able to find a way to equalize the opportunity for all patients waiting for an organ, regardless of their financial situation.

A question that is likely to become even more important in the future is that of xenotransplants. Body parts from some animals have long been used in some forms of human surgery. For example, heart valves from pigs have been used to replaced damaged heart valves in humans. But researchers are now exploring the possibility of using complete organs from animals such as pigs for transplantation into humans. Some scientists are even trying to engineer pigs genetically so that their organs are even more like those of humans, thus increasing the likelihood of a successful transplantation.

Some critics wonder whether humans should be put at risk by having such transplants. Even if the transplant is successful, they say, it is not clear what the long-term effects of introducing a foreign body part into a human might be. Perhaps the greatest of these concerns is that microorganisms unique to the donor animal will also be introduced into the human. Without a natural immunity to those microorganisms, the human might develop an infection that could be untreatable.

Finally, some animal rights activists object to the use of animals for xenotransplants. They argue that animals other than humans have rights also and that they should not be bred and raised solely for the purpose of providing organs to humans.

References

Caplan, Arthur. "Organ Procurement and Transplantation: Ethical and Practical Issues." http://www.upenn.edu/ldi/issuebrief2_5.html. 4 August 1998.

Carnell, Brian. "The Moral, Legal and Economic Issues of Organ Sales." http://www.carnell.com/organs.html. 4 August 1998.

"Ethical Issues." http://www.lhsc.on.ca/transplant/ethic.htm. 4 August 1998.

"Frequently Asked Questions about Xenotransplants." http://www.transweb.org/qa/qa_txp/faq_xeno.html. 4 August 1998.

"Issues in Health Care: Organ Transplantation." http://www.geocities.com/HotSprings/3872/organ.htm. 4 August 1998.

Meckler, Laura. "Lawmakers Press Shalala about Organ Transplants." *Oregonian*, 19 June 1998, A23.

"Organ Procurement and Transplantation Network: Final Rule." *Federal Register*, 2 April 1998, 16295–16370; also at http://www.hrsa.dhhs.gov/News-PA/ruletxt.htm. 4 August 1998.

Petechuk, David. "Transplant, Surgical." In *Gale Encyclopedia of Science*, edited by Bridget Travers, 2711–3716. Detroit: Gale Research, 1996.

Reinert, Patty. "Final Decisions." *Houston Chronicle*, 27 November 1997, A1+ (special report); also at http://www.chron.com/content/chronicle/special/transplant/html. 7 April 1998.

"U.S. Drafts Rules for Cross-species Organ Transplants." *San Francisco Chronicle*, 21 September 1996, A1+.

Oxy Gas

See Ethanol as a Fuel; MTBE

Ozone Depletion

Ozone is an allotrope (form) of oxygen with molecules that consist of three atoms per molecule (O_3) rather than the two atoms per molecule (O_2) found in the most common form of oxygen. The level of ozone in the Earth's stratosphere is significantly higher than it is in the regions of the atmosphere higher and lower than the stratosphere. This "ozone layer" is important to the survival of life on Earth because it protects plants and animals from the harmful effects of the sun's ultraviolet radiation. Most scientists now believe that certain chemicals produced by human activities can result in the destruction of ozone molecules, causing a depletion of the ozone layer.

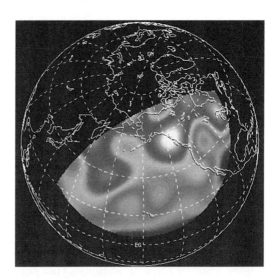

Efforts by the world's nations to reduce the production and release of CFCs appears to have had some success in reducing the risk of ozone depletion in the stratosphere. (Corbis)

Ozone in the Stratosphere

Ozone is produced in the stratosphere as a result of the action of solar energy (sunlight) on ordinary oxygen molecules. Individual oxygen atoms formed in the dissociation of oxygen molecules react with other oxygen molecules to form ozone:

$$O_2 - \text{solar energy} \longrightarrow O + O$$
$$O + O_2 \rightarrow O_3$$

This reaction is reversible. Ozone molecules can break apart to form oxygen atoms and molecules:

$$O_3 \rightarrow O + O_2$$
$$O + O \rightarrow O_2$$

Over very long periods of time, these two reactions have come into equilibrium, and a relatively constant amount of ozone remains in the stratosphere. The absolute amount of ozone is actually relatively small. If all of that ozone could be transported to the Earth's surface, it would form a layer no more than 3 millimeters thick. But in the very low density characteristic of the stratosphere, ozone constitutes a significant fraction of the gases present. Its concentration is estimated to be about 10–20 ppm (parts per million) or 5×10^{12} molecules per cubic centimeter at a maximum.

Solar Radiation and Its Effects on Living Organisms

From the standpoint of life on Earth, the most important property of ozone is its tendency to absorb radiation in the range between 250 and 350 nanometers.

Solar energy consists of radiation with a wide variety of wavelengths, from about 200 to 3,600 nanometers, or between the mid-ultraviolet (UV) to the far infrared regions. Some of this radiation reflects off the Earth's atmosphere and is returned to outer space, some is absorbed by gases in the atmosphere (such as ozone), and some passes through the atmosphere to the Earth's surface.

Radiation with wavelengths between about 100 and 400 nanometers is referred to as UV radiation. UV radiation with maximum wavelengths up to about 250 nanometers is called far-UV radiation, while that with wavelengths greater than 250 nanometers is known as near-UV radiation.

Finally, near-UV radiation is further categorized as UV-A radiation (wavelengths of 320–380 nanometers), UV-B radiation (wavelengths of 290–320 nanometers), and UV-C radiation (wavelengths of 250–290 nanometers). These forms of near-UV radiation are distinguished from each other largely because of their effects on human life and their fate in the atmosphere. UV-A radiation, for example, is able to penetrate the atmosphere almost completely and is thought to be relatively harmless to life on Earth. It may, however, cause long-term health effects, such as damage to the immune system of organisms, that are not well understood.

UV-C radiation, by contrast, is extremely hazardous to most forms of life on Earth. In fact, it is sometimes used in medical situations as a sterilizing agent because it kills microorganisms. Fortunately, almost no UV-C radiation reaches the Earth's surface because it is efficiently absorbed by ozone in the stratosphere.

Finally, UV-B radiation is known to cause a variety of deleterious effects on plants and animals. These effects include skin cancers, premature aging of skin, cataracts and other eye disorders, and damage to deoxyribonucleic acid (DNA). UV-B radiation is largely absorbed by ozone in the stratosphere, although a small amount does reach the Earth's surface.

The Ozone "Hole"

As early as the 1960s, scientists began to feel concern about the possible loss of ozone in the stratosphere because of various human activities. At first, these concerns were based on the possible use of aircraft capable of flying in the stratosphere, such as the supersonic transport (SST) and the space shuttle. Scientists worried that exhaust gases from these aircraft might react with and destroy ozone molecules.

These concerns became more serious in 1985 when British meteorologist James Farmer announced that his research team had found a "hole" in the ozone layer over the Antarctic. The term *hole* was used to describe a vertical column of the atmosphere in which the concentration of ozone was significantly less than normal. The concentration of ozone in 1984, Farmer reported, was about 200 du (Dobson units) compared to a normal level of about 300 du.

Later research showed two disturbing trends. First, earlier records showed that the ozone hole was not a new phenomenon but had been developing for nearly two decades, since the late 1960s. Second, research after 1985 showed that the Antarctic hole was becoming larger in size and the level of ozone depletion even greater in subsequent years. At one point, scientists announced that certain regions of the hole showed an almost complete absence of ozone.

Chlorofluorocarbons

For a brief period of time, scientists debated the possible causes of stratospheric ozone depletion. Some physicists argued that the phenomenon was a natural event caused by changes in patterns of air movements above the Antarctic. Before long, however, a consensus developed that the cause of this phenomenon was the escape into the atmosphere of certain synthetic chemicals known as chlorofluorocarbons (CFCs). CFCs had been invented in the 1920s by Thomas Midgley Jr., an employee of the Frigidaire Corporation. They had rapidly become extremely popular for use in a wide variety of applications, including dry cleaning, air-conditioning and refrigeration systems, fire extinguishers, foaming materials, and aerosol sprays. Production of CFCs mushroomed from about 8.3 million kilograms in 1945 to 283.5 kilograms in 1965 to 715.0 million kilograms in 1975.

One of the properties that made CFCs so desirable commercially was their stability and inertness. They do not decompose readily and so can be used over and over again for long periods of time. Ironically, CFCs behave quite differently in the upper atmosphere, where they are exposed to the energetic effects of solar radiation. Indeed, as early as the 1970s, Mario Molina and F. Sherwood Rowland had demonstrated that CFCs exposed to solar radiation will break down with the formation of free chlorine atoms:

$$CFC - solar\ radiation \longrightarrow CFC^* + Cl$$

These chlorine atoms, in turn, are capable of attacking and destroying ozone molecules in a complex sequence of reactions. Briefly, these reactions involve the formation of an intermediary product, chlorine oxide (ClO), with the eventual regeneration of the chlorine atom:

$$Cl + O_3 \rightarrow ClO + O_2$$
$$ClO + O \rightarrow Cl + O_2$$

These equations illustrate two key points. First, the free chlorine atom is capable of converting an ozone molecule (O_3) into an oxygen molecule (O_2). Second, the chlorine atom is not lost in the reaction but is regenerated as a result of the second step. That is, a single chlorine atom is capable of destroying thousands of ozone molecules before it is lost by some other chemical mechanism. By the mid-1980s, many scientists had become convinced that CFCs escaping from the Earth's surface were causing the destruction of ozone in the stratosphere by processes such as those described here.

Political Action

In a remarkably short period of time, the world's political leaders recognized the severity of the ozone problem. They agreed to meet and negotiate a treaty calling for drastic limitations on the production of CFCs. That treaty, the Montreal Protocol on Substances that Deplete the Ozone Layer, was signed in 1987. It called for a reduction in the production of certain CFCs by 20 percent of their 1986 levels by 1995 and by 50 percent by 2000.

Ink on the treaty was scarcely dry before new reports of ozone depletion appeared. The Antarctic hole was growing larger, a similar hole was being predicted over the Arctic, and reduced ozone levels over lower latitudes were also being detected. Motivated by this news, two more meetings were held on the Montreal Protocol, one in London in 1990 and the other in Copenhagen in 1992. At these two meetings, amendments to the Montreal Protocol were adopted accelerating the rate at which the originally designated CFCs were to be eliminated and adding additional ozone depleting chemicals to the list of banned chemicals.

Disputes over Theory and Action

In some ways, the Montreal Protocol represents an amazing accomplishment in science and politics. The world's nations became convinced of the seriousness of the ozone crisis in a remarkably short time, and they responded quickly and efficiently with actions that were to have enormous financial consequences for some large and important industries.

It is hardly surprising to note, therefore, that the road to the Montreal Protocol was hardly a smooth one. Although most scientists were rather quickly convinced as to the facts of ozone depletion and the role played by CFCs, a few scientists withheld judgment on the issue. They argued that data were incomplete or unconvincing and that the Molina/Rowland theory was inadequate to explain what, they thought, might be nothing other than a natural phenomenon of ozone fluctuation in the atmosphere.

For example, a persistent critic of the Molina/Rowland theory, Fred Singer, testified as follows at a 1995 hearing before the U.S. House of Representative's Subcommittee on Energy and Environment: "The bottom line is this: Currently available scientific evidence does not support a ban on the production of chlorofluorocarbons (CFCs or freons), halons, and especially methyl bromide. . . . Yet because of the absence of full scientific debate of the evidence, relying instead on unproven theories, we now have an international treaty that will conservatively cost the U.S. economy some $100 billion dollars [*sic*]" (U.S. House Committee on Science 1995, 57–58).

The manufacturers of CFCs and other ozone depleting substances were, quite naturally, especially skeptical. Still, when faced with mounting evidence, they gradually acknowledged the connection between their products and ozone depletion. In 1990, for example, only five years after Farmer's initial reports on ozone depletion, E. I. du Pont de Nemours and Company, the world's largest producer of CFCs, announced that it would eliminate production of the chemicals by the year 2000. Only three years later, the company accelerated these plans, announcing that it would stop producing CFCs by 1995.

Latest Developments

By 1997, the first effects of the Montreal Protocol were becoming readily apparent. Scientists at the U.S. National Oceanic and Atmospheric Administration (NOAA) announced that the concentration of chlorine oxide in the atmosphere had begun to decrease. Because of the persistence of such chemicals in the atmosphere, no decrease in ozone depletion itself would occur immediately even if all ozone depleting chemicals disappeared tomorrow. However, scientists felt confident enough in the NOAA announcement to predict that levels of ozone in the atmosphere might once more begin to increase in the period between 2005 and 2010.

A number of natural and human factors must be taken into consideration, however, in predicting how long it will take for complete recovery of the ozone layer. For example, some of the most troublesome of ozone depleting chemicals remain in the atmosphere for dozens or hundreds of years. Second, an increase in the concentration of carbon dioxide in the troposphere (the lowest level of the atmosphere) is producing a cooling effect in the stratosphere, the layer above the troposphere. This decrease in stratospheric temperatures is likely to lead, in turn, to the formation of more clouds, on which the decomposition of CFCs and other ozone depleting chemicals occurs most readily. Finally, smuggling of CFCs across national borders has resulted in a greater availability of these compounds and, therefore, a greater release of them into the atmosphere than had been permitted or anticipated by the Montreal Protocol.

See also: Chlorofluorocarbons; Hydrofluorocarbons

References

"The Accelerated Phaseout of Class I Ozone-depleting Substances." http://www.epa.gov/docs/ozone/title6/phaseout/accfact.html. 23 June 19.

Freedman, Bill. "Ozone Layer Depletion." *Gale Encyclopedia of Science*, edited by Bridget Travers. Detroit: Gale Research, 1996, 2655–2659.

Hileman, Bette. "Ozone Treaty: Successful, but Pitfalls Remain." *Chemical and Engineering News*, 15 September 1997, 24.

Lyman, F. "As the Ozone Thins, the Plot Thickens." *Amicus Journal*, Summer 1991, 20–28+.

Newton, David E. *The Ozone Dilemma*. Santa Barbara, CA: ABC-CLIO, 1995.

"Ozone at Halley, Rothera and Vernadsky/Faraday." http://www.nbs.ac.uk/public/icd/jds/ozone/. This site also contains links to other sites with useful information on ozone depletion.

Recer, Paul. "Good News for Earth: 'Bad Chemicals' Decline." *Oregonian*, 31 May 1996, A1+.

U.S. House Committee on Science. *Scientific Integrity and Public Trust: The Science behind Federal Policies and Mandates: Case Study 1—Stratospheric Ozone: Myths and Realities: Hearing before the Subcommittee on Energy and Environment*. 104th Cong., 1st sess., 20 September 1995.

Zurer, Pamela. "Ozone Depletion's Recurring Surprises Challenge Atmospheric Scientists." *Chemical and Engineering News*, 24 May 1993, 8–18.

The U.S. Environmental Protection Agency maintains a web site on ozone depletion at http://www.epa.gov/ozone/.

Ozone Depletion Potential

Ozone depletion potential (ODP) is a measure of the potential impact of any given substance on ozone in the Earth's atmosphere. It is defined as the ratio between (1) the amount of ozone destroyed by a unit mass (say, one metric ton) of that substance compared to (2) the amount of ozone destroyed by a comparable amount of the standard compound, CFC-11. The ODP of a chemical depends primarily on two factors: the efficiency with which it breaks down to release chlorine or bromine atoms in the atmosphere and its natural residency rate (how long it stays) in the atmosphere. ODP rates range from 0.05 for HCFC-142b and 0.09 for HCFC-141b to 0.9 for CFC-12 and CFC-113. The designations "HCFC" and "CFC" stand, respectively, for members of the hydrochlorofluorocarbon and chlorofluorocarbon families. The highest measured ODP for chemicals found commonly in the atmosphere is 1.1 for carbon tetrachloride (CCl_4).

See also: Chlorofluorocarbons; Hydrochlorofluorocarbons; Ozone Depletion

Partial-Birth Abortions

See Intact Dilation Evacuation

Particle Accelerators

A particle accelerator is a device used to make tiny particles of matter move at very high speeds, often close to the speed of light. Those fast-moving particles are then caused to collide with other rapidly moving particles or with stationary targets, such as sheets of gold, aluminum, or other materials. The energy released in such collisions is greater than that produced by any other artificial means available to humans. In such environments, events occur that cannot be observed under any other conditions on Earth. For example, particles are formed that do not otherwise exist on Earth, and interactions can be observed among particles that are unknown under normal terrestrial conditions.

Discoveries with Particle Accelerators

The purpose of such experiments is to explore the fundamental nature of matter. Experiments with particle accelerators have already revealed that matter is much more complex than had once been imagined. For example, scientists once thought that the basic units of matter were the proton, the neutron, and the electron. These units, they thought, combined in a variety of ways to produce atoms of all the elements of which matter is made.

Experiments with particle accelerators have shown, however, that particles even simpler than protons and neutrons exist. These particles are known as *quarks*. In addition, particles responsible for the transmission of various forces have been discovered.

For more than sixty years, new discoveries in particle physics have been associated with improvements in particle technology. Early machines were able to generate only modest amounts of energy, measured in the range of a few million electron volts. An *electron volt* is a unit used to measure the energy produced by a particle accelerator. With the earliest machines, only modest breakthroughs in our understanding of the complexity of matter were possible. As particle technology improved, however, so did the energy capability of those machines. Today, the most powerful particle accelerators produce energies in the range of a few billion electron volts or even a few trillion electron volts. As particle accelerators become more powerful, they make possible ever more detailed analysis of the complexity of matter.

Types of Particle Accelerators

Particle accelerators are of two general types, linear and circular. A linear accelerator *(linac)* consists of a series of tin can–like tubes arranged in a straight line. Subatomic particles (particles smaller than atoms) are introduced at one end of the machine. These particles are usually electrons or positrons (positive electrons). Then the particles are accelerated through the tubes by means of a changing electric current that pushes them out of one tube and pulls them into the next tube in sequence.

The case of the Superconducting Super Collider (SSC) raises the question as to whether the United States will ever again be able to afford to build the kind of machines to make it the world's leader in particle physics. (Corbis/Kevin Fleming)

With each successive push and pull, the particles gain speed. The longer the linac, the greater the speed the particles can attain. The longest existing linac is located at the Stanford Linear Accelerator Center (SLAC) in Stanford, California. That machine is three kilometers (two miles) in length. It generates an electron beam with an energy of about 32 gigaelectron volts, or billion electron volts.

One problem with linear accelerators is their physical size. To make a powerful linac, it is necessary to make the machine as long as possible. To do so involves solving a number of difficult technical problems. For example, in building the Stanford linac, engineers had to take into consideration the curvature of the Earth's surface to keep the accelerating tubes exactly in line with each other.

One solution to this problem is to use a circular accelerator that accelerates particles in a circle rather than a straight line. The first machine of this type was built by U.S. physicist E. O. Lawrence in 1932. Lawrence's machine was called a *cyclotron*. The cyclotron has gone through a number of adaptations, improvements, and modifications over the past seven decades. Machines that have evolved from the cyclotron have been the synchrotrons, synchrocylcotrons, self-focusing cyclotrons, and colliding beam accelerators.

Circular accelerators all operate on a common principle. A moving charged particle follows a curved path when it moves through a magnetic field. That is, suppose that a pair of large magnets were placed above and below the accelerating tubes in a linac. The path of particles traveling through those tubes would be bent, up or down, to the left or right, depending on the orientation of the magnetic field. In principle, then, a circular particle accelerator is only a linac surrounded by magnets with accelerating tubes arranged in a circle rather than a straight line.

Particles are accelerated to higher and higher speeds in a circular accelerator by being made to travel around the machine more and more times. The controlling factor in a circular accelerator is not just its physical size, as with a linac. It is the size of the magnets. As the particles in the machine accelerate to greater and greater speeds, the magnetic field must be made stronger and stronger to keep the particles moving in a circle.

The problems that particle accelerator engineers face are many and complex. Fundamentally, however, the two biggest issues are (1) how long one can make a linac and (2) how powerful one can make the magnets needed for a circular accelerator.

Since completion of the SLAC linac in 1966, most of the effort to produce more powerful accelerators has focused on circular accelerators. The latest development in circular accelerator technology has been "ring" machines in which two beams of particles, traveling in opposite directions, are caused to collide with each other. In the most powerful of these machines, these collisions release energies in the trillions of electron volts.

The Superconducting Super Collider

In the mid-1980s, an important scientific breakthrough changed the future of particle accelerator technology. Scientists discovered how to make powerful magnets that weigh much less than traditional magnets. These new magnets are made of *superconducting* materials, which conduct an electric current without resistance. That is, an electrical current that passes through such a material experiences no resistance to its flow and continues to flow through the material essentially forever.

The discovery of superconducting materials was an enormous boon to particle accelerator designers because all magnets used in accelerators are electromagnets. However, the metals of which they are made are resistant to the flow of electric current. They can be made to generate powerful magnetic fields only if they are made large enough to overcome resistance to the flow of electricity sufficiently to maintain such fields. With superconducting materials, the size of the magnets needed for the accelerators was dramatically smaller than in traditional designs. Studies began almost immediately on the design of a powerful new accelerator with magnets made of superconducting materials. That machine became known as the superconducting super collider (SSC).

Plans called for a circular particle accelerator 82.944 kilometers (51.539 miles) in circumference buried at least 6 meters (20 feet) under ground. The accelerating rings were to be built inside concrete tunnels about 3 meters (10 feet) in diameter. The machine was designed to produce particles with an energy of about 20 trillion electron volts. Physicists hoped and believed that the SSC would bring particle physics to a new plateau, answering some basic questions about the nature of matter and, undoubtedly, raising new ones.

Politics of the SSC Proposal

The SSC proposal became the focus of an intense political debate. On the one hand, proponents of the machine argued that it was essential if the United States were to remain a leader in the field of particle physics. Without the machine, they pointed out, young physicists would leave the United States to go where better technology was available. In addition, construction and operation of the machine would be an economic boon to both local and national economies.

Critics felt that the SSC was a waste of money. It would do nothing to solve poverty, health problems, or other problems of national significance. It was simply throwing money away on the pet projects of a handful of elite physicists. The debate that developed was a repeat of the one that almost always accompanies proposals for large and expensive projects in basic research, such as, for example, the Apollo Project's plan to land a human on the Moon.

Eventually the pro-SSC camp won out. In 1987, President Ronald Reagan announced his support for the construction of the SSC. In 1998, energy secretary John Herrington announced that a site outside Waxahachie, Texas, had been selected for the machine. A few months later, construction on the SSC began. The project did not go well, however. More and more technical problems arose. Costs continued to rise. Eventually, the estimated cost of completing the project reached three times the original estimate of $4 billion. In 1990, the U.S. Congress decided that it had had enough of the SSC. It canceled the project, leaving a partially excavated tunnel in Texas as the primary legacy of the project.

Particle Accelerators after the SSC

The failure of the U.S. project did not mean the end of all new particle accelerator construction, however. In Europe, the Centre Européen pour la Recherche Nucléaire (CERN, or European Laboratory for Nuclear Research) had long been planning a powerful circular accelerator, the large hadron collider (LHC).

The LHC was planned as the latest of a series of powerful accelerators located at CERN outside Geneva, near the Swiss-French border. It was designed to consist of a central ring 36.3 kilometers (16.5 miles) in circumference, capable of accelerating protons to energies of 14 trillion electron volts. That would make the LHC the most powerful accelerator in the world. It is scheduled for completion in 2005 at an estimated cost of $6 billion.

In late 1997, the U.S. government took the unusual step of agreeing to help fund the construction of the LHC. The Department of Energy announced plans to provide equipment and services worth more than $500 million to CERN. The collapse of the SSC project had made it clear that that kind of project had no immediate future in the United States. The best way to ensure access to LHC by U.S. scientists, therefore, was for the United States to become a working partner in the CERN project.

U.S. particle physicists were not entirely satisfied with this decision. They hoped to recover for the United States some of the prestige of being home to a powerful new particle accelerator. Some began to talk about a new generation of linacs, more powerful than the SLAC machine. As one physicist said, "The U.S. needs good science like this, from accelerators. Without them, we (physicists) are going to lose a whole generation of brilliant students. They will go into biology or something" (Petit 1998, A6).

The linac envisioned by these physicists would be about 30 kilometers (20 miles) long with a capacity to accelerate electrons to an energy of 1–1.5 trillion electron volts. This energy level would be less than that of the LHC. But it would still be high enough to generate groundbreaking discoveries in particle physicists. Under the most favorable circumstances, the new linac would probably not be completed until 2009. Between now and then, a debate similar to the one over the SSC and the cost of basic research is likely to be replayed over this new issue.

References

Brown, Malcolm W, "Powerful Accelerator to Be Completed Early." *New York Times*, 24 December 1996, B12.

Glashow, Sheldon L., and Leon M. Lederman. "The SSC: A Machine for the Nineties." *Physics Today*, March 1985, 28–37.

Newton, David E. "Accelerators." In *The Gale Encyclopedia of Science*, edited by Bridget Travers, 8–11. Detroit: Gale Research, 1996.

———. *Particle Accelerators: From the Cyclotron to the Superconducting Super Collider*. New York: Franklin Watts, 1989.

O'Leary, Hazel, Secretary of Energy, and U.S. Representative George E. Brown Jr. "The Future of the Superconducting Super Collider." http://www.lbl.gov/Science-Articles/Archive/ssc-and-future.html. 28 June 1998.

Petit, Charles. "Physicists Dream Big Dreams about Smashing Tiny Particles." *San Francisco Chronicle*, 9 March 1998, A6+.

"The Superconducting Super Collider History." http://www.hep.net/ssc/new/history/. 28 June 1998.

U.S. House Committee on Science, Space, and Technology. *Termination of the Superconducting Super Collider Project: Hearing before the Subcommittee on Science.* 103d Cong., 2d sess., 15 March 1994.

Pesticides

See DDT; Mediterranean Fruit Fly

Pollutant Standard Index

The Pollutant Standard Index (PSI) is a system for determining air quality in a given area that, in turn, can be used by governmental agencies in issuing advisories about recommended limitations on human activities for that area. The PSI is defined as the ratio of the concentration of any given air pollutant (such as carbon monoxide) to the recommended concentration of that pollutant times 100. Recommended levels of pollutants come from the U.S. Environmental Protection Agency (EPA) or from state or local agencies. The most important of these agencies is the California Department of Environmental Quality. Because of its special problems of air pollution, California standards may be different from, and usually more rigorous than, those of the federal government.

Calculating a PSI

For example, suppose that the concentration of ozone over Oakland, California, on a particular day is determined to be 0.18 ppm (parts per million) over a one-hour period. Federal standards for ozone are 0.12 ppm and for California, 0.09 ppm. The PSIs for the Oakland measurement, then, are:

Federal: $\frac{0.18 \text{ppm}}{0.12 \text{ppm}} \times 100 = 150$

State: $\frac{0.018 \text{ppm}}{0.09 \text{ppm}} \times 100 = 200$

The PSI scale ranges from 0 (complete absence of a pollutant) to 500 (the most extreme concentration of a pollutant). A PSI rating of 100 represents "moderate" conditions, in which the concentration of a pollutant corresponds exactly to the federal or California standard. PSI ratings of 50 or less are defined as a "good" condition for the pollutant. PSI ratings of greater than 100 are regarded as "unhealthful" to various degrees. Above a PSI of 200, the concentration of a pollutant is regarded as high enough to present a serous risk to human health. At that level, a pollution "alert" is usually issued.

Pollution Alerts

A Stage 1 pollution alert (PSI: 200–275) calls on citizens to avoid strenuous activities, such as jogging and sports. People, especially younger children, the elderly, and those with respiratory disorders, are warned to remain indoors. A Stage 2 alert (PSI: 275–400) leads to recommendations for a complete restriction of physical activity, again especially for more sensitive groups. A Stage 3 alert (PSI: greater than 400) is the most serious warning given. Individuals are required to remain at home and all polluting industries and other activities are required to close down. Stage 3 alerts are extremely rare.

References

"Measuring Air Quality: The Pollutant Standards Index" (EPA Publication 451/K-94-001); also at http://www.epa.gov/oar/oaqps/psi.html. 21 June 1998.

Pollution

See Feedlot Pollution; Formaldehyde; Halons; Indoor Air Pollution; Methyl Bromide; Noise Pollution; Radon; Salton Sea; Secondhand Smoke; Superfund

Polybrominated Biphenyls

Polybrominated biphenyls (PBBs) are organic compounds whose molecular structure consists of two benzene rings joined to each other and two or more bromine atoms substituted for hydrogen atoms in the benzene rings. A typical structure of the PBBs is shown below:

The compounds were first prepared by the German chemical company Chemische Fabrik Kalb, who received a patent on their manufacture in 1964. A U.S. patent for the production of PBBs was awarded to the Dow Chemical Company in 1973.

A particularly popular PBB is the one that contains six bromine atoms and is known, therefore, as hexabrominated biphenyl. This compound was found to be a highly effective flame retardant when added to certain plastics. The commercial name of the product containing hexabrominated biphenyl is Firemaster BP-6. During the 1970s, Firemaster BP-6 became very popular as an additive in the plastics industry used for telephones, calculators, television cabinets, automobile parts, and hair dryers. When such products are exposed to flames, the presence of Firemaster BP-6 causes them to melt rather than to burst into flame.

Like other organic compounds that contain chlorine and bromine, the PBBs are highly toxic. In fact, they are chemical relatives of compounds used as pesticides and herbicides. These compounds are also highly persistent in the environment. That is, once released to the environment, they tend to remain in water and soil for very long periods of time. The half life of some PBBs, for example, may be as high as three years. *Half life* is the period of time it takes for one half of a sample of the compound to break down.

PBBs have a special place in environmental history because of an event that occurred in Michigan in 1973. The company that manufactured PBBs accidentally mislabeled a number of bags of the product. They were incorrectly marked as contain-

ing magnesium oxide, a chemical added to livestock feed for nutritional purposes. Eventually thousands of cattle in Michigan were fed PBBs in their feed rather than magnesium oxide. Since the two compounds look similar to each other physically, farmers were not aware of the erroneous substitution. Over time, a number of cattle became sick and died. It was some time before researchers were able to determine the cause of this problem. By the time PBBs were finally identified as the culprits, the compounds had been ingested by millions of humans through the beef and milk they ate and drank. Hundreds of Michigan residents eventually became ill with symptoms ranging from stomachaches and nausea to severe joint pain and damage to the immune system.

PBB is still used as a flame retardant although it is now known to be a serious health hazard to humans and other animals and is considered to be a likely carcinogen.

References

Carter, Luther. "Michigan's PBB Incident: Chemical Mix-up leads to Disaster." *Science,* 16 April 1976. 240–243.

Chen, Edwin. "Michigan's PBB Disaster: An Update." *The Atlantic,* October 1979, 16–26.

"Polybrominated Biphenyls (PBBs)." Go to http://atsdr1.atsdr.cdc.gov/search, and ask for polybrominated biphenyls. 21 May 1999.

"Polybrominated Biphenyls (PBBs)." In *Gale Encyclopedia of Science,* edited by Bridget Travers, 2879–2880. Detroit: Gale Research, 1996.

"TR-398: Toxicology and Carcinogenesis Studies of Polybrominated Biphenyls (CAS No. 67774-32-7)(Firemaster FF-1®) in F344/N rats and B6C3F$_1$ mice (feed studies)." http://ntp-server.niehs.nih. gov/htdocs/LT-Studies/TR398.html. 4 July 1998.

Polychlorinated Biphenyls

Polychlorinated biphenyls (PCBs) are organic compounds that contain two benzene rings with two or more chlorine atoms substituted for hydrogen atoms in the ring. A typical polychlorinated biphenyl has the structure shown below.

The PCBs were first prepared synthetically by the German chemists H. Schmidt and G. Schulz in 1881. They were of very little commercial interest, however, until the 1930s. They then became popular for a number of industrial applications because of certain highly desirable physical and chemical properties. For example, they are very stable compounds (that is, they do not break down easily), they do not evaporate readily, and they transmit an electrical charge easily. Because of these properties, they became widely used as insulating fluids in electrical transformers, as additives to certain kinds of paints, for the treatment of cotton and asbestos products, and in the manufacture of plastics.

In the 1960s, reports began to appear of PCBs in the environment. These reports were troublesome because the very properties that made PCBs desirable in industry also made them dangerous as environmental pollutants. That is, they are highly toxic compounds and once released to the environment, they do not easily break down. They are ingested by organisms low in the food chain and then are accumulated as those organisms are eaten by others higher up in the food chain.

Perhaps the most dramatic accident involving the PCBs occurred in Japan in 1968. Cooking oil was contaminated by PCBs, and thousands of individuals experienced damage to their livers and intestinal and lymphatic systems. Health problems associated with exposure to PCBs were also reported in the Hudson River Valley of New

York State. A plant manufacturing electrical components was found to have discharged wastes containing PCBs into the Hudson River for more than twenty-five years. Concentrations of PCBs ten to 100 times that recommended were recorded in regions downstream of the factory.

By the early 1970s, manufacturers of PCBs had begun to cut back on or stop production of these compounds voluntarily. In 1976, the Toxic Substances Control Act banned all production of the compounds in the United States and severely limited the uses to which they could be put.

That action did not entirely solve the problem of PCB pollution. The compounds were still escaping into the environment from their few remaining permitted uses and from old and discarded equipment. In addition, PCBs discarded into the environment years earlier still remained in soil and water. As late as 1981, duck hunters in New York State and Montana were being warned to limit their consumption of killed waterfowl because of high levels of PCBs discovered in the birds.

References

Ahmed, A. K. "PCBs in the Environment" *Environment,* March 1976: 6–11.

Chan, Ian, et al. "The Environmental Aspects of Polychlorinated Biphenyls (PCBs)." http://bordeaux.uwaterloo.ca/biol447/assignment1/pcbs.html. 4 July 1998.

Layne, E. N. "The ABCs of PCBs." *Audubon* 74, no. 1 (1972): 116–118.

Manahan, Stanley E. *Environmental Chemistry.* 6th ed. Boca Raton, FL: Lewis Publishers, 1994, 211–214.

"Mitch's PCB Info." http://www.indepen.com/pcbs/index.htm. 21 May 1999.

Moran, Joseph M., Michael D. Moran, and James H. Wiersma. *Introduction to Environmental Science.* 2d ed. New York: W. H. Freeman and Company, 1986, 226–229.

"Polychlorinated Biphenyls (PCBs)." http://www.atsdr.cdc.gov/tfacts17.html. 21 May 1999.

"Proceedings of the Expert Panel Workshop to Evaluate the Public Health Implications for the Treatment and

Disposal of Polychlorinated Biphenyls-Contaminated Waste." Available from National Technical Information Service, Springfield, VA, 22161, (703) 487–4650.

Polygraph

A polygraph is a device used to determine the truth or falsity of statements made by a person. The polygraph is also known as a lie detector.

Historical Background

The principle behind the polygraph is that human physiological traits change when a person makes a statement that is not true. Most people are at least subconsciously aware of that principle. A person who is obviously nervous while making a statement may be nervous because he or she knows that statement to be less than the truth.

Primitive methods for using this principle in determining the truth and falsity of statements go back hundreds of years. For example, English law once decreed that a person accused of telling a lie be forced to eat a "trial slice" of cheese and bread. If the person was unable to swallow the "trial slice," he or she was determined to be guilty of telling a lie. The rationale was that nervousness can result in one's mouth becoming dry, making it difficult to swallow a dry food.

The earliest form of a polygraph can be traced to Italian criminologist Cesare Lombroso. Lombroso is regarded as being one of the earliest scientific criminologists. He attempted to find physical characteristics by which one could recognize a criminal. He believed that such characteristics must exist since he was convinced that criminal behavior was genetically determined. In some of his most famous research, Lombroso measured the size and shape of people's heads and tried to correlate those measurements with criminal or noncriminal behavior. Lombroso's ideas are no longer regarded as having any scientific validity.

The lie detecting device invented by Lombroso was called a *hydrosphygmograph,*

a "water blood-pressure testing device." It consisted of a tank filled with water into which a person placed his or her hand. The tank was then covered with a rubber membrane. Changes in the person's blood pressure and pulse could be detected as changes in pressure within the tank. Such changes, Lombroso believed, indicated that a person was not telling the truth.

The Modern Polygraph

The earliest practical polygraph was invented in 1921 by John A. Larson, a medical student at the University of California and later a police officer in California. Larson's device measured two attributes: blood pressure and respiration. About a decade later, another police officer, Leonarde Keeler, added a third component, a galvanometer. The galvanometer was used to measure changes in electrical conductivity of the skin, known as the galvanic skin reflex (GSR). Such changes had been known for some years to accompany lying.

A polygraph test today consists of three phases. In the first phase, the polygraph operator talks with the subject of the test, attempting to put him or her at ease. The operator also tries to develop a rapport with the subject so that the actual test can be conducted under the most positive circumstances.

In phase two of the test, the actual physical measurements are made. The operator asks questions that can be answered "yes" or "no." The subject's blood pressure, respiration rate, and GSR are measured and recorded during each response that he or she makes. The operator includes three kinds of questions in the interview. Some questions are thought to be certain to elicit honest answers, such as the subject's name and home address. Other questions are designed to evoke answers that are almost certainly lies. One might ask, for example, if one has ever lied to a good friend. The first two question types provide baseline responses for the operator, recordings that indicate the subject's physiological responses during truth telling and lie telling.

The third type of question is that relating to the crime for which the subject is being examined. If the subject is thought to have taken materials from a factory, for example, he or she will be asked directly whether those materials were taken. By comparing the physiological responses to these questions to the baseline responses, the operator is expected to be able to determine whether the subject is lying or not.

The third phase of the test involves the operator's analysis of the subject's responses. If those responses appear to indicate that the subject is telling the truth, the procedure is terminated. If the recordings seem to suggest that the subject is lying, the operator may share that conclusion with the subject and attempt to get more honest responses from the subject.

Problems with Polygraph Analyses

The use of polygraph tests has long been a matter of serious controversy among law enforcement officials, civil libertarians, and the general public. One concern is that such tests are an unwarranted intrusion on an individual who may have little or no choice in deciding how to respond. Of greater concern, however, is the reliability of the polygraph itself. Critics point out that subjects of polygraph tests are often very nervous and excited. Their physiological responses tend to be abnormal even when they are telling the truth.

A number of studies have been conducted on the accuracy of polygraph tests. Those studies show that the rate of false positives may range from 0 to 75 percent. A false positive is a decision that a person has been lying when, in fact, he or she was actually telling the truth. The best polygraph operators consistently make correct judgments 90 percent of the time or better. The average rate of correct judgments for operators with minimal training, however, may be closer to 75 percent.

In response to the concerns over polygraph testing, the U.S. Congress passed a law in 1988 prohibiting private businesses from requiring such tests of workers and job applicants. The law provided for certain exceptions, including people who work in security services and employees suspected of financial crimes. In general, the courts do not allow polygraph tests to be admitted as evidence in trials unless the prosecution, defense, and judge all agree.

In recent years, a move to make polygraph tests admissible on behalf of defendants developed. That is, rather than using the tests to show that someone is guilty, attorneys wanted to use them to show that their clients were innocent. Such clients often asked for and were given polygraph tests to prove that they were telling the truth about some matter.

The case in this field that finally reached the U.S. Supreme Court involved an air force enlisted man who had been accused of using illegal drugs. That man took a polygraph test that showed he was telling the truth when he said he had not taken those drugs. The military court prohibited introduction of the test results, and the man was convicted, discharged, and sentenced to thirty months in prison. He sued to have those tests included in his testimony, which would have forced a retrial on his case. In March 1998, the Court ruled against the air force man. It said that there was still insufficient evidence that polygraph test results were reliable and accurate. Justice Clarence Thomas acknowledged that, in general, polygraph testing often produced the correct result. But, he added, "there is simply no way to know in a particular case whether a polygraph examiner's conclusion is accurate" (Farrell 1998, A1).

References

Abrams, Stan. *The Complete Polygraph Handbook.* Lexington, MA: Lexington Books, 1989.

Block, Eugene B. *Lie Detectors: Their History and Use.* New York: D. McKay Company, 1977.

Farrell, John Aloysius. "Supreme Court Calls Polygraph Unreliable." *Boston Globe,* 1 April 1998, A1.

Gale, Anthony, ed. *The Polygraph Test: Lies, Truth, and Science.* London: Sage Publications, 1988.

"Lie Detector and Polygraph Tests." http://www.angelfire.com/pa/blmresearch/index.html. 28 June 1998.

Matté, James Allen. *Forensic Psychophysiology Using the Polygraph.* Williamsville, NY: J.A.M. Publications, 1996.

Mauro, Tony. "Supreme Court Justices Take Polygraph Test." *USA Today,* 3 November 1997, 6A.

"Polygraph Bill/Voluntary Tests." http://www.senate.gov/~rpc/rva/1002/100237.htm. 28 June 1998.

"Polygraph Testing." http://www.fbi.gov/kids/poly/poly.htm. 28 June 1998.

Home page for the American Polygraph Association on the Internet is at http://www.polygraph.org:83/.

Population

To some extent, the most fundamental issue underlying many socioscientific problems is population. For example, an important factor—some say *the* most important factor—responsible for problems of pollution is population. The more people living in an area, the argument goes, the more air, water, noise, and other forms of pollution. One cannot begin to think of solving pollution problems, then, without first dealing with problems of population growth.

One might extend that argument to other areas of social concern, such as the loss of land due to erosion; food shortages; diminished supplies of timber, fish, metals, and other natural resources; insufficiency of water supplies; and a generally low standard of living and health care for much of the world's population.

The 1960s were a period of great awakening about population issues. A few especially powerful individuals wrote and spoke about the threat posed to the world by overpopulation. One of the most powerful of these works was Paul Ehrlich's *The Population Bomb.* In his book, Ehrlich argued that "the cancer of population growth must be cut out" (Ehrlich 1968, 311). Failing

to do so, he warned, would mean that humans would "breed ourselves into oblivion" (Ehrlich 1968, 311).

Unsurprisingly, there have always been critics of Ehrlich and the views on population he represents, those who feel somewhat less pessimistic about the dangers of population growth. These critics have argued in essence that humans have developed an extraordinary capability of dealing with societal problems in the past. It seems likely, therefore, that they will also be able to find ways of dealing with population issues, they have argued.

Fundamentals of Demography

The population of a nation, state, city, or other region depends on four factors: birthrate, death rate, rate of immigration, and rate of emigration. A simple way to express population growth in an area is the following equation:

population growth = number of live births + number of immigrants − number of deaths − number of emigrants

The fundamental factors of population growth are affected to some extent by natural courses. The spread of an epidemic, for example, is likely to increase the death rate in a region and to reduce, therefore, the overall population. The failure of an essential food crop, as in the Irish potato famines of 1846–1848, can dramatically affect emigration rates out of one country (Ireland, in this example) and into another (the United States, for example).

The four population-determining factors are, however, often more strongly affected by human actions than might at first be obvious. For example, improvements in medical technology tend to decrease death rates and, thus, increase rates of population growth. Some other human factors that affect population growth are the following:

1. *Sex education:* The more individuals know about sexual behavior, presumably, the more likely they will be able to make informed

Some authorities believe that the best way to solve the world's population problems is to find ways to improve the standard of living of its poorest nations. (WHO/J. P. Revel/CRED)

judgments about having children. Conversely, the less people know about sexuality, the greater the rate of accidental births is likely to be. Thus, societies that encourage sex education may tend to have lower birthrates while those that discourage sex education may tend to have higher birthrates. Interestingly enough, the population effects of sex education policies, as with other practices outlined below, are likely to be incidental to the primary reason(s) for a given policy on the practice. That is, a society may be opposed to sex education and may prohibit sex education for its children for religious or some other reason(s). The population effects of this policy may not be apparent to the majority of people in the society. Indeed, the effects on population may conceivably even be counter to those the society would choose for itself if population issues were the main criterion for making social decisions.

2. *Abortion:* In the same vein, most people probably tend to base their views on abortion on religious, ethical, or other grounds. However, it is apparent that proabortion views tend to contribute to decreases in population growth while antiabortion views tend to lead to increases in population growth.

3. *Homosexual behavior:* The drive that encourages people to marry and have children is very likely to be greater in a society that views homosexual behavior as immoral, mentally aberrant, or "wrong" in some other sense than it is in a society that is more accepting of such behavior. Homophobic views tend, therefore, to contribute to population growth while more

accepting views of homosexual behavior tend to contribute to reduced growth.

4. *Views toward women:* In some societies, women may be assigned a limited role that emphasizes motherhood. In such societies, the birthrate and rate of population growth are likely to be relatively high. As more options outside the home become available to women, they may place relatively less value on motherhood and having children. In such situations, the birthrate and rate of population growth may tend to decrease.

5. *Economic incentives and disincentives:* Governments can make economic decisions that may influence the rate of population growth. For example, a government may decide to place a tax on any family with more than some number of children. More commonly, it may provide a tax benefit for having children. In such cases, the more children a family has, the less tax it pays. These two tax policies are likely to discourage and encourage, respectively, the rate of population growth in a nation.

Demographic Transition

One of the most basic issues among demographers (those who study population issues) is the *demographic transition.* This term refers to a stage in the development of a society characterized by a declining birth and death rate and a reduced rate of population growth. This pattern is often found in more industrialized nations where medical science makes it possible for people to live longer and where women have a greater variety of out-of-home career options.

The question among demographers with regard to the demographic transition is: Which comes first, the chicken or the egg? That is, is it necessary for a society to reduce its birthrate before it can achieve significant improvements in economic and social conditions? Or do improvements in economic and social conditions lead to a situation in which people choose or are able to reduce birthrates?

No one can say what the answer to that question is. But the two possible choices lead to two dramatically different views on population policies. According to one view, nations should do whatever they can to reduce birthrates. (It is almost always assumed that nations will *want* to reduce death rates.) As birthrates go down, according to this view, the share of economic goods and services per person in the society increases and the standard of living for everyone improves.

According to the second view, governments should stay out of the business of family planning. They should focus, instead, on improving their economies. As these economies grow, the total amount of goods and services increases, and the amount available to each individual citizen increases. With increasing wealth, families will choose to have fewer children, and the birthrate and rate of population growth will decrease.

These two views on the demographic transition have been in conflict in the United States at the level of national policymaking for many years. Some government officials believe that the United States should have an aggressive program of making birth control information and materials available to nations throughout the world with high rates of population growth. These officials argue that bringing down birthrates and rates of population growth may be the single most important step the United States can take in helping these nations improve their economies.

On the other side of this debate are those officials who argue that the United States should leave population issues to individual nations. It may be appropriate for Americans to provide economic aid to less developed nations, they say, because that is more likely to bring prosperity and, as a consequence, reduced birthrates and rates of population growth.

Population Today

In some ways, the study of population patterns today is a fascinating survey of an enormous unplanned (or poorly planned) experiment in population growth. For more than thirty years, demographers have been watching and documenting population changes and trying to sort out the factors that might have caused these changes. Today, some patterns of population growth in the world are clear, but the reasons for those patterns are in as much dispute as they were in the 1960s.

For example, birthrates and rates of population growth have fallen worldwide. In the United States, for example, the total fertility rate (TFR) has fallen continuously since 1790, when it was 7.7. The TFR is the average number of children born per woman in her lifetime. The TFR in the United States is now 1.9, less than the replacement rate. The replacement rate is the TFR needed to maintain a population of constant size. One might expect the replacement rate to be 2.0. That is, it would seem to be necessary for each couple to have two children, on average, to keep the population constant. Infant death rates must be factored into this calculation, however, since individuals who die young do not live long enough to reproduce themselves. In actual practice, the replacement rate tends to be about 2.1.

When the TFR falls below the replacement rate, population does not begin to decrease immediately. When people live sixty, seventy, or more years, it takes decades for changes in the rate of population growth to have their effects. Thus, the *rate* of population growth in the United States has been falling for some time. However, the absolute size of the U.S. population will continue to increase for a number of years.

Some nations have experienced even more dramatic changes in TFR than the United States. In Italy, for example, the TFR in 1997 was 1.17. This number means that each Italian couple is producing, on average, slightly more than one child. If maintained in the future, this rate would mean that the Italian population would reach zero by the year 2200.

That possibility is, of course, quite unlikely. Any society aware of that kind of threat to its existence will almost certainly take action to reverse TFR trends. What is of more concern to demographers is the change in composition of the Italian population. As birthrates go down, the average age of a population increases. In 1997, for example, there were more Italians over the age of sixty than under the age of twenty. In theory, the average age of an Italian worker in 2050 would be seventy. No retirement system would be able to function in a population with that age structure. Clearly, the economic and social consequences of Italy's declining birthrate are enormous.

But Italy is not the only nation with declining birthrates. Those rates have fallen to less than the replacement rate in the rest of the European Union (1.43), Japan (1.46), and even China (1.82). Overall, most of the world's nations appear to have "turned the corner" in terms of population growth. They have adopted, or allowed to develop, population policies that have led to decreases in birthrates.

These patterns have not eliminated many of the concerns about the effects of overpopulation. Many demographers continue to write and speak about the dangers of population growth, even as birthrates continue to fall. The demographers point out the long lag time between declining birthrates and declining populations and stress the threats to the environment and natural resources as populations continue to grow.

In fact, many demographers have begun to emphasize a new category of problems created by growing populations. They feel that reducing birthrates, while important, may be only one population issue with which societies have to deal. Perhaps an equally important issue involves the distribution of wealth in societies. They argue that nations should begin to find ways to shift resources from the small number of very wealthy individuals to the much larger number of very poor individuals. Similar changes may be necessary on the level of nations, with the very rich industrialized countries having to find ways of making greater contributions to the development of the world's poorest nations.

See also: *The Limits to Growth*

References

Abernethy, Virginia D., and Garrett Hardin. *Population Politics: The Choices that Shape Our Future.* New York: Plenum Publishing, 1993.

Bouvier, Leon. "Planet Earth 1984–2034: A Demographic Vision." *Population Bulletin* 39, no. 1 (1984).

Daugherty, Helen Ginn, and Kenneth C. W. Kammeyer. *An Introduction to Population.* 2d ed. New York: The Guilford Press, 1995.

Durning, Alan Thein, and Christopher D. Crowther. *Misplaced Blame: The Real Roots of Population Growth.* Seattle, WA: Northwest Environment Watch, 1997.

Ehrlich, Paul, and Anne Ehrlich. *The Population Explosion.* New York: Doubleday, 1990.

Fyson, Nance Lui. *World Population.* New York: Franklin Watts, 1998.

Lutz, Wolfgang. *The Future of World Population.* Washington, DC: Population Reference Bureau, 1994.

Markley, O. W., ed. *Twenty-First Century Earth.* San Diego: Greenhaven Press, 1996.

Mazur, Laurie Ann, and Timothy Wirth. *Beyond the Numbers: A Reader on Population, Consumption, and the Environment.* Washington, DC: Island Press, 1994.

McFalls, Joseph A., Jr. *Population: A Lively Introduction.* Washington, DC: Population Reference Bureau, 1995.

Miller, G. Tyler, Jr. *Environmental Science*, 3d ed. Belmont, CA: Wadsworth Publishing, 1991, Chapter 6.

Newton, David E. *Population: Too Many People?* Hillside, NJ: Enslow Publishers, 1992.

Southwick, Charles H. *Global Ecology in Human Perspective.* New York: Oxford University Press, 1996, Chapters 14 and 15.

Wattenberg, Ben "The Population Explosion Fizzles." *Sunday Oregonian*, 7 December 1997, E3.

An extensive list of books, pamphlets, data sheets, and other materials on population is available from the Population Reference Bureau, 1875 Connecticut Ave. N.W., Suite 540, Washington, D.C., 20009–5728. A valuable source of information on population issues is the Population Council, whose web site is at http://www.popcouncil.org. The *Population Index* contains annotated bibliographies of books, journal articles, working papers, and other materials on population topics. It can be accessed at http://popindex.princeton.edu/index.html.

Prescribed Burn

A prescribed burn is a fire that is set intentionally in a forest, on a prairie, or at some other location. The primary purpose of a prescribed burn is to prevent the outbreak of a much larger and more disastrous natural fire in the same area. Prescribed burns are also known as controlled burns.

For most of U.S. history, government agencies have devoted huge amounts of financial resources, personnel training, and educational efforts to preventing forest fires. Smokey the Bear may be one of the best known advertising figures in the nation. Smokey's message has always been to stop forest fires before they break out.

There can be little doubt that these efforts have had some very positive effects. Many forests, grasslands, prairies, and other natural resources would have been destroyed had humans been less careful about starting fires. But fire prevention has also had some unexpected and negative results.

Benefits of Natural Fires

Ecologists now know that fire is a natural and necessary part of any ecosystem. Rather than thinking of fire as a natural enemy, many individuals and agencies have come to regard fire as something that every ecosystem needs to experience at some point in its evolution.

One function of fire is to clear away dead vegetation and live plants that would oth-

Prescribed burns are an ongoing environmental issue because political pressures are sometimes at odds with scientific evidence about how such burns should be used. (Corbis/Michael S. Yamashita)

erwise smother the growth of other plants. Fire also results in the formation of ash, which becomes part of soil. Ash contains nitrogen, phosphorus, potassium, calcium, and other minerals from dead plants that are needed to replenish the fertility of soils. Fire is also needed by some kinds of plants in order for reproduction to occur. Some seeds will not germinate, for example, unless they are first broken apart by the heat of fires. An area swept by fire may also be more congenial to certain kinds of animal life.

Overall, forest and land managers are now generally convinced of the value of and need for prescribed burns. In 1997, for example, the U.S. Forest Service carried out prescribed burns on 325,000 hectares (805,000 acres) of woodlands in the western United States. Managers expected to increase that amount to nearly half a million hectares (more than a million acres) in 1998. These totals were more than twice the amount of land burned intentionally over the period between 1986 and 1996.

Objections to Prescribed Burns

Not everyone is enthusiastic about the use of prescribed burns as a method of land management. For example, lumber companies suggest that they should have the opportunity to remove dead trees rather than having them burned off. People who live in the vicinity of prescribed burns may object to the smoke produced by such fires. People with asthma, allergies, and respiratory disorders are especially likely to be inconvenienced by prescribed burns. Other observers worry that intentional fires may actually threaten the habitat of some animals and will almost certainly kill some of them.

Finally, the effectiveness of Smokey the Bear's message cannot be discounted. Most adult Americans have grown up with the "Prevent Forest Fires" message. They may now find it difficult to understand how intentionally setting forest fires is a good practice.

References

Becker, Rachel, and Liz Stack. "Do People Purposely Set Prairie Fires?" http://www.highlands.w-cook.k12.il.us/Prairie/BURNING/prairiefires%201. 5 July 19/98.

"Fire Management." http://www.nps.gov/htdocs2/planning/mngmtplc/nrmfm.html. 5 July 1998.

Freedman, Bill. "Prescribed Burn." In *The Gale Encyclopedia of Science,* edited by Bridget Travers, 2921–2923. Detroit: Gale Research, 1996.

Koehler, John T. "Comprehensive Training: The Use of Prescribed Burning as a Wildfire Prevention Tool." *Fire Report Newsletter,* 21 July 1997, 1; also at http://www.firefighting.com/FRN/fr2–12–1.htm. 5 July 1998.

Matthews, Mark. "In Western Woodlands, Forest Service Begins Setting Fires to Prevent Fires." *Washington Post,* 19 April 1997; also at http://forests.org/gopher/america/sintburn.txt. 5 July 1998.

"Native Plants Require Fire." http://www.epa.gov/glnpo/greenacres/ga-fire.html. 5 July 1998.

"The Role of Fire in Ecosystems." http://www.blm.gov/education/fire/fire1.html. 5 July 1998.

Sonner, Scott. "Foresters Defend Controlled Burning." *Oregonian,* 5 July 1997, C1+.

Privacy and the Internet

Americans hold a wide variety of political views, ranging from the far Left to the extreme Right. One issue on which an overwhelming number of people agree, however, is the right to privacy. Indeed, some writers have argued that the right to be left alone, free of governmental intrusion, may be *the* most important right to which a U.S. citizen can lay claim.

Like all rights, however, the right to privacy is not absolute. Indeed, it is not uncommon for privacy rights to come into conflict with other kinds of rights. For example, employers have certain property rights that allow them to control and pro-tect their assets from employee theft. What happens when an employer decides to monitor the private behavior of employees to make sure they are not stealing from the company? Are the employer's property rights or the employee's privacy rights predominant?

Electronic Devices and Privacy

As it happens, the conflict between privacy rights and property rights is a common topic of disagreement. That issue has been made more complex by the growing availability of electronic devices for monitoring personal behavior. For example, some companies have installed video cameras—usually with the knowledge or consent of employees—in rest rooms, changing rooms, and other facilities in which a person might reasonably expect to enjoy privacy. These cameras allow a company to make sure that employees are not walking off with company property, engaging in illegal acts (such as smoking marijuana), or behaving in other ways the employer deems to be unacceptable.

The explosion of computer technology has dramatically expanded the options available to companies as well as to governmental agencies and others interested in "listening in" on the private activities of ordinary citizens. For example, companies can collect all kinds of information about employees who use computers tied in to a central network. This kind of data collection can range across the spectrum, from reading the employee's electronic mail to calculating the amount of time the employee spends away from his or her desk. The problem with such practices, from a privacy standpoint, is that the employee may never know about the electronic surveillance that is taking place.

Surveillance issues can be even more complex when they involve children and adolescents. Do people who have not reached some "legal" age, such as eighteen or twenty-one, have the same rights to privacy as adult citizens? For some civil libertarians, the answer is easy. Constitutional rights, they argue, apply to *all* Americans, no matter their age. Other scholars are not so sure. They believe that at least some rights under some circumstances are reserved for adults.

In this debate, U.S. courts have tended to lean toward the second of these views. They have allowed unannounced searches of student lockers, for example, when comparable searches of an adult's private property would never be permitted. Concerns about "the greater good" achieved as the result of such a search or even a belief that the state knows better than a child as to what is "good" for him or her has sometimes motivated these decisions.

Sex and Privacy

Debates over privacy issues are often muddied when they involve sexual topics. Some of the most contentious debates over privacy rights relate to a person's right to buy or view sexually explicit material and/or the right of someone to offer for sale or viewing such materials. People who would have no objection to the private showing of a grisly war movie might, by contrast, have violent objections to a similar private showing that involved naked bodies.

In this regard, there is probably less disagreement about the privacy rights of children than in any other areas. Generally speaking, sexually explicit materials are widely thought to be unsuitable for children and, usually, adolescents. Even when such materials are permitted to be sold to or seen by adults, they are regarded as being "off-limits" to anyone under the age of eighteen (or some other approved age). Legislation to limit a child's access to sexually explicit material on the television, in theaters, in books, on the Internet, and elsewhere typically draws nearly unanimous approval, even when similar legislation for adults earns much less support.

The Internet

The birth of the Internet has arguably raised issues of personal privacy to a level never seen or imagined before. At the touch of a computer key, a person can have instant access to information from millions of web sites throughout the world. In the privacy of one's own home, a person can read on-line some of the greatest literature ever written, find out the scores of last night's soccer matches in Germany, order a new lawn mower from Canada, or see pictures of naked men and women. As long as a user pays certain access fees, that information is available to anyone who knows how to operate a computer, from a six-year-old child to an eighty-year-old retiree.

The Internet had hardly escaped from its cocoon in the mid-1990s before questions began to appear about the appearance of "inappropriate" material on web sites. What was the government's role in shielding its citizens, especially its children, from offensive information on the World Wide Web? As one might expect, that question was most commonly framed in terms of offensive sexual information, in contrast to possibly offensive or even dangerous political, philosophical, military, or other information. Also, the question was most commonly raised in terms of protecting children, although the potential harm to viewing adults was by no means ignored.

Communications Decency Act of 1996

The first effort to regulate commerce on the Internet was the Communications Decency Act (CDA) of 1996, originally introduced by Representative Mike Oxley (R-Ohio). This act prohibited the distribution of "obscene or indecent" material via the Internet to anyone under the age of eighteen. President Bill Clinton signed the bill on 8 February 1996. Three weeks later, a group calling itself the Citizens Internet Empowerment Coalition filed suit to prevent enactment of the law. The suit was joined by a wide variety of groups that included the American Booksellers Association, the Freedom to Read Foundation, the Society of Professional Journalists, Apple, Microsoft, America Online, Wired, and HotWired.

The plaintiffs' argument was that information transmitted electronically on the Internet is no different from that available on the printed page. All such information is protected by the right of free speech guaranteed in the First Amendment to the Constitution. People of all ages have the right to have access to that information in the privacy of their own homes, the plaintiffs argued.

The case worked its way through the courts and was decided by the U.S. Supreme Court on 26 June 1997. The Court held in favor of the plaintiffs, claiming that "[t]he interest in encouraging freedom of expression in a democratic society outweighs any theoretical but unproved benefit of censorship."

Child Online Protection Act of 1998

Congress was not deterred by the Court's decision, however. Before the ink on the Court's ruling was barely dry, efforts were under way to pass similar legislation. That legislation was incorporated in the Child Online Protection Act (COPA) of 1998, passed by Congress and signed by President Clinton in October 1998. The COPA made it a crime to communicate materials "harmful to minors" via the Internet. Web sites were required to obtain proof of a user's age by collecting a valid credit card number, driver's license number, or other form of identification before permitting access to its information. Penalty for failing to do so was a fine of up to $150,000 and/or six months in jail.

Once again, a suit was filed against implementation of the COPA by the American Civil Liberties Union and other groups. The arguments offered by the plaintiffs were similar to those in the case against the CDA. On 21 December 1998, U.S. District Judge Lowell Reed issued a temporary injunction against COPA. Six weeks later, Judge Reed continued that injunction, saying that "[p]erhaps we do the minors of this country harm if First Amendment protections, which they will with age inherit fully, are chipped away in the name of their pro-

tection." While this case still has the potential for traveling the long road to the U.S. Supreme Court, there seems to be little reason to believe its ultimate fate will be different from that of the CDA.

Other Issues

One of the ongoing concerns about both the CDA and COPA has nothing to do with privacy issues per se. It has to do with their limited reach. The Internet is, after all, a worldwide system. Laws passed by the U.S. Congress have, under the best of circumstances, only limited impact on the residents of foreign countries. If these residents choose to continue providing sexually explicit and other "offensive" materials on their web sites, there is not much the U.S. government can do about it. Even if the CDA and/or COPA *had* been upheld in the courts, it is not clear how effectively they would have protected children in the United States from objectionable materials from other countries.

Efforts by the federal government to limit access to information on the Internet have been the focus of much of the debate over privacy issues. But those issues go beyond governmental policies. Many individuals and companies are very much concerned about the kinds of material available to Internet users of all ages and have attempted to devise ways of controlling that material. The mammoth Internet service, America Online (AOL), for example, has been criticized for developing policies to limit the kinds of speech transmitted on its service. Those policies are summarized in legal agreements between AOL and its members called Terms of Service (TOS).

AOL's TOS deal largely with issues such as licensing, billing, and the like. But Section 2.5 of the TOS concerns on-line conduct. It outlines behaviors that the company regards as inappropriate and that can lead to termination of a member's service. These behaviors include the transmission through AOL of any "unlawful, harmful, profane, hateful, racially, ethnically or otherwise objectionable material of any kind." This policy is said to be enforced by the company's Community Action Team that searches for and deletes "forbidden" content.

One of the least attractive by-products of the Internet has been the process known as *spamming*. Spamming is the widespread distribution of electronic mail messages to users who, in most instances, have not requested information and are not interested in receiving the transmissions. In the majority of cases, spamming is the Internet equivalent of "junk mail" that arrives at private homes every day.

Spamming has another function, however. It can also be the means by which some individual attempts to inform large numbers of people about his or her own political, social, economic, or other views on some subject. In such cases, the question may arise as to the right to free speech being exercised by the sender versus the expectation of privacy by the receivers of the spamming.

An interesting legal development in this area took place in late 1998. The computer chip manufacturer Intel Corporation filed suit against a former employee named Ken Hamidi. After losing his job at Intel, Hamidi began a crusade to discredit the company. Over a period of two years, he sent out seven spam mailings to as many as 30,000 Intel employees. He reputedly included incorrect and damaging information about the company in these mailings. In its lawsuit, the company claimed that it had suffered "immeasurable" and "irreparable" harm as a result of Hamidi's spamming. Hamidi's attorney, by contrast, argued that sending out a batch of electronic messages was not fundamentally different from mailing out the same number of letters through the U.S. mails. And the latter activity is certainly not illegal.

Where this case will end is not clear. Many legal scholars are not convinced that precedents based on the use of the mail service are comparable to spamming on the World Wide Web. Constitutional scholar

Laurence Tribe was quoted as saying in connection with this case that "the rules that apply to the mail don't apply to e-mail. The terrain of cyberlaw is being constructed as we speak" (Abate 1998, B1).

References

Abate, Tom. "Corporate America Is Eyeing the Outcome of Intel's E-Mail Case." *San Francisco Chronicle*, 14 December 1998, B1+.

Atkinson, Craig. "Censorship and the Internet." http://www.golden.net/~craig/censor/essay.html. 11 December 1998.

"Can Congress Censor the Internet?" Ford Marrin Esposito Witmeyer and Gleser. http://www.fmew.com/archive/censor/. 11 December 1998.

"Congress Agrees to Internet Censorship Legislation." http://www.cdt.org/publications/pp_4.26.html. 11 December 1998.

Eglehof, James. "Censorship on America Online." http://www.aolsucks.org/censor/. 11 December 1998.

Miller, Leslie. "Judge Blocks Net Porn Law." *USA Today,* 2 February 1999, 1A.

The American Civil Liberties Union publishes a number of booklets dealing with privacy issues. They can be contacted at 125 Broad Street, New York, NY, 10004. Some titles include "Take Back Your Data Info Kit!" "Surveillance Incorporated," and the ACLU Handbook *Your Right to Privacy* as well as the video "Through the Keyhole: Privacy in the Workplace—An Endangered Right."

Psychosurgery

Psychosurgery is any medical procedure performed on the brain in order to modify behavior. It is, therefore, a form of brain surgery. Brain surgery itself is also done for a broader set of reasons, such as the relief of pain or the removal of a tumor. The term *trephination* is often used to describe any form of brain surgery.

History of Psychosurgery

Brain surgery has a very long history indeed. It was practiced by the Egyptians about 2,000 years ago, and, more recently, it was developed to a rather high level of sophistication in the Western Hemisphere by the Incas.

The first reports of psychosurgery in modern times can be traced to the work of a Swiss physician, Gottlieb Burkhardt. In 1892, Burkhardt removed parts of the cerebral cortex from six patients who had been diagnosed as being aggressive and demented. His work did not seem to help the patients, and objections from colleagues and the general public forced him to discontinue his experiments.

Credit for the popularization of psychosurgery is usually given to Portuguese physician Antonio C. de A. F. Egas Moniz. Moniz had apparently become interested in this procedure upon hearing of research on chimpanzees carried out by two American scientists, John F. Fulton and Caryle G. Jacobson, in 1935. These researchers reported that removal of the animals' frontal lobes had apparently calmed them down and made them less anxious after failing to learn new tasks. The chimpanzees also, however, appeared to become less adept at those tasks.

Moniz's first procedures in 1935 were rather crude. He operated on patients institutionalized for psychotic behavior by drilling two holes into their foreheads. He then injected absolute alcohol into the patients' brains, causing death of the brain tissue. In later procedures, he inserted a wire loop into the holes and twisted it to physically destroy brain tissue.

Moniz conducted his research on patients who had not been helped by any other standard procedures, such as counseling, drug therapy, or shock therapy. His technique seemed to reduce a patient's aggressive behavior and anxiety, but it also led to fairly dramatic changes in the patient's personality and intellectual abilities.

His work became widely known as perhaps the only procedure available for dealing with severely psychotic disorders. In retrospect, more careful assessments of the overall effects of psychosurgery on these individuals might have been appropriate.

But standards for the care and follow-up of patients like those with whom Moniz dealt were not very rigorous. The esteem with which he was held by his colleagues is evidenced by the fact that he was awarded the 1949 Nobel Prize in Medicine or Physiology for his work.

Later Developments

For about two decades after Moniz began his work, psychosurgery became increasingly popular among those working with psychotic patients. One reason for this popularity was improvements made in surgical techniques by U.S. physicians Walter J. Freeman, James W. Watts, and J. G. Lyerly. Freeman and Watts also suggested a new name for the surgery carried out by Moniz: *prefrontal lobotomy*, or just *lobotomy*. By some estimates, more than 50,000 lobotomies were performed in the 1940s and 1950s.

Gradually, the popularity of lobotomies to treat mental disorders began to decline. One reason for this decline was the serious side effects experienced by patients. Some became apathetic and unresponsive to their environment. In many cases, they were little more than "human vegetables." At other times, the operation resulted in wild, uncontrolled behavior, accompanied by fits, seizures, and loss of control over bodily functions.

A second reason for the decline in use of lobotomies to treat mental disorders was the appearance of certain chemicals that achieved similar results with more benign side effects. Chlorpromazine was one of the most successful of these new antipsychotic drugs.

Techniques used in brain surgery have improved immensely over the past half century. The earliest work of this kind could very nearly be characterized as butchery compared to the methods used today. One of the most advanced surgical procedures is known as *stereotactic neurosurgery*. In this process, the precise location of a problem area in the brain can be located on a three-dimensional grid. A surgeon needs to make only one very small hole in the patient's skull. Then, a thin probe can be inserted into the brain and the problem area removed or cut.

In fact, some critics of psychosurgery complain that surgeons today use technical language to hide the kind of procedures they do. The procedures may, in fact, be more elegant, sophisticated, and precise, but their effects on a patient's brain are often not much different from what they once were, is the objection.

Controversies over the Use of Psychosurgery

Although the number of prefrontal lobotomies was drastically reduced after 1955, the controversy surrounding this procedure did not disappear. Animal research and a small number of operations on humans continued. A number of critics urged a complete ban on all psychosurgery in the United States. One of the most prominent spokespersons in this movement was Peter Breggin, a psychiatrist from Washington, D.C. Partly as a result of Breggin's efforts, a special study by the National Commission for the Protection of Human Subjects of Biomedical and Behavioral Research was ordered in 1975.

The report of the commission was issued in the fall of 1976. It found that techniques for psychosurgery had become much more sophisticated. The rather crude methods of the 1940s and 1950s were no longer advised or used by most physicians. Techniques used in the 1970s were found generally to bring relief from pain or severe depression without the terrible side effects of earlier operations. Even members of the commission who began with negative attitudes about psychosurgery ended up supporting its uses for certain purposes.

The commission's report certainly did not end the controversy about psychosurgery, however. For one thing, it found that no one quite knows how the procedure works, that is, how it affects the brain. Some commission members felt that the procedure's success might have been brought about by special personal factors in the doctor-patient relationship. The lack

of research on the methods used by surgeons was also a source of concern among commission members.

The availability of psychosurgical techniques also raises fundamental and profound questions about the nature of mental illness itself. There is no question that throughout history, certain groups of people have been defined as "mentally ill" for political, social, or other nonmedical reasons. Even today, it is not uncommon to hear someone say, "He's crazy" or "She's nuts," suggesting a mental disorder when the only problem may be a difference of opinion with the speaker.

From time to time, the dangers of using psychosurgery as a tool of social control become frighteningly obvious. For example, in the late 1960s, a group of three physicians recommended the use of psychosurgery on leaders of the civil rights riots of those years. The possibility of defining civil disobedience as an emotional problem and then "curing" it by psychosurgery is certainly a real one.

Finally, the question of "informed consent" is a particularly difficult one in the case of psychosurgery. *Informed consent* means that any person is thought to have the right to understand any medical procedure that is planned for him or her and that that person then has the right to disagree or agree to submit to that procedure. But some individuals who might need the kind of help offered by psychosurgery are those least able to understand the procedure or agree to its use. Should an operation of this kind be performed without a person's knowing what is going to happen and agreeing to allow the procedure?

Prefrontal lobotomies are still considered to be a viable treatment option for patients with certain kinds of mental disorders. According to one standard, only those patients with "severe, chronic, disabling and treatment refractory [resistant] psychiatric illness should be considered for surgical intervention" (Sabbatini 1998). Such patients generally fall into two large categories, as defined by the *Diagnostic and Statistical*

Manual, fouth edition (DSM-IV) of the American Psychiatric Association. Those categories include patients with (1) obsessive compulsive disorders and (2) major affective disorders, such as major depression or bipolar (manic-depressive) disorders.

References

"Brief History of the Lobotomy." http://public.carleton.edu/~vestc/pages/brief.html. 20 June 1998.

Doorey, Marie. "Psychosurgery." In *The Gale Encyclopedia of Science,* edited by Bridget Travers, 2968–2971. Detroit: Gale Research, 1996.

Earp, J. D. "Psychosurgery." [Position Paper of the Canadian Psychiatric Association.] http://cpa.medical.org/pubs/papers/psy_surgery.html. 20 June 1998.

Herbert, Wray. "Psychosurgery Redux." *U.S. News and World Report,* 3 November 1997; also at http://www.usnews.com/usnews/issue/971103/3surg.htm. 20 June 1998.

Sabbatini, M. E. "The History of Psychosurgery." *Brain and Mind,* June/August 1997; also at http://www.epub.org./br/cm/n02/historia/pioneers.htm. 21 May 1999.

Vertoscki, Frank T., Jr. "Lobotomy's Back." *Discover,* October 1977; also at http://public.carleton.edu/~vestc/text/loboback.html. 20 June 1998.

R

Radon

Radon is a naturally occurring radioactive gas with a significant potential as a hazard to human health under certain specific and relatively limited conditions. Radon is formed when uranium and other radioactive elements decay. Radioactive decay is the process by which an atomic nucleus breaks apart with the release of some form of radiation.

As an example, uranium-238 decays with the emission of an alpha particle to produce thorium-234. This process is the first step in a series of decay reactions that result in the formation of thorium-234, protactinium-234, uranium-234, thorium-230, radium-226, and then radon-222. That sequence continues beyond radon-222 until a stable isotope, lead-206, is formed.

The decay of radon-222 occurs with the emission of a beta particle and a half life of 3.8 days. A sister isotope, radon-220, is formed during a second radioactive decay scheme that begins with thorium-232. Radon-220 also decays with the emission of a beta particle, but with a shorter half life of fifty-five seconds.

Health Effects of Radon

As with any radioactive material, exposure to radon poses certain types of health risks for humans. The most important of those risks is lung cancer. By some estimates, anywhere from 5,000 to 20,000 deaths occur annually in the United States as a result of exposure to radon. This makes radon the second leading cause of lung cancer after smoking.

Since radon is a gas, it is inhaled and passes through the respiratory system. Beta radiation released during the decay of radon within this system is presumed to produce the cell damage observed after long-term exposure to the element. According to the best estimates available, the additional risk for lung cancer for a person exposed to normal concentrations of radon (about one picocurie per liter) over a lifetime of seventy years is three to thirteen cases per thousand individuals. The "safe" level of radon exposure established by the U.S. Environmental Protection Agency (EPA) is higher than that amount, four picocuries per liter. (A picocurie is a unit of measurement of exposure to radiation.)

Conditions of Risk

Exposure to dangerous levels of radon is typically a problem under only two circumstances: (1) among individuals who work with uranium ore, mine tailings, or other uranium-containing wastes; and (2) inside homes and other structures built on ground with unusually high levels of radioactive materials. The risk in the first of these two cases can be dramatic. It is said, for example, that up to 30 percent of all workers in uranium mines in some parts of the world eventually die of lung cancer. In the United States, the Uranium Mill Tailings Radiation Control Act of 1978 includes cleanup provisions designed to protect workers from such risks.

Concerns about indoor radon pollution are of somewhat more recent vintage. In the mid-1980s, studies of indoor air quality showed the level of radon to be unexpectedly high in certain buildings. Researchers were eventually able to construct a scenario for circumstances in which radon might pose a threat to human health. Uranium and other radioactive elements in the soil beneath a building produce radon gas. That gas seeps through the floors of the building into its interior spaces. Radon may also come from radioactive materials in bedrock and groundwater beneath the building. Once inside the building, radon tends to accumulate in concentrations significantly higher than it would outside the building, where it would rapidly diffuse away.

By some estimates, more than 8 million homes in the United States have radon levels exceeding the EPA standard of four picocuries per liter. The homes in which radon is a special problem are those in which one or more smokers also live. In such cases, the two most important causes of lung cancer—smoking and radon—interact synergistically. Experts believe that about three-quarters of all deaths caused by radon occur among smokers.

The risks posed by radon in buildings are relatively easy to control. Inexpensive radon test kits are now available and can be used by anyone to measure the level of radon in his or her home or office. Structures with high radon levels can be modified—sometimes as easily as by improving air circulation and ventilation—to reduce the risk posed by radon.

References

Brenner, David J. *Radon: Risk and Remedy.* Salt Lake City, UT: W. H. Freeman, 1989.

Brookins, Douglas G. *The Indoor Radon Problem.* Irvington, NY: Columbia University Press, 1990.

"A Citizen's Guide to Radon." 2d ed. (EPA Document #402-K92-001) Washington, DC: U.S. Environmental Protection Agency, 1992.

"Consumer's Guide to Radon Reduction." (EPA Document #402-K92-003) Washington, DC: U.S. Environmental Protection Agency, 1992.

"Radon Health Risks: Frequently Asked Questions." http://www.epa.gov/iaq/radon/radonaq1.html. 21 June 19/98.

The U.S. Environmental Protection Agency maintains a web site on radon at http://www.epa.gov/iaq/radon/.

Rain Forests

A rain forest is a geographic region characterized by annual rainfalls of at least 250 centimeters (100 inches). Such forests are characterized by the presence of very large, often very old, trees with large leaves, producing an extended canopy throughout the region. Rain forests maintain distinctive animal populations adapted to the physical characteristics provided by such trees.

Tropical and Temperate Rain Forests

Rain forests are often subdivided into *tropical rain forests* and *temperate rain forests* depending on the annual average temperature of the region. Tropical and temperate rain forests differ from each other in some important ways. For example, temperate rain forests tend to be dominated by coniferous ("pine") trees, while tropical rain forests contain a much broader diversity of large trees. The animal life found in each of these biomes is, obviously, strongly determined by this fact and other characteristics of the two kinds of rain forests.

On the other hand, the two types of rain forests have many characteristics in common. For example, there is generally no sustained dry period in either biome. As a result, fires tend to be rare, and trees tend to live to very old ages. As they die of natural causes, they are replaced by younger trees that will pass through a similar life cycle. The net biological productivity of rain forests tends, therefore, to be relatively constant.

Most rain forests also tend to be very complex ecosystems. Although they may be dominated by large species of trees, they also harbor a large variety of other plants, ranging from the smallest mosses to medium-size trees. Indeed, tropical rain forests are regarded as one of the most bio-

Loss of rain forests is widely believed to be one of the world's most serious environmental issues, but it has become an issue partly because that loss results from improved living conditions for some people who live in the tropics. (Corbis/Wolfgang Kaehler)

logically diverse and rich ecosystems found anywhere on the planet.

Benefits of Rain Forests

Rain forests are important for a number of reasons. First of all, they are home to a bewildering variety of plant and animal species. By some estimates, they contain well over half of all species found on the Earth. Millions of humans must be included in this census also.

Rain forests are also the source of many of the foods that make up the everyday diet of humans. These foods include coffee, tea, cocoa, palm and coconut oils, bananas and citrus fruits, many different vegetables, peanuts and cashew nuts, and beans and grains.

Some of the world's finest woods also are produced in the rain forests. These woods include balsa, mahogany, redwood, sandalwood, and teak. These woods are used not only for routine construction of homes and other buildings but also for the manufacture of some of the finest wood products in the world.

The rain forests are also a seemingly endless source of drugs and pharmaceuticals. At least 1,400 tropical plants, for example, are being studied as possible treatments for cancer. Many other rain forest products have been found that can be used to treat a variety of other diseases, including diabetes, arthritis, glaucoma, heart disorders, rheumatism, and inflammations. So far, less than 1 percent of 250,000 plants found in rain forests have been studied for such effects. Scientists believe that many of the unstudied plants will provide other medical products of use to humans.

Rain forests are also an important part of global cycles, such as the hydrological (water) and carbon dioxide cycles. For example, the forests take carbon dioxide out of the atmosphere and convert it to carbohydrate through the process of photosynthesis. When rain forest plants die and decay, they return that carbon dioxide to the atmosphere. The rain forests obviously will have an important role in dealing with the increasing levels of carbon dioxide in the atmosphere that have led to concerns about global warming.

Finally, the rain forests are home to many human societies. In many cases, these societies consist of only small numbers of individuals who survive under primitive conditions. Neither of these factors is relevant, of course, to the right to retain their native homelands in the rain forests.

Threats to the Rain Forests

Over the past three decades, the world's rain forests have been exposed to increasing pressures from human activities. Large corporations and national governments have begun to appreciate what a "gold mine" of natural resources the rain forests represent. They have begun to exploit these rich biomes at an astonishing pace. The Rainforest Action Network (RAN) estimates that 1 hectare (2.5 acres) of rain forest is being destroyed every second. That amount is equivalent to the loss of an area the size of New York City every day and the loss of an area the size of Poland every year. RAN estimates that, on average, 137 species of organisms are being driven into extinction every day. At that rate, the world is losing a total of about 50,000 species every year.

A major threat to rain forests today is logging. In virtually every rain forest, large corporations have begun to remove trees at an ecologically alarming rate. In the Malaysian state of Sarawak, for example, half of all land area has been declared suitable for logging. Of that amount, 8 percent has been set aside for permanent protection, while the rest has been declared open for sale to lumber companies. Natives in Sarawak and the rest of Malaysia have been protesting the denuding of their land, as

have a number of international environmental groups. But thus far, those complaints have fallen on deaf ears. Both national and state governments in Malaysia regard rain forest trees as "green gold" and see the utilization of this resource as an important factor in the development of their own economies.

That story has been told over and over again, in nearly every rain forest. In the late 1990s, for example, logging was begun in the Columbia River Forest Reserve of Belize. The area had long been regarded as one of the last untouched tropical rain forests. Again, native Mayan people have protested the loss of their native habitat and have sued to have logging stopped. But Belize is a small country with a large foreign debt. Sale of logging rights is one of the easiest and most profitable ways of dealing with this debt.

Native communities in the rain forests may themselves be presenting a threat to their natural homelands. Many are experiencing explosive population growths that place heavy demands on the forests for wood to burn and with which to build homes. The traditional farming techniques used by rain forest dwellers—slash-and-burn agriculture—may also be an increasing threat to the regions. At one time, the number of rain forest dwellers was very small compared to the size of the land on which they lived. The effects of slash-and-burn agriculture were largely insignificant. As the amount of rain forest decreases in size and populations increase, slash-and-burn may become a relatively more important factor in loss of forest lands.

The potential risks posed by slash-and-burn techniques were illustrated in the late 1990s. The arrival of an El Niño weather pattern in 1996 and 1997 changed typical rainfall patterns in many parts of the world. Regions that normally received hundreds of centimeters of rain each year experienced unusual dry spells. As a result, fires set by slash-and-burn farmers burned out of control. Very large regions of rain forests in Mexico, Indonesia, Malaysia, and other parts of the world were destroyed by these fires.

Development presents another threat to rain forests in many parts of the world. In South America, for example, a number of nations have signed political agreements to promote trade within the region. In order to carry out the terms of those agreements, many new roads, airports, work areas, and housing developments will have to be built, many within the rain forests.

Agricultural development poses another risk for rain forests. Today, many farmers living in or near rain forests find it profitable to cut down trees and convert an area to a farm or pasture. It makes more economic sense for them to sell fruits, vegetables, and beef to large corporations than to just "get by" growing their own food on a piece of land. As an example, as recently as 1950, about an eighth of the land area of Costa Rica was used for cattle ranching. By 1985, more than a third of the national land area had been converted to pasture land for the raising of cattle. The majority of that land had come from tropical rain forests.

In some parts of the world, oil exploration creates yet one more threat to the rain forests. In parts of the Ecuadorian Amazon, for example, the discovery of oil has led to a drilling "boom." Companies are moving into parts of the rain forests, drilling for oil, and building an infrastructure to be used in moving the oil to distribution points. A number of the world's most spectacular rain forest preserves are threatened by this development, even though the drilling may not take place within them. Oil spills at drilling sites and along the distribution routes are expected to damage plant and animal life in those preserves.

Solutions to Rain Forest Loss

Environmental groups, some governmental agencies, and indigenous people have been searching for methods to protect rain forests. The problem is an overwhelming one because of the enormous economic profit that can be made by exploitation of the rain forests. Appeals to the "better in-stincts" of large corporations and impoverished national governments tend to fall on deaf ears.

One of the few successful approaches thus far has been to pay nations, states, individuals, tribes, and other entities to take rain forest land out of production. For example, it is now possible for environmentally concerned individuals to "buy an acre" of land in a rain forest by making a contribution to one or another environmental group. These efforts may seem modest in comparison with the programs of large multinational corporations. But they do provide a concrete way of salvaging at least some small portion of one of the world's great resources.

See also: Headwaters Forest; Northwest Forest Plan; Slash-and-Burn

References

Freedman, Bill. "Rain Forest," in *Gale Encyclopedia of Science,* edited by Bridget Travers, 3048–3051. Detroit: Gale Research, 1996.

Gray, Denis D. "Malaysia Puts Blade to Virgin Wilderness." *San Francisco Chronicle,* 20 May 1996, A8+.

McConahay, Mary Jo. "On the Chopping Block in Belize." *San Francisco Examiner,* 2 February 1997, C10.

Pickering, Kevin T., and Lewis A. Owen. *An Introduction to Global Environmental Issues.* London: Routledge, 1994, 251–256

Rainforest Information Centre. http://forests.org/ric/welcome.htm. 21 May 1999.

Revkin, A. *The Burning Season: The Murder of Chico Mendes and the Fight for the Amazon Rain Forest.* London: Collins, 1990.

Turk, Jonathan, and Amos Turk. *Environmental Science.* 4th ed. Philadelphia: Saunders College Publishing, 1988, 284–293.

A good source of information about the rain forests is the Rainforest Action Network, whose web page is located at http://www.ran.org.

Religion and Science

See Creationism; Science and Religion

Research, Basic and Applied

The research conducted by scientists can often be divided into two general categories: basic and applied. Basic research is sometimes also called pure research.

Basic Research

Basic research is research carried out primarily to answer some question of interest to a scientist without concern as to any practical applications it may have. That research *could* ultimately lead to some practical applications, although those applications are of little or no interest in the design and conduct of the original research. Attempts by chemists to make entirely new compounds just to satisfy their curiosity can be an example of basic research. Literally millions of compounds that do not exist in the natural world have been synthesized by chemists just to find out about the physical, chemical, and other properties of those compounds. Probably well over 95 percent of those compounds are never produced a second time after they are originally made and studied.

Basic research contains a certain element of competition between humans and nature, a battle by researchers to uncover new facts about the natural world. It is, to some extent, a game or puzzle of enormous overall magnitude. In fact, interviews with scientists often reveal the sense of joy and excitement they find in extracting new information from science that is not entirely unlike the triumph of one sports team or competitor over another.

Applied Research

That sense of elation is often expressed by those who work in applied research also. *Applied research* is research designed to solve some specific problem, such as the design of a new kind of film or the search for a cure for a disease. The excitement in applied research is less likely to involve the discovery of some entirely new fact about the natural world than it is to involve finding an answer to a puzzle on which one has been working.

The Relation between Basic and Applied Research

The division between basic and applied research is by no means a clear one. Scientists usually have no reason to declare that their work falls into one category or the other. But it would not be unusual for one working in the field of basic research to have an inkling of some potential application of that research years into the future. Also, those who are working to solve specific problems in science may find that they need to conduct additional basic research on some aspect of a topic before moving forward on their main line of study.

In addition, there is obviously a line that connects *some* basic research with *all* applied research. That is, the knowledge used to solve practical problems must have come at some point in the past from basic research. It is often very difficult to trace that line backward into history, but that line certainly must exist. In fact, some historians of science have actually tried to trace the parentage in basic research of some important modern technical discoveries, such as the transistor and atomic energy.

Issues of Funding

Questions about basic and applied research might be no more than interesting intellectual exercises were it not for one additional factor: money. At one time, not so long ago, most research projects in science were relatively inexpensive. They were funded by colleges and universities, by business and industry, by philanthropic agencies, and, occasionally, by individual scientists themselves.

Since World War II, however, that scenario has changed dramatically. Today, many research projects are large and complex endeavors, requiring the participation of large groups of scientists and the use of expensive equipment. Many scientists now spend a significant portion of their time writing research proposals, requests for funding for their research.

Collecting money for applied research is relatively easier than it is for basic research. There is often some interested group willing to pay for research that will lead to a cure for spinal bifida or to a more efficient gasoline engine or to a longer-lasting plastic or to some other practical outcome. It is much more difficult to get funding for a project whose goal it is to study the formation of volcanoes on the ocean floor or the nature of triple-bonded dysprosium complexes or the shapes of ice crystals or some other topic of basic research. Without knowing the potential application(s) for such research, who will provide for its funding?

It is for this reason that basic researchers may try to suggest possible applications for the results of their work. They may have to "stretch" in order to find a practical application but doing so can greatly increase their ability to obtain funding. In fact, to the extent that they can get funding agencies to think about those applications, their chances for funding are likely to improve.

Tax Dollars for Basic Research

Today, funds for basic research still come from a variety of sources, including colleges and universities, corporations, philanthropic foundations, wealthy individuals, and governmental agencies. But the more costly a project in basic research is, the more likely it is that funding will be available from only one source: the federal government.

As an example, consider the Apollo space program of the 1960s and 1970s. That program was created with the goal of placing a human on the Moon by the year 1970. The program was initiated in 1961 by President John F. Kennedy for a variety of reasons, some of them political and some, scientific. The program had little or no practical value, however, except for a number of "spin-off" benefits that developed as an incidental consequence of the program's main thrust.

The task facing Kennedy and the rest of the federal government for many years was to convince legislators and the general public that spending billions of dollars in tax revenues on the Apollo program was a wise investment of funds. The Apollo appeal had to be based on national pride and the excitement of exploration and discovery with, to a lesser extent, arguments for practical fringe benefits the program might bring the ordinary U.S. citizen. The overwhelming thrill of exploring outer space was sufficient to maintain public and legislative enthusiasm for the Apollo program for many years. Indeed, the fact that the National Aeronautics and Space Administration (NASA) has been able to maintain a program of space discovery for nearly three decades is a tribute to the U.S. public's willingness to support basic research at a very significant level for many years.

Throughout this period, however, Congress has had to deal with the ongoing question as to how funds for basic research could be justified in terms of the practical needs of society, such as unemployment, housing, health issues, poverty, and crime. Every year, legislators have to decide whether a tax dollar is better spent on learning more about the moons of Saturn or building more public housing.

In some cases, it has been simply too difficult to convince Congress to provide support for basic research. The massive multibillion-dollar plan to build the superconducting super collider (SSC) is an example. That machine was designed to learn more about the fundamental nature of matter, a subject for which it was extremely difficult to find practical applications. President Ronald Reagan and Congress were eventually convinced of the value of this project and approved construction of the SSC. But legislators bailed out of the project and discontinued funding only two years after construction began, when cost overruns on the machine more than doubled.

For many years, Congress has followed a general "10 percent" rule of thumb for funding projects in basic research. That is, it tends to provide about one dollar for basic research projects for every nine dollars spent on applied research. Those funds provide a significant source of funding for basic research in the United States, although the total amount available is still small enough to ensure that projects in basic research from many fields of science will constantly be in competition with each other.

See also: Human Genome Project; Particle Accelerators; Search for Extraterrestrial Intelligence; Space Station

References

Basic Research and U.S. Prosperity." http://www.ced.org/docs/Br_rel.htm. 24 June 1998.

"Basic vs. Applied Research." http://www.lbl.gov/Education/ELSI/Frames/research-main-f.html. 24 June 1998.

"DOD Basic Research Remains Flat in FY 1998 Spending Bill." http://webster.aip.org/enews/fyi/1997/fyi97.117.htm. 21 May 1999.

Dutton, John A., and Lawson Crowe. "Setting Priorities among Scientific Initiatives." *American Scientist,* November/December 1988, 599–603.

"Future Course of Basic Science Research Is Debated at Conference Series." http://www.cc.columbia.edu/cu/record/record2014.37.html. 24 June 1998.

Greenberg, Daniel S. *The Politics of Pure Science.* New York: The New American Library, 1967.

Henig, Robin Marantz. "Basic Research Has a Place, Says the Man with the Golden Fleece." *BioScience,* March 1980, 149–152.

Kingburg, John M. "One Man's Poison." *BioScience,* March 1980, 171–176.

Lederman, Leon M. "The Value of Fundamental Science." *Scientific American,* November 1984, 40–47.

"Nobel Laureates Press for Science Funding." http://www.ripn.net:8081/infomag/dbase/B006E/961028–009.txt. 24 June 1998. http://www.aip.org/enews/fyi/1996/fyi96.150.htm. 21 May 1996.

Sarzin, Anne. "ARC Chair Warns Against Funding Shift from Basic to Applied Science." http://www.usyd.edu.au/su/exterel/news/970501News/1.5.ARC.html. 24 June 1998.

Weaver, Warren. "Basic Research and the Common Good." *Saturday Review,* 9 August 1969, 17–18.

Wong, Eugene. "An Economic Case for Basic Research." *Nature,* 16 May 1996, 187–189.

Right to Die

The Right to Die movement is also known by other names, such as Death with Dignity, Assisted Suicide, and active euthanasia. These terms refer to the ending of a person's life, based on his or her expressed wish, usually with the assistance of a qualified medical worker.

Choosing to Die

Does a person have a "right" to end his or her own life? That question is a complex and difficult one that has troubled Western civilization for 2,000 years. The question can be subdivided into two large parts, concerning those who want to die—to commit suicide—for emotional reasons and those who want to die for physical reasons. In the first category are those individuals who have lost hope for a happier, more fulfilling life and who choose, therefore, to die. In the second category are those who suffer from a terminal illness that may be very painful. These people may want to hasten a death that they feel will come soon anyway, but not soon enough for them.

In most Judeo-Christian societies, suicide under any circumstances has been considered to be immoral. The basic argument is that God has given life, and only he has the right to decide when that life shall end. Thus, suicide is regarded as a mortal sin, and the act has traditionally been illegal in most Western nations.

More to the point, perhaps, *assisting* another person to commit suicide has been and is illegal in most countries and in most states of the United States. In the United States, for example, forty-four states (as of 1998) had laws specifically banning assisted suicide. Five other states had no law dealing with assisted suicide, and one, Oregon, permitted assisted suicide.

Progress in Medical Technology

The simple moral principle that "the Lord giveth, and the Lord taketh away" long ago became muddied, however. Humans have long fought to preserve life even when it might appear that God had decided to end that life. Medical science could perhaps be defined as an organized system to thwart God's will in this respect.

Progress in medical technology over the past few decades has made life-and-death issues even more complicated. Organ transplants, life-sustaining systems, and other advances have made it possible to sustain life almost indefinitely in many cases. The quality of that life may be in question, but a physician's ability to maintain respiration and brain wave function continues to improve. In the Australian debate over assisted suicide, one member of parliament pointed out that "in 20 years' time we will be able to keep virtually everyone alive indefinitely. More and more, we are going to die when someone makes the decision that we are going to die" (Mydans 1997, A1).

Legal Steps

The 1990s saw a flurry of activity over the issue of assisted suicide. It is not clear exactly what produced this activity, although the constantly improving quality of medical care was almost certainly a factor. Another factor may well have been the human immunodeficiency virus (HIV) epidemic. For more than a decade, thousands of HIV-infected individuals were faced with the prospect of a long and painful death experience. Many of those individuals felt that taking their own lives should be an option to a long and horrible death. They sought out assistance in bringing about their own deaths.

In general, the response by various governmental bodies to the question of assisted suicide has varied. In 1993, for example, the Dutch parliament adopted a law that prevents the prosecution of a physician who carries out the carefully detailed instructions for aiding a terminal patient to commit suicide. The Dutch experience was

eventually to become an international test case for both those who favor and those who oppose assisted suicide. It seems that elements could be found in that experience to support either view of assisted suicide.

In 1996, the world's first voluntary euthanasia law was adopted in Australia's Northern Territory. That law permits any person over the age of eighteen who is mentally and physically competent to request his or her own death. That request must be approved by three doctors, one of whom is a specialist who can confirm that the patient is terminally ill and one of whom is a psychiatrist who can attest that the person is not clinically depressed. The use of this law in the assisted suicide of at least three individuals prompted Australia's national parliament to review the legislation in 1997.

Activity in the United States

In the United States, the first state to adopt similar legislation was Oregon. In 1994, state voters approved a Right to Die initiative permitting physicians to assist in an individual's suicide under carefully described provisions. That law was challenged in the courts and declared unconstitutional by a lower court. While that decision was being appealed, opponents of the Right to Die law filed a second initiative petition, this one designed to overturn the original law. That initiative was defeated in 1997. The vote against the initiative (in favor of the Right to Die law) passed by an even larger margin than did the original initiative of 1994.

During the same period, suits were brought in two states, New York and Washington, challenging laws that prohibit assisted suicide. These suits eventually worked their way through the court system to the U.S. Supreme Court, which issued its decision on 26 June 1997. The decision was an interesting one that was hailed as a victory by both sides of the Right to Die debate.

The Court ruled 9 to 0, first of all, that no constitutional "right to die" existed. It pointed out that suicide and assisted suicide have both been considered to be immoral and illegal throughout Judeo-Christian history and that the Constitution contained nothing to change that history. It upheld, therefore, the right of Washington and New York to pass laws banning assisted suicide.

But the Court also acknowledged, thereby, the right of individual states to pass legislation on this issue, presumably legislation that might permit as well as prohibit assisted suicide. In a concurring opinion, in fact, Justice Stephen G. Breyer even acknowledged that physicians had the right to prescribe medications to control pain "despite the risk that those drugs themselves will kill." Other justices pointed out that the question of assisted suicide was one that had to be dealt with legislatively rather than judicially and that decisions in future cases might take very different directions were circumstances different from those of the New York and Washington cases.

As of 1998, then, the visible and legal status of the Right to Die movement remained in flux. On a different level, however, the debate had taken quite a different direction. Many individuals inside and outside the medical community had apparently concluded that progress in medical technology had made some forms of assisted suicide inevitable. A highly sophisticated network of assisted-suicide providers had formed within the community of acquired immunodeficiency syndrome (AIDS) patients, for example, to help the terminally ill in that community to commit suicide.

In addition, one physician, Dr. Jack Kevorkian, had chosen to make a personal campaign to aid terminally ill patients to commit suicide. Kervorkian is thought to have assisted at least four dozen individuals to end their own lives. He was prosecuted a number of times in his home state of Michigan for murder, assisting in a suicide, and delivering a controlled substance to a patient. In the first four cases, either one or more charges were dropped against him, or Kevorkian was found innocent of the charges.

That situation changed, however, when the television program *60 Minutes* broadcast a video showing Kevorkian aiding in the suicide of a patient. Following that episode, Kevorkian was again tried for murder, assisting in a suicide, and delivering a controlled substance. On 13 April 1999 Kevorkian was found guilty of second-degree murder and the controlled substance charge and was sentenced to ten to twenty-five years in prison.

In some ways, the development of the issue of assisted suicide can be compared in its development to the pre–*Roe v. Wade* debates over abortion. The law, public opinion, and the way people actually behave are still at odds with each other.

References

Barnett, Jim, and Dave Hogan. "Assisted-suicide Foes Turn to Congress." *Oregonian,* 6 June 1998, A1+.

———. "Suicide Law Passes U.S. Review." *Oregonian,* 5 June 1998, A1+.

Carter, Stephen L. "Whose Death Is It?" *Sunday Oregonian,* 4 August 1996, B1+.

Cassel, Christine K. "Physician-assisted Suicide: Progress or Peril?" In *Birth to Death,* edited by David C. Thomasma and Thomasine Kushner, 218–230. Cambridge: Cambridge University Press, 1996.

Hoover, Erin, and Gail Kinsey Hill. "Two Die Using Suicide Law." *Oregonian,* 26 March 1998, A1+.

Humber, James M., Robert F. Almeder, and Gregg A. Kasting. *Physician-Assisted Death.* Totowa, NJ: Humana Press, 1994.

Kass, Leon R. "'I Will Give No Deadly Drug': Why Doctors Must Not Kill." In *Birth to Death,* edited by David C. Thomasma and Thomasine Kushner, 231–246. Cambridge: Cambridge University Press, 1996.

Knox, Richard A. "Poll: Many U.S. Doctors Would Honor Assisted-suicide Request." *Oregonian,* 23 April 1998, D12.

Mydans, Seth. "Australia Poised to Overturn Its Euthanasia Law." *San Francisco Examiner,* 2 February 1997, A7.

O'Keefe, Mark. "Court Says No Right to Die Exists." *Oregonian,* 27 June 1997, A1.

Price, Richard, and Tony Mauro. "Advocates Promise to Press the Fight." *USA Today*, 27 June 1997, 4A.

"6% of Doctors Say They've Aided Suicides." *San Francisco Chronicle*, 23 April 1998, A3.

Right to Privacy

See Privacy and the Internet

Ritalin

Some children are nervous, excitable, hyperactive, and disruptive in school. Almost any parent, teacher, child—almost *anyone*—recognizes that fact. Nearly everyone can remember a classmate who was "antsy" and could not sit still for a whole class period or someone who constantly caused disruptions in the classrooms. Why are some children that way?

That question would probably have seemed absurd to an earlier generation. Most people were probably willing to accept the fact that children who are growing up are filled with energy. Their bodies automatically and naturally look for outlets for that excess energy. Hyperactivity, from that standpoint, is hardly abnormal. What *would* be abnormal would be to expect young, growing children to sit passively, hour after hour and day after day, in school classrooms.

Childhood Hyperactivity as a Medical Problem

That view of childhood hyperactivity began to change in the 1970s. Scientific studies had begun to show that many forms of behavior are controlled by biochemical changes in the body. Such behaviors could no longer be regarded as "just the way kids are" but as actions caused specifically by the presence, absence, excess, or deficiency of particular chemicals in the brain.

Did childhood hyperactivity belong in this category? Some experts thought it did. First, these experts invented a clinical name for the condition: *attention deficit disorder* (ADD) or *attention deficit hyperactivity disorder* (ADHD). Then they began looking for drugs with which to treat the condition.

The most effective drug discovered for the treatment of ADD was Ritalin. Ironically, Ritalin is a member of the amphetamine family of drugs. Most amphetamines are stimulants. Yet, when used with ADD patients, Ritalin generally has the opposite effect. It tends to calm such individuals down and allow them to concentrate on learning tasks.

For many parents and teachers, Ritalin was a miracle drug. It solved discipline and control problems among children who had been resistant to any and all other approaches. Parents who had spent thousands of dollars on counseling, medical tests, and a variety of therapies found the answer to their problems in a pill. Ritalin became so popular that by 1997, more than one in thirty children between the ages of five and nineteen in the United States had a prescription for the drug.

Questions about the Use of Ritalin

Not everyone saw Ritalin as the ideal answer for childhood hyperactivity, however. Some authorities wondered about the long-term effects of the drug. It had been tested and approved by the U.S. Food and Drug Administration, of course, but it became an "overnight sensation" before there was any evidence about possible long-term effects.

Other critics wondered about the use of pills to deal with behavior problems. Ritalin was certainly a quick and easy solution to classroom and home discipline problems. But did all those children to whom it was being given really have a *biological* disorder? Or were at least some of them simply children with a lot of energy that could not be suppressed in a classroom?

Only recently has there been some evidence that children diagnosed with ADD may have brain structures different from those not diagnosed with the condition. But these findings are still preliminary. It is not clear what fraction of those who use Ritalin do have nontypical brain structures or any other biological basis for their behavior.

Finally, a new issue has arisen in the 1990s with regard to Ritalin. The drug has now become popular among college students for whom it was prescribed many years earlier in their elementary and high school days. The drug is used not primarily for recreational purposes, apparently, as is marijuana or LSD. Instead, grade-conscious students use it to "stay focused" on their class work.

Health officials are concerned about this growing popularity of Ritalin. In the first place, continued use of the drug requires larger and larger doses for it to have an effect. Some students report having to take twenty-five tablets a day to maintain its effect. Also, health officials know that doses this large can have harmful effects on the body, including sleeplessness, loss of appetite, and fatigue.

References

Bower, B. "Kids Talk about the 'Good Pill.'" *Science News*, 27 May 1989, 332.

Breggin, Peter. *Talking Back to Ritalin: What Doctors Aren't Telling You about Stimulants for Children.* Monroe, ME: Common Courage Press, 1998. Also, see a review of this book in Marilyn Elias, "Book Puts Ritalin Through the Wringer." *USA Today,* 6 March 1998, 11D.

Chacon, Richard, "Drug for Fidgety Schoolchildren Turning up on College Campuses." *San Francisco Chronicle,* 23 February 1998, A9.

Cowart, Virginia S. "The Ritalin Controversy: What's Made This Drug's Opponents Hyperactive?" *Journal of the American Medical Association,* 6 May 1998, 2521–2523, with response in Letters to the Editor in 21 October 1988, 2219.

Divoky, Diane. "Ritalin: Education's Fix-it Drug?" *Phi Delta Kappan,* April 1989, 599–605.

Grandparents and Parents Against Ritalin, Inc. http://www.chesapeake.net/vparker/moz3.htm. 17 June 1998.

Hadley, James. "Facts about Childhood Hyperactivity." *Children Today,* July-August 1984, 8–13.

"Little Evidence Found of Incorrect Diagnosis or Overprescription for ADHD." http://www.add.org/content/treatment/jama.htm. 17 June 1998.

"Ritalin." http://www.add.org/content/treat1.htm. 17 June 1998.

RU486

Most abortions performed today are surgical procedures. They involve the use of some physical technique for removing a fetus from a woman's uterus. As an example, one of the most common abortion techniques is known as dilation and curettage (or D and C). In this procedure, a woman's cervix is first dilated (expanded), and her uterus is then scraped to remove a developing fetus. As with any surgical procedure, abortions of this kind can be very traumatic for a woman. They also carry some medical risk, although that risk tends to be small.

Some health authorities believe that chemical abortions may be preferable to surgical abortions. They have the potential of reducing or eliminating both emotional and medical problems involved with surgical abortions. Researchers have been attempting for many years to develop chemicals that could be used safely for such abortions.

Action of RU486

In the early 1980s, such a chemical was developed in Europe. That chemical is mifepristone, better known as RU486. The initials *RU* designate the developer and manufacturer of the drug, the French firm of Roussel Uclaf. RU486 is classified as an antiprogestin, a substance that blocks the action of progesterone. Progesterone is a hormone needed to maintain the early stages of pregnancy. In the absence of progesterone, a fertilized egg does not implant on the uterine wall and develop properly.

RU486 must be taken during the first seven weeks of pregnancy. It is then followed two days later with a second drug, misoprostol, that induces uterine contractions. These contractions then expel the fetus. A third visit is required in order to ensure that loss of the fetus has occurred properly. If the fetus is not expelled, it may be born with defects caused by mifepristone and/or misoprostol.

As is the case with any new drug, RU486 was tested extensively for both efficacy and safety. Efficacy tests ensure that a drug performs as anticipated, while safety tests ensure that it does not cause unacceptable medical complications in patients. Nearly 5,000 women were involved in clinical trials of the drug. The drug passed both tests and was made available for use in Europe in 1988. Since that time, an estimated 200,000 women in Europe have taken RU486 to end pregnancies.

Issues Raised by RU486

Strong objections have been raised by antiabortion groups to the use of RU486. One matter of concern has been long-term effects of the drug. This question is often raised about new drugs that have been available for only a few years. It is generally not possible to know with certainty what long-term effects, such as cancer, may be associated with the use of such drugs.

Probably more important to antiabortion groups, however, is the ease with which abortions can be obtained using RU486. Pregnant women do not need to visit special clinics in order to obtain surgical abortions. Any physician can prescribe RU486, and the series of three visits associated with the drug's use can be accomplished easily and with virtually no public awareness that an abortion has taken place.

The privacy for abortions afforded by RU486 is both an advantage for women who choose to have an abortion and a tactical challenge for antiabortion groups. In the past, these groups have focused attention on clinics where abortions are performed, have picketed and harassed workers at these clinics, and have inspired bombings and killings at such clinics. The availability of RU486 promises to shift the focus of abortions away from such public clinics and into the private offices of physicians.

Approval of RU486 in the United States

By the late 1980s, Roussel Uclaf was being bombarded by opponents of abortion rights because of its involvement with the production and sale of RU486. The company decided that it had no interest in battling these critics in the United States. As a result, in 1994 it devoted all rights to use of the drug in the United States to the Population Council. The Population Council is a nongovernmental agency interested in a variety of population issues, including birth control.

The council's first task was to obtain approval for sale of the drug in the United States. It assembled and presented to the U.S. Food and Drug Administration (FDA) scientific evidence about the safety and efficacy of the drug. In September 1996, the agency announced its approval for the manufacture and distribution of the drug. However, it withheld final approval for the drug's sale pending additional information as to how the drug would be manufactured and labeled.

This final stage of drug approval is typically a routine step that takes relatively little time. With a drug as controversial as RU486, however, nothing is likely to be routine. Further delays in the approval process were announced by the FDA in 1997 and 1998.

One problem that developed in November 1996 involved Joseph Pike, the businessman who was licensed by the FDA to raise money for distribution of the drug. In November 1996, Pike admitted that he had concealed prior criminal convictions for fraud and forgery. The FDA withdrew Pike's approval and then asked the Population Council for additional scientific information about the drug. It had, apparently, had second thoughts about its original decision to approve the drug.

The Population Council complied with the FDA's request and, in late 1998, provided the agency with four more volumes of research data on the drug. It also submitted a plan for labeling and manufacturing the drug. In early 1999, the Population Council was expecting that final FDA approval for manufacture and sale of the drug would occur soon.

References

"Abortion Pill Would Change the Abortion Debate." *USA Today,* 20 July 1996; also at http://www.usatoday.com/life/health/lhs711.htm. 20 June 1998.

"Accord Reached to Allow Sales of Abortion Pill." *San Francisco Chronicle,* 14 February 1997, A7.

"Bibliography of Medical Articles on RU 486 and Breast Cancer." http://www.feminist.org/other/bc/bcru486.html. 20 June 1998.

"FDA Backs Abortion Pill for U.S. Use." *San Francisco Chronicle,* 19 September 1996, A1+.

"Feminist Majority Foundation Reports on Mifepristone." http://www.feminist.org/gateway/ru486one.html. 20 June 1998.

"The Fight to Make RU 486 Available to U.S. Women: A Chronology in Brief." http://www.feminist.org/gateway/ru486two/html. 20 June 1998.

Haney, Daniel Q. "U.S. Tests Show Abortion Pill 92% Effective." *San Francisco Chronicle,* 30 April 1998, A3.

MacDonald, Annette. "RU-486: A Dangerous Drug." *Vancouver Sun,* 18 September 1992; also at http://www.ru486.org.ru1.htm. 20 June 1998.

"The RU-486 Files." http://www.ru486.org/. 20 June 1998.

Zitner, Aaron. "The Saga of RU-486." *Boston Globe Magazine,* 23 November 1997; also at http://www.boston.com/globe/magazine/11–23/ru486/. 20 June 1998.

S

Saccharin

Saccharin was first prepared in 1878 by Constantine Fahlberg, a chemist at Johns Hopkins University. Saccharin is about 500 times as sweet as an equal amount of sucrose (table sugar), against which the sweetness of all compounds is compared. Fahlberg recognized the commercial value of saccharin and, after conducting tests on its safety in the human diet, applied for a patent on the compound in 1894.

Health Issues Posed by Saccharin

At the time, there were very few controls over the foods sold in the marketplace or over substances added to them. Scientists knew that saccharin passed through the human body without being digested. It contributed no calories, therefore, to a person's diet. Such a product held considerable appeal to companies that manufacture food products. Saccharin could be used in place of sucrose to add a sweet taste to foods but without adding any calories to the food products.

Saccharin's popularity as a food additive increased dramatically in the 1960s as people began to become more health conscious about the foods they ate. "Low-cal" and "diet" foods and drinks began to appear on the market and rapidly became very popular. Saccharin was also sold in pure form as a sugar-substitute sweetener. The best known form of pure saccharin is probably Sweet 'N Low®.

Concerns about possible health effects of saccharin appeared as early as the 1950s. Studies conducted by the U.S. Food and Drug Administration (FDA) suggested that saccharin could cause structural damage to the kidneys of experimental animals, but it was not toxic to the animals nor did it cause cancer. No regulatory action was taken because saccharin was then, and for the next twenty years continued to be, the only low-calorie sweetener available for use in food products.

Then, in 1970, a study by Canadian researchers found that very large amounts of saccharin in the diet of laboratory rats were correlated with unusually high rates of bladder cancer in the rats. Some observers took these results to mean that saccharin might be a carcinogen in humans also.

Governmental Response

The response by the U.S. government to the Canadian study has been a classic example of the issues confronting regulatory agencies with regard to a number of food additives. In 1977, the FDA adopted a new rule that severely limited the use of saccharin in consumer products. The new regulation was written largely in response to concerns raised by the Canadian research.

The FDA regulation had hardly taken effect when the U.S. Congress stepped into the controversy. Influenced at least partly by demands of the food industry, Congress revoked the ban on saccharin and decided instead that products containing saccharin would have to carry the following notice: "Use of this product may be hazardous to your health. This product contains saccharin which has been determined to cause cancer in laboratory animals."

The food industry's position was that, first, no satisfactory low-calorie substitute for saccharin was then available. Second, no evidence existed to suggest that saccharin causes cancer *in humans.*

Indeed, saccharin has never been defined as a "known carcinogen" by the FDA, as have a number of other substances. It has, however, generally been regarded as a "suspected carcinogen."

Current Status

The battle over the status of saccharin has now gone on for more than twenty years. In late 1997, the debate moved one step further. A panel of scientific experts was appointed by the National Toxicology Program of the National Institute of Environmental Health Sciences (NIEHS) to review the scientific data on fourteen substances then listed as "suspected carcinogens." Representatives on both sides of the debate appeared before the panel to present their views.

As expected, industry experts along with a number of independent scientists called for removing saccharin from the list of "suspected carcinogens." They argued that even after twenty years of research, there was no convincing evidence that saccharin caused cancer in humans. On the other side of the case were a number of public health advocates and independent scientists. They argued that the available animal studies were sufficiently clear to require that saccharin continue to be listed as a "suspected carcinogen."

On 31 October 1997, the scientific panel voted, four to three, to retain saccharin on the list of "suspected carcinogens." The panel's decision was then passed on to the executive committee on the National Toxicology Program. That committee voted in late 1998, six to three, to reverse the scientific panel's decision and to remove saccharin from the list of suspected carcinogens. The executive committee's decision was still not the last word, however. Its recommendations were to be passed on to the director of the NIEHS and, eventually, to the Secretary of Health and Human Services. It is conceivable that a decision made even at that level will not mark the end of the long controversy about saccharin.

References

"Low-calorie Sweeteners: Saccharin." http://caloriecontrol.org/sacchar.html. 19 July 1998.

McGinley, Laurie. "Saccharin May Be Delisted from NIH's Carcinogen List." http://www.junkscience.com/news/saccharin.html. 19 July 1998.

"New Federal Saccharin Review Planned: Could Lead to Removal from List of Carcinogens." http://www.nih.gov/news/pr/jul97/niehs–18.htm. 19 July 1998.

"Saccharin Still Poses Cancer Risk, Scientists Tell Federal Agency." http://www.cspinet.org/new/saccharn.htm. 19 July 1998.

Schulte, Brigid. "Panel Lists Saccharin as Carcinogen." *Oregonian,* 1 November 1997, A12.

Sagebrush Rebellion

Sagebrush Rebellion is a term used to describe efforts by some Western states and municipalities within those states to regain control over land within their borders currently under the ownership of the U.S. government.

Background

The federal government owns large stretches of land in every state of the union. These lands have been set aside as national parks, national monuments, national historic sites, national wildlife refuges, national wilderness areas, and many other nationally controlled areas. In addition, other land areas are owned or controlled by a variety of governmental agencies, most notably the U.S. Department of Defense.

In some states, the federal government owns or controls a significant fraction of land within the state's boundaries. In Nevada, for example, about 87 percent of all land area is owned or managed by the Bureau of Land Management (BLM), the U.S. Fish and

Wildlife Service, the U.S. Forest Service, the U.S. military, and other agencies.

Federal control of such large segments of state lands has long been a sore point for the states and the municipalities located within them. Whatever justification there may have been at one time for ceding control of these regions to the federal government, they say, those reasons are no longer valid. The federal government should return control of all or most of its land holdings to the states and other governmental units in which they lie.

A mitigating factor in leading states to increase pressure on the federal government for return of those lands was passage by the U.S. Congress in 1976 of the Federal Land Policy and Management Act (FLPMA). This act directed the BLM to take a more aggressive role in monitoring the uses to which land under its control was put.

Prior to adoption of the FLPMA, the BLM had little or no legislative authority to determine how federal lands would be used. As a consequence, ranchers, miners, lumber companies, and the general public used many of these lands in just about any way they chose. Passage of the FLPMA suddenly forced these groups to deal with rigorous new standards for the use of lands they had long taken for granted.

States' Rights, Conservation, Preservation, and Development

The opening salvo in the Sagebrush Rebellion was fired in June 1979 when then-governor Bob List of Nevada signed Nevada Assembly Bill 413. This bill decreed that 20 million hectares (49 million acres) of land then claimed by the federal government was to be returned to the state of Nevada.

Although the bill was the first broad statement by a state on the issue of land use, the Sagebrush Rebellion had been brewing for some time. At least five other western states had considered legislation similar to that of Nevada's, and hosts of private citizens had begun to make their concerns known in the lobbies of state and federal legislatures.

The so-called rebellion seemed to gain important momentum in 1980 with the election of Ronald Reagan as president. Reagan had long supported the principles of the rebellion, and the cabinet choices he made reinforced his announced intention to reduce the role of federal government in all phases of U.S. life, including the control of western lands. His choice for secretary of the interior, James Watt, almost immediately announced plans to open federal lands for exploration of fossil fuels and other resources, to cut back on the acquisition of new lands for federal use, and to begin selling off federal lands to private groups.

An additional element in the Sagebrush Rebellion was the theme of states' rights versus the rights of the federal government. This debate had had its beginning in the early history of this nation, as the drafters of the Constitution battled over the rights to be allocated to the central government versus those to be reserved for individual states. The final compromise in that debate was the Tenth Amendment, which says that rights not specifically granted to the federal government or specifically denied to the states were to be assigned to the states.

Many Sagebrush Rebels argued that the Tenth Amendment made it clear that lands currently held by the federal government were held illegally. There was no provision anywhere in the Constitution, they said, that permitted the government to take large stretches of state land without the state's express permission. And many states no longer agreed to this kind of arrangement.

Later Developments

Environmentalists were galvanized by the Reagan administration's efforts to give away federal lands. They took their case to Congress and the U.S. people. Although some shift in policies about federal land use did occur, few of the most extreme plans of Watt and other Reagan officials were ever carried out. According to some experts, the rebellion lost ground and was no longer a major political force by the mid-1980s.

An important factor in that trend was a series of lawsuits that were decided consistently in opposition to the rebels. Court after court ruled that the federal government had the legal authority to hold the lands it did and that states had no authority to pass laws transferring ownership of federal lands to themselves.

The notion that the rebellion is over is, however, vastly oversimplified. The fundamental issue on which the rebellion was based had been around for decades before the 1970s and did not disappear when state authorities ran out of options in their fight with the federal government. In fact, as the Sagebrush Rebellion was apparently winding down, a second movement was already under way, a movement sometimes known as Sagebrush Rebellion II. Sagebrush Rebellion II is characterized by efforts by cities, towns, counties, and other municipalities to adopt laws, ordinances, and other legislation invalidating federal ownership of property within their boundaries.

As an example, Boundary County, Idaho, adopted a land use plan in 1994 claiming authority over all decisions made with regard to federal and state land within its borders. The plan was based on a model developed two years earlier that had already been adopted in at least forty western counties. The plan was described as a "wise use" ordinance. *Wise use* was a term used for land use plans inspired by principles of the Sagebrush Rebellion. The federal district court ruled that Boundary County's "wise use" plan was unconstitutional and that local governmental agencies had no control over state and federal lands within their geographic region.

The Sagebrush Rebellion, wise use land use plans, and environmental reactions are not, of course, phenomena unique to the 1970s and 1980s. They represent an ongoing disagreement as to the way large stretches of the United States should be utilized, whether for development, conservation, or preservation. This disagreement will almost certainly continue in the future.

References

Larmer, Paul. "Wise-use Ordinances Suffer Legal Setback." *High Country News,* 21 February 1994; also at http://www.hcn.org/1994/feb21/dir/wr5.html. 9 August 1998.

Miller, Steve. "The East's 132-year Fight against Nevada's Ranchers." http://www.zianet.com/wblase/endtimes/rebel2a.htm. 9 August 1998.

"Minnesota 'Wise Use' Group." http://www.ewg.org/pub/home/clear/by_clear/Fifty-States/Minnesota.html. 9 August 1998.

"States Rights Attacked." http://192.148.252.38/pub/tezcat.Constitution/Feds_Attack.txt. 9 August 1998.

Salton Sea

The Salton Sea is a body of water in the central region of southern California, about 40 kilometers (25 miles) north of the Mexican border. The sea was formed between 1905 and 1907 when the Colorado River broke through irrigation dikes near Yuma, Arizona, and flowed westward into California. The damaged dikes were not repaired for nearly two years. By the time they had been restored, a new lake 65 kilometers long and 21 kilometers wide (40 miles long by 13 miles)—the Salton Sea—had been formed. Today, the Salton Sea covers about 47,000 hectares, or 240,000 acres, and has an average depth of about 10 meters (30 feet).

For a period of time, the Salton Sea was a popular tourist attraction. It was by far the largest body of inland water readily accessible from Los Angeles, San Diego, and other populated regions of southern California. At one time, it attracted more tourists than any other resort area in California and more visitors than Yellowstone National Park.

Ecological Problems

The Salton Sea's topography and geology have doomed it to a horrible fate, however. The lake has no natural outlet. Water flows

Birds have been dying because of pollution at the Salton Sea for more than two decades, but an effective constituency interested in solving this problem has only recently begun to exert its influence to change this situation. (Corbis/George Lepp)

into the lake from streams and from runoffs from farms and irrigation systems. But once in the lake, the water has nowhere to go, except to escape by evaporation. That is, the water cycle in the lake is the same as it is in oceans. In such a case, water in the lake constantly becomes more and more salty. At the present, the lake is twenty-five times as salty as the Pacific Ocean.

Under these conditions, most aquatic organisms are unable to survive. Fish, and the birds who feed off them, have begun to die off in very large numbers. Their rotting carcasses line the shore of the lake and produce a stench that has discouraged visiting tourists. The lake has experienced other ecological problems also. In 1996, a rare bacterium, *Vibrio alginolyticus,* spread through fish in the lake, killing thousands of fish and more than 14,000 birds that fed on them. Among the dead birds were 1,600 brown pelicans, listed as an endangered species.

In May 1998, another extensive fish kill occurred. Again birds who fed on the infected and dead fish were also killed. Among these were more than 6,000 double-crested cormorants. Authorities were unsure as to the exact cause of this disaster but thought that a toxic mixture of agricultural runoffs was at least partly to blame for the disaster. Included in the mix were pesticides, fertilizer residues, and selenium. Selenium is an element needed by plants and animals in very small amounts but toxic to animals in much larger amounts. Experts believed that the elevated levels of selenium, not ordinarily present in bodies of water, were at least partly responsible for the death of fish and birds.

Solving the Problem

Solving the pollution problems in the Salton Sea is made difficult partly because of the complex mix of factors that lead to the pollution. In addition, no one has yet been able to suggest a mechanism for reducing the lake's salinity (salt content). One proposal is for the construction of an evaporation pond at one end of the lake. Lake water would be pumped into that pond, keeping salinity in the rest of the lake at a constant level. But no one has determined what to do with the mass of toxic chemicals that would collect as water evaporated from the pond itself.

In 1998, legislators from California pleaded with the Department of Interior to begin studies on the causes of and solution for Salton Sea problems. Meanwhile, tourism at the Salton Sea National Wildlife Refuge had plummeted as visitors no longer found the site the beautiful natural wonder it had once been.

References

"Another Ecological Disaster Strikes Dying Salton Sea." *San Francisco Chronicle,* 27 May 1998, A15+.

Fordahl, Matthew. "California's Salton Sea Becomes a Hazard." *Oregonian,* 25 October 1996, A26.

Kopytoff, Verne. "In Spring, Birds Return to the Salton Sea and Die in Droves." *New York Times,* 24 March 1998, E4.

"The Salton Sea." http://www.desertusa. com/salton/salton.html. 11 June 1998.

"Senator Feinstein Says CAL-FED Type Approach Needed to Stop the Environmental Crisis at the Salton Sea." http://www.senate.gov/~feinstein/ releases98/salton.html. 11 June 1998.

Savage Rapids Dam

The Savage Rapids Dam is located on the Rogue River in south-central Oregon. It has become the focus of an ongoing dispute between dam owners—the Grants Pass Irrigation District (GPID)—and state and federal conservation authorities. This dispute is of more than local interest since it highlights many of the issues that arise when efforts are made to prevent the decline of fish populations in many parts of the Western states.

History of the Dam

The Savage Rapids Dam was built in 1921 for the purpose of providing water to an extensive irrigation system in the area around Grants Pass, Oregon. Construction of the dam prevented access by coho and chinook salmon, steelhead, and sea-run cutthroat to their traditional spawning grounds. In recognition of this fact, the GPID made a number of modifications to the dam to make upstream parts of the Rogue River and its tributaries more accessible to these fish.

For example, it built fish ladders on both the north side (in 1923) of the dam and the south side (about a decade later). In 1934, screens were installed to prevent fish from being swept into the irrigation system. These modifications never worked as well as planned. By some estimates, a quarter of all coho salmon are killed while trying to navigate the dam.

Decline of Salmon Populations

The Savage Rapids Dam issue is only one of many similar cases that have arisen since the mid-1980s in the Pacific Northwest. As early as 1981, environmentalists had begun to warn about the declining number of salmon, steelhead, and other food and game fish. A number of factors were identified as contributing to this decline: the construction of dams, such as the Savage Rapids Dam; intensive logging and cattle grazing along river banks that results in erosion and water unfit for the survival of fish; and industrial development that results in the dumping of toxic materials in rivers and streams.

Over a period of nearly two decades, the federal government and state and provincial governments in the Pacific Northwest invested more than $3 billion in a variety of plans to reverse the decline in fish populations. Most authorities began those campaigns with high hopes. They felt that salmon and other fish could be saved from extinction if sufficient goodwill, planning, and financial resources were available. By the late 1990s, it had become obvious that this optimism was unjustified. Salmon numbers had continued to decline.

For example, various governmental agencies had built a number of hatcheries in which to raise young salmon and other fish. The young fish were then returned to their native rivers. In these experiments, however, the majority of fish never returned to spawn after their first year. At one hatchery, for example, only 6 percent of the fish released ever returned to spawn.

The Cost of Saving the Salmon

At that point, some people began to ask whether the cause was a hopeless one. After the expenditure of so much money, with so few results, it seemed to those observers that certain fish species should simply be allowed to become extinct. The chairman of the Northwest Power Planning Council, a group organized to help save the salmon, eventually declared that "we shouldn't spend a lot of money just to slow down (the salmon's) rate of extinction" (Brinckman 1997).

Many environmentalists were unwilling to give up on these species quite that readily. They lobbied for having some of the fish species most at risk declared as endangered species under the Endangered Species Act (ESA). This listing, they felt, would provide them with a strong weapon to do whatever was necessary at whatever cost to save the salmon.

Current Issues at Savage Rapids Dam

Many features of this debate were being repeated on a smaller and more local scale in the Grants Pass area. Over a period of nearly two decades, the GPID and the Oregon Water Resources Commission (WRC) sparred with each other over efforts for protecting fish on the Rogue River. In 1994, the GPID agreed to remove the dam. But two years later, it had not only not taken any action in that direction but had actually changed its mind. It claimed that public opinion in the district was too strongly opposed to dam removal.

In April 1997, environmentalists got the weapon they needed in the Grants Pass area when the coho salmon was declared an endangered species by the U.S. Environmental Protection Agency. All of the relevant governmental agencies, including the U.S. Fish and Wildlife Service and the Bureau of Reclamation, as well as private conservation groups, such as Trout Unlimited and Water Watch of Oregon, argued that dam removal was clearly the most economical and efficient way of meeting the requirements of the ESA. As of late 1998, however, the irrigation district was still opposed to this action and continued looking for other options that would save the dam and avoid any further damage to the coho salmon population.

As the 1990s drew to a close, there was every indication that the Savage Rapids Dam issue would be replayed over and over again in the Pacific Northwest. In mid-1998, for example, Senator Slade Gorton (R-Wash.) introduced an amendment to a Department of the Interior appropriations act that called for specific congressional approval before any dam anywhere could be demolished. Although business and agricultural leaders were pleased by the Gorton amendment, environmentalists were aghast. The director of the Seattle office of the National Marine Fisheries Service said that the amendment "would completely and fundamentally disrupt salmon recovery in the Columbia Basin" (Barnett and Hogan 1998, C1).

References

Barnett, Jim, and Dave Hogan. "Gorton Plan Angers NW Rivers Advocates." *Sunday Oregonian*, 19 July 1998, C1+.

Brinckman, Jonathan. "After 16 Years and $3 Billion, Salmon Still in Peril." *San Francisco Examiner*, 17 August 1997, A19.

Cox, J. "Seeking a Solution for the Dam." http://www.rogueriverpress.com/seeking.htm. 8 August 1998.

"Endangered Species Act Issues." http://cru1.cahe.wsu.edu/whatcom/environ/wildlife/wildlifl.htm. 8 August 1998.

"Facts about Savage Rapids Dam." http://www.nwr.noaa.gov/ipress/042298_1.htm. 8 August 1998.

"Fight Over Future of Savage Rapids Dam Continues." Radio broadcast on "Oregon Considered." Available at http://www.opb.org/nwnews/savagerapids2-orcon.asp. 8 August 1998.

"Fish and Wildlife Issues." http://www.nwppc.org/fw_issu.htm. 8 August 1998.

Sato, Mike. "Salmon Recovery Will Benefit Puget Sound, Citizens Group Asserts." http://www.pugetsound.org/releases/r1998/release0226.html. 8 August 1998.

"Water Resources Commission Agrees to Contested Case Hearing with Grants Pass Irrigation District." Access via http://www.wrd.state.or.us/search/search.shtml. 22 May 1999.

Science and Religion

Science and religion are two systems developed by humans for understanding the natural world and, to one extent or another, the place of humans in that world. Conflicts between these two worldviews have existed throughout human history but especially over the last five centuries.

Nonscientific Views of the Natural World

Modern science can be said to have arisen during the late sixteenth century, largely as a result of the work of the great Italian scientist, Galileo Galilei. Prior to that time, a number of thinkers had used quasi-scientific methods for describing the world. The Golden Age of Greek civilization, for example, was characterized by efforts to view the natural world in a logical and rational, although not a truly scientific, manner. The element lacking from the Greek approach to their system of thought was experimentation.

Such versions of *natural philosophy* (in contrast to true scientific thought) were relatively rare prior to Galileo's elucidation of the modern scientific method. Most attempts to understand and explain the natural world were framed within mythological, mystical, or religious philosophies. For example, every human culture of which we know has some explanation as to how the world began and how humans first appeared in the world. In most cases, the act of creation is carried out by some supreme being, or some group of supreme beings. For example, aboriginal Australians tell of a great god who "travels across the face of the Earth," creating trees, hills, witchity grubs, ponds, and humans.

Early Jewish and Christian writings also contained explanations as to how the world was created and how humans first appeared on the Earth. Much of that description is contained in the Bible in the book of Genesis where, in fact, two slightly different versions of the creation can be found. The Old Testament also contains other descriptions and/or explanations of natural phenomena, such as the parting of the Red Sea by Moses and the halting of the Sun in its movement across the sky by Elijah.

As the Christian church grew over the centuries, its leaders promulgated a number of explanations as to the nature of the physical world and of humans. One of the best-known examples involves the physical structure of the universe. It was believed that the Earth was at the center of the physical universe, with the other planets, the Moon, and the Sun orbiting around it. Beyond the orbits of these bodies was a celestial sphere, beyond which could be found God, his angels, and the heavens. The stars were nothing other than tiny holes in the celestial sphere through which shown celestial light.

This view of the physical universe was developed, of course, not through any form of scientific experimentation. It grew out of the beliefs of early Christian leaders that the Earth must be the center of the universe because it was created by God for humans. One accepts this view not because of empirical evidence but because of the position of authority of those who teach it.

In fact, this point forms the basis of how "truth" is determined in religion. One believes a statement to be true because one has faith that the individual who makes the statement has been endowed with some supernatural authority to make that statement. The pope, for example, is regarded as the authoritative voice on many matters for those who regard themselves as Roman Catholics. Statements that the pope makes are to be taken as true *on faith* because he has been given the special authority to make those statements.

In many cases, religious statements are also regarded as "true" because they can be found in a holy book (such as the Bible or the Koran) or because they can be inferred from statements that appear in such books.

Truth in Science

Other systems of thought make use of other criteria for deciding which statements are true and which are not. In the visual arts, for example, an artist's description of a tree may look nothing at all like a tree to another person. A painting by Picasso or Dali is a vivid illustration of this fact. But the description still has a valid claim to be "true" if it represents some kind of reality to the artist himself or herself.

The reason that Galileo is so great in the history of human civilization is that he defined, almost single-handedly, a way of thinking about the natural world that was virtually unknown before his time. The "ground rules" that Galileo laid down for science contain a number of important features. One of those was the specification as to how one knows a statement is true in science. A statement can be regarded as "true," Galileo said, if it has been tested by experiment over and over again and found *not* to be "not-true."

One of the intriguing consequences of this position is that it is never really possible to "prove" for eternity that some given statement is "true." One can only show, as it is tested over and over again, that it is more and more unlikely to be false and, therefore, more and more likely to be "true."

Differences between Science and Religion

Perhaps the most fundamental difference between science and religion, then, is the standards by which truth is judged. In religion, statements are accepted as being true based on one's faith in the promulgations of a superior person or book. In science, statements are accepted as being more and more probably true based on their ability to withstand repeated experimental tests.

Science and religion differ on another important basis: the subjects with which they deal. Historically, religions have not hesitated from making statements about every aspect of the physical world and about every fact of human nature. The early Christian church had views on every aspect of nature, not just the structure of the universe. For example, it had explanations as to how the Earth was formed, why it had the shape it did, how humans got here, where comets came from, what the source of fossils was, how various metals were different from each other, what the causes of various weather conditions were, and so on. As virtually the sole source of knowledge in prescientific days, religion had an answer for every question that humans might ask.

By contrast, Galileo very specifically limited the range of scientific questions. Only those problems can be studied, he said, that allowed a person to physically manipulate objects and make measurements on those objects. Thus, science could legitimately study the motion of a falling object, but could never attack the question as to what love or fear is. Humans emotions have, or at least have generally been thought to have, no physical component that can be manipulated and measured.

Conflicts between Science and Religion

It is virtually impossible to imagine how conflicts between science and religion would *not* have arisen over time. As soon as Galileo, Copernicus, and other early scientists began their studies, they found information that conflicted with traditional (religious) information. For example, Galileo observed the Moon and found that its surface was covered with spots. Members of the church denied that the Moon could have spots. It was part of the celestial sphere and, therefore, had to be perfect.

Galileo suggested that people look through the telescope and see the spots he had seen. But his opponents replied that they had no need to look through the telescope. They claimed to already know what the Moon looked like.

As science developed, it produced more and more statements about the natural world that were in conflict with traditional (religious) views. At first, the way that organized religion dealt with this problem was by force of authority. It prohibited Galileo, for example, from conducting further experiments or from revealing to the general public the information he had gained. The Catholic Church simply banned the publication of books that contained information at conflict with its own views on various features of the natural world.

Some of the debates between science and religion have been more dramatic and more prominent than others. The publication of Charles Darwin's works on the evolution of plants and animals and of humans, in particular, may be the best known of these debates.

In sum, it would appear that science has usually had the better part of the conflict between these two great worldviews when the subject matter is the physical world itself (which it always is when science is involved). Time after time, explanations of natural phenomena promulgated by formal religion on the basis of faith alone have not been able to survive against the scientific method of repeated testing. It seems likely that an overwhelming number of the statements that people accept today as being "true" about the natural world are those obtained from scientific research, not from some religious dogma.

Science and Religion, Today and Tomorrow

It is easy to write about the conflict between science and religion as if it were something that occurred long ago in history. It is true that theologians in the modern world are much less likely to take on explanations of the physical world—such as the cause of earthquakes, for example—

as problems in rational philosophy or religious thought. In that regard, they tend to be less likely to come into conflict with scientists—such as seismologists—than was the case centuries ago.

And yet, to a considerable extent, that conflict has not disappeared. Most Americans tend to regard themselves as religious and to believe in God. By that admission, they take on at least some portion or shading of a supernatural view of the world. If one believes in God, one may also tend to believe that that God has the ability to work any wonders anywhere he wants at any time he wants.

For example, religious broadcaster Pat Robertson was reported as warning the city of Orlando, Florida, in 1998 about the possibility that hurricanes would strike their city because the city government had approved a "Gay Days" holiday there. He had earlier reported saving Virginia Beach, Virginia, from Hurricane Gloria in 1985 by praying for protection. Robertson reflects in these comments the view that natural phenomena, such as hurricanes and earthquakes, occur not solely because of natural causes but as acts of retribution by a punishing God.

Professional scientists are probably in one of the most difficult positions of anyone when science comes into conflict with religion. Scientists have to learn to apply one standard of truth in one part of their lives and a different standard of truth in another part of their lives. Many scientists regard themselves as good Christians, good Jews, good Muslims, or faithful members of some other religions and have obviously found ways to make this accommodation.

In fact, one can imagine that the historical debate between science and religion has not really subsided at all but may soon start to heat up again. One reason for this suspicion is that science has now begun to move into areas of research that were historically regarded as "off limits." For example, the statement was made above that love and fear are not subjects of scientific research because they have no physical components.

Yet, that statement is probably not true. Some researchers are looking for physical manifestations—changes in biochemistry, for example—that accompany these and other emotions.

In addition, the field of sociobiology has begun to investigate the chemical and genetic basis of certain human characteristics, such as altruism, that would once have been thought to be beyond the scope of scientific inquiry. As science continues to expand the subject matter in which it is interested, therefore, one might anticipate that it will cross over into a field that was once regarded as the sole province of philosophy and theology. When that happens, one cannot but wonder how science and religion themselves will change, if at all, and what new accommodations will be made between them.

See also: Creationism

References
"Activists' Corner: Pat Robertson Warns Orlando of God's Judgement for Hosting "Gay Day." http://www.infidels.org/activist/current/wire/stories/pat_and_gayDays.html. 22 July 1998.

Atkins, Peter. "Religion—the Antithesis to Science." *Chemistry and Industry*, 20 January 1997. Also at http://ci.mond.org/9702/8.html. 25 July 1998.

Barbour, Ian G. *Religion and Science: Historical and Contemporary Issues.* New York: HarperCollins, 1997.

Dawkins, Richard. "When Religion Steps on Science's Turf." *Free Inquiry*, Spring 1998. Also at http://www.secularhumanism.org/library/fi/dawkins_18_2.html. 25 July 1998.

Hoffman, Roald, and Shira Leibowitz Schmidt. *Old Wine, New Flasks: Reflections on Science and Jewish Tradition.* New York: W. H. Freeman, 1997.

Lindberg, David C., and Ronald L. Numbers, eds. *God and Nature: Historical Essays on the Encounter Between Christianity and Science.* Berkeley: University of California Press, 1986.

Polkinghorne, John C. *Belief in God in an Age of Science.* New Haven, CT: Yale University Press, 1998.

Price, Deb. "As a Kingmaker, Pat Robertson Would Make Most of Us Prey." *Detroit News,* 29 September 1995. Also at http://www.detnews.com/menu/stories/18098.htm. 22 July 1998.

Russell, Bertrand. *Religion and Science.* London: Oxford University Press, 1960.

Stenger, Victor. "Essays on Science and Religion." http://www.infidels.org/library/modern/science/. 22 July 1998.

White, Andrew Dickson. *A History of the Warfare of Science with Theology in Christendom.* New York: Dover Publications, 1896, 1960.

Search for Extraterrestrial Intelligence

The question as to whether intelligent life on Earth is unique has fascinated humans for centuries. Is our planet the only one in the universe on which intelligent beings live? Or are there others like Earth? And, if so, how many and where are they?

Origins of the Search for Extraterrestrial Intelligence

Until the middle of the twentieth century, such questions were purely speculative, the subject of science fiction stories. But then, a number of serious astronomers began to ask how such questions could be asked from a scientific standpoint. Improved technology, such as powerful telescopes that can detect radiation in the whole range of the electromagnetic spectrum, made a scientific study of life beyond the Earth a real possibility. And with those technological advances was born a new field of scientific inquiry, the Search for Extraterrestrial Intelligence (SETI).

SETI is actually a subdivision of the more general field of astrobiology. Astrobiology deals with the study of all life forms, intelligent or not, in regions beyond the Earth's atmosphere. Astrobiologists are as interested in one-cell organisms on the surface of Mars as they are in intelligent beings on planets elsewhere in our galaxy.

SETI research has been encouraged not only by new technologies but also by a dawning realization of the immense scope of the universe. Astronomers now estimate

Financial support for research on the Search for Extraterrestrial Intelligence (SETI) has been difficult to obtain because many people do not see any practical benefit to be derived from such research. (Corbis/Roger Ressmeyer)

that the universe contains hundreds of billions of stars. Many of those stars have the potential for hosting at least one Earthlike planet. SETI enthusiasts have guessed that our own Milky Way Galaxy could contain anywhere from one to ten billion Earthlike planets.

How many of those planets might contain life of any kind, let alone intelligent life, is, of course, completely unknown. Scientists still know very little about the origins of life on our own planet. So they are not well equipped to guess about the progress of a comparable event on other planets.

Goals of SETI Research

SETI research tends to focus on two main objectives. One is finding Earthlike planets that have environments in which intelligent life, as we know it, could survive. An important breakthrough in that area occurred in 1996 when planets were found circling stars in our own galaxy. At least one of

these planets was judged to be sufficiently distant from its own star to support Earthlike conditions.

The second focus of SETI research is the search for electromagnetic signals from other planets and the intentional transmission of such signals from Earth into space. The adjective *intentional* is important here because humans have been transmitting signals randomly into space since the invention of radio and television. Many of the signals transmitted for use on our own planet—radio and television programs—have escaped into space. They have the ability to travel very great distances across space where, if intelligent civilizations do exist, they can be intercepted.

A major part of SETI research has been simply to listen for incoming messages from intelligent beings on other planets. The assumption has always been that such beings might try to communicate with Earth by sending out detectable messages, such as the same set of impulses repeated in an identifiable pattern. In order to listen for such messages, astronomers have pointed their radio telescopes to various parts of the sky and listened for radio signals that appear to be anything other than random noises. Thus far, they have failed to detect such signals.

Financial Support for SETI

For those interested in SETI, the search for intelligent beings is a legitimate field of scientific research that has legitimate claims for research dollars. However, not all scientists agree with this view. After all, using radio telescopes to listen for messages from outer space and constructing equipment with which to send messages detracts funds and equipment for other astronomical projects, they say. Someone who truly believes in the objectives of SETI has to make a commitment to spending time on SETI rather than some other scientific project. To date, relatively few astronomers have chosen to do so.

In 1991, the U.S. Congress voted to give $100 million to the National Aeronautics and Space Administration (NASA) for the support of a ten-year project for the search for intelligent life in the universe. Part of that money was to be channeled through the SETI Institute, located near Stanford University. The SETI Institute was created by Dr. Frank Drake, one of the "grand old men" of SETI. Congress's commitment to SETI research lasted only two years, however. Dissatisfied with the project's lack of success, funding was canceled in 1993. The SETI Institute has continued to function on its own, however, largely through the financial support of David Packard and William Hewlett, cofounders of the Hewlett-Packard Company; Gordon Moore, chairman of the Intel Corporation; and Paul Allen, cofounder of Microsoft, Inc.

In addition to the SETI Institute, research on extraterrestrial intelligence is being conducted at four other locations. They are Project Phoenix, in Palo Alto, California; Project Serendip, sponsored by the University of California at Berkeley; Projects Meta and Beta, sponsored by the Planetary Society and Harvard University; and Project Big Ear, sponsored by Ohio State University.

Prospects for government support of SETI research began to improve once more toward the end of the 1990s. Dan Goldin, administrator of NASA, decided to make SETI a priority for NASA-funded research. He announced plans for a new institute of astrobiology within NASA, increased support for NASA's Origins program that deals with the origin and distribution of life in the universe, and plans to construct a telescope to orbit beyond Jupiter and search for Earthlike planets outside the solar system.

References

Achenbach, Joel. "Search for Aliens No Longer on Edge of Science Galaxy." *Sunday Oregonian,* 15 March 1998, G3.

Boyd, Robert S. "Is Anyone Out There?" *Oregonian,* 23 May 1996, B11+.

Bracewell, Ronald N. *Intelligent Life in Outer Space.* San Francisco: W. H. Freeman and Company, 1975.

Davidson, Keay. "Search for Alien Life Forms Expands." *San Francisco Examiner,* 12 April 1998, D1+.

Gallant, Roy A. *Beyond Earth: The Search for Extraterrestrial Life.* New York: Four Winds Press, 1977.

Sagan, Carl. *The Cosmic Connection: An Extraterrestrial Perspective.* New York: Dell Publishing, 1973.

Shklovskii, I. S., and Carl Sagan. *Intelligent Life in the Universe.* San Francisco: Holden-Day, 1966.

Secondhand Smoke

Secondhand smoke is smoke produced by burning tobacco products, such as those in cigarettes, cigars, and pipes, that is inhaled by someone other than the smoker himself or herself. For example, a person seated next to a smoker in an office is exposed to the products of smoking from the second worker. Secondhand smoke is also referred to as *passive smoke,* since it is taken in by a person without any effort on his or her part. By contrast, smoke inhaled by a smoker herself or himself is referred to as *active smoke.*

Smoking as a Health and Social Issue

In recent years, the dispute over the possible health effects of smoking in general has become one of the most acrimonious of all social issues. Public health officials have long suspected that smoking is a major cause of lung cancer, emphysema, and other respiratory problems. Over a period of at least two decades, regulatory agencies have slowly expanded the number of restrictions on the advertisement and use of tobacco products. For example, all tobacco products for sale in the United States now carry some kind of warning label outlining the possible health effects of tobacco on smokers.

During the late 1990s, it became increasingly evident that tobacco companies, contrary to their earlier protestations, have long known about the health effects of tobacco and have misrepresented those dangers to smokers. In some cases, companies appear to have actually modified natural tobacco products in one way or another to increase the addictive properties of those products.

Disclosures of that evidence have led to very large legal actions taken by individual states, cities, and other municipalities against tobacco companies. Those suits have aimed at collecting from those companies some portion of the costs in treating tobacco-related health problems. The federal government has also become involved in an effort to adopt legislation that will punish tobacco companies financially and severely restrict the promotion and sale of their products in the future.

The Dangers of Secondhand Smoke
Scientists have also long been interested in the question as to what risk, if any, is posed by secondhand smoke. That is, suppose that mother and father in a family both smoke. Are children in the family at risk for health problems because of the secondhand smoke they inhale?

Again, many health scientists have believed for some time that they know the answer to that question. They have become convinced that passive smoke poses many of the same problems for nonsmokers that active smoke does for smokers. In 1993, the U.S. Environmental Protection Agency (EPA) released an important report on the health effects of secondhand smoke. The EPA report was based on a survey of thirty earlier studies conducted worldwide on passive smoke, eleven of those in the United States.

In their report, EPA officials estimated that passive smoke is responsible for about 3,000 cases of lung cancer in the United States each year. The report also suggested that infants and young children of smoking parents were also at risk because of secondhand smoke. The report estimated that between 150,000 and 300,000 lower respiratory tract infections each year could be traced to secondhand smoke, about 7,500 to 15,000 of which resulted in hospitalization of the infant or child. The report summa-

rized a number of other possible health effects of passive smoke. For example, it said that between 200,000 and 1,000,000 children with asthma had their condition made worse as the result of exposure to secondhand smoke.

Influence of the EPA Report
Many people think that the EPA report on secondhand smoke was a turning point in the campaign against smoking. After the report was issued, governmental agencies became more aggressive about limiting the use of tobacco products in enclosed areas. In 1994, for example, the U.S. Congress banned smoking in public schools except in certain areas where children are not permitted. Amtrak banned smoking on short- and medium-distance trips and provided separate smoking areas on long-distance trips. Many cities, counties, and states banned smoking in all public places, such as offices and other workplaces. In 1998, the state of California even adopted a ban on smoking in bars, taverns, and gaming clubs. This action was regarded by many as the most severe smoking ban yet to be adopted in the nation.

Other Views of Secondhand Smoke
Tobacco companies had long held a different view about secondhand smoke. They argued that the evidence linking passive smoking with health problems was weak or nonexistent. In this regard, they took essentially the same position that they had held for decades about active smoking. Even as the companies began to admit that some health problems might be associated with active smoking, however, they continued to deny that such an association existed between passive smoke and health problems. They used this argument in speaking out against governmental bans on smoking that began to appear throughout the nation.

A new element appeared in this debate in July 1998, however. A federal district judge in North Carolina, William L. Osteen, ruled that the 1993 EPA report on secondhand smoke was flawed. He decided that the report overstated the health effects of passive smoke and that the EPA had used questionable scientific and regulatory practices in reaching its decision about those effects.

Reactions to Judge Osteen's ruling were mixed. Tobacco companies were expected to use that ruling to fight against additional smoking bans that were being planned for public facilities. It seemed possible that they might even use the ruling to have earlier bans overturned. Other observers felt that the issue of secondhand smoke had already been fought and lost by smokers and tobacco companies. Too many Americans had already become too convinced about the health effect of smoke to change their attitudes about the place of smoking in life in the United States.

The issue of secondhand smoke is likely to take its place among a host of other tobacco-related questions with which Americans will have to deal in the coming decade. The basic problem is that tobacco smoke contains a number of products known or believed to be carcinogenic. Public health officials are concerned, therefore, about the problems that tobacco smoking causes for the nation. On the other hand, smoking is not illegal. No matter how many harmful substances tobacco contains, the federal government has never decided that people should be prevented from using tobacco. How these two issues—public health and the right to pursue a legal hobby—will be resolved has yet to be determined.

References

Davis, Robert, and Wendy Koch. "As Society Turned, Smokers Learned to "Just Live with It." USA Today, 22 July 1998, 1A+.

"Lung Cancer: Risk from Second-hand Smoke." http://pharminfo.com/pubs/msb/sec_smke.html. 2 August 1998.

"Research Shows Second-hand Smoke and High Cholesterol Damage Heart." http://www.pslgroup.com/dg/6fa2.htm. 2 August 1998.

"Secondhand Smoke." http://www.epa.gov/iaq/pubs/etsbro.html. 2 August 1998.

Sick Building Syndrome

See Indoor Air Pollution

Slash-and-Burn

Slash-and-burn is a land use technique employed primarily by nomadic people living in tropical areas. The technique involves the cutting of trees in a limited area, usually no more than an acre in size, and then setting fire to those trees. The resulting fire usually clears all vegetation in the area.

The technique is an effective, if primitive, form of land use technology. It provides nomadic people with a clear space on which to plant crops and assures that the space will be fertile because of ashes left from burning vegetation. The fertility of land produced by slash-and-burn is not persistent, however, and the land can seldom sustain crops for more than one or two years. The tribe then moves on to a new location and repeats the process.

Environmental Consequences of Slash-and-Burn

Traditional slash-and-burn techniques seldom have serious long-term environmental consequences. They have been used for many years and affect only small sections of a forested area. They become a serious environmental hazard only when used on a much larger scale, as when food or logging businesses clear tens or hundreds of thousands of acres for the conduct of their operations.

An example of the problems that can arise occurred in 1997 in Indonesia. A number of agricultural and lumbering companies had been practicing slash-and-burn technologies in remote areas of Borneo and Sumatra since 1983. Their method was to log an area and then spray the area with gasoline. When the gasoline was ignited, all vegetation was burned away, leaving huge sections where new crops could be planted. The annual monsoon rains that arrive in July and August could be counted on to put out the massive fires that were produced.

Slash-and-burn agriculture has been a widely used and generally effective method of farming in many parts of the world for centuries but has become a serious environmental problem recently as the extent to which it has been used has greatly increased. (Corbis/Chris Hellier)

In 1997, however, the monsoon rains did not appear as expected. The year was an El Niño year, in which weather patterns changed dramatically in many parts of the world. The slash-and-burn fires continued to rage out of control throughout the summer of 1997. Huge clouds of smoke spread across Indonesia and into other parts of Southeast Asia. Businesses and schools closed for days at a time, and people were often able to travel out of doors only while wearing masks.

Some experts placed the cost of the incident at tens of billions of dollars in lost economic production, timber, and tourism. They saw the disaster as a "wake-up call" to Indonesia and other Southeast Asian governments about their insensitivity to land use and other environmental practices.

References

"Amazon Interactive" http://www.eduweb.com/amazon.html. 22 May 1999.

"Banning Slash-and-Burn Is Impractical and Impossible, Says Agricultural Research Consortium." http://www.cgiar.org/icraf/pre_rele/press_1.htm. 24 June 1998.

Miller, G. Tyler, Jr. *Environmental Science*. 3d ed. Belmont, CA: Wadsworth Publishing, 1991, 23–25.

Schmetzer, Uli. "A Burning in the Lungs." *Oregonian*, 30 September 1997, A11.

Southwick, Charles H. *Global Ecology in Human Perspective*. New York: Oxford University Press, 137–140.

Sociobiology

Sociobiology is a field of study that attempts to search for a biological basis for all social behavior. The term was used as early as 1946, but it became well known only thirty years later. It then reappeared in connection with research carried out by U.S. entomologist Edward O. Wilson.

Sociobiology in Animals Other than Humans

Wilson's interests in sociobiology grew out of his own research and that of others in insects, primarily ants. He discovered a number of behaviors common among these animals that could not be explained on the basis of traditional evolutionary theory.

An example of a behavior that *can* be explained in terms of evolutionary theory is cooperation among individuals of a species. When organisms work together to obtain food, fight off an enemy, or promote their well-being in some other way, the community as a whole is more likely to survive. And the key to survival, according to evolutionary theory, is for the community's gene pool—not its individual members—to survive.

But other kinds of behavior are more difficult to explain. One such behavior is *altruism,* which refers to an animal's sacrificing itself so that one or more members of its community can survive. For example, a firefighter who risks certain death to rescue a child from a burning building is demonstrating altruism. The firefighter does nothing at all to ensure his or her own survival and, thus, appears to make no contribution whatsoever to the continuation of his or her own genes in the community.

Wilson argued that a trait like altruism could be explained genetically. For example, an ant that exhibits altruism—is willing to sacrifice itself—shares genes with other members of its own community. The individual ant may die and thus lose any chance of transmitting its own genes to the next generation. But by its own sacrifice, it may increase the likelihood that other members of its population—those with whom it shares genes—*will* be able to survive and pass on their genes.

One of the examples Wilson used in developing his theory of sociobiology was a kind of Malaysian ant that literally "blows itself up" when invaded by an enemy species. The ant ruptures its own glands, releasing a toxic material that kills invaders and acts as a chemical warning signal to other members of its community.

Behavior and Genes

Wilson argued that social behaviors, such as altruism, were determined by the genetic makeup of an organism, just as were physical characteristics. That is, we know that an organism's hair color, eye color, body shape, and dozens of other traits are determined by its genes. Why is it not possible, Wilson asked, that social behaviors, such as aggressiveness, curiosity, and compassion, might not also be coded for by genes?

In other words, biologists might begin to search for a gene or set of genes that was responsible for the appearance of altruism in an organism. Some individuals might have those genes, just as some individuals have genes for red hair, and others might not.

Wilson outlined his ideas about sociobiology in a book by that name, published in 1975. In the last chapter of that book, he speculated on the applications of sociobiological principles to human beings. He outlined a future in which all of our knowledge of psychology and sociology could be expressed in terms of genes and, ultimately, nerve cells in the human brain. He acknowledged that "when we have progressed enough to explain ourselves in these mechanistic terms . . . the results might be hard to accept" (Wilson 1975, 575).

Three years later, Wilson expanded on these thoughts in another book, *On Human Nature.* In that book, Wilson examined in detail what might be implied by the concept that human behaviors are controlled by a person's genetic makeup.

Objections to Wilson's Theories

Wilson's ideas met with very strong and sometimes violent reaction. At a 1978 meeting of the American Association for the Advancement of Science, for example, protesters disrupted a meeting at which Wilson spoke and poured a pitcher of water over his head.

Critics have raised relatively little complaints about Wilson's work, and that of others, on sociobiology in lower animals. His analysis of ant behavior, for example, has been the focus of a considerable body of additional and related work on the connection between social behaviors and evolutionary biology. It is Wilson's attempts to extend his theories to human beings that have been extensively criticized. The fundamental issue involved in this debate is one that has troubled biologists for centuries. It is fundamentally the question of "nature versus nurture." That is, do humans behave the way they do because of genetic characteristics with which they are born ("nature") or because of the way they are raised and education ("nurture")?

Most people would suspect that the answer to this question is "a little of each." That is, we can hardly deny that some human behaviors are clearly determined or at least strongly influenced by genetic factors, such as the tendency to be right- or left-handed. Other behaviors may have both a genetic and a learning basis. Still others may be completely learned, without any genetic basis at all. Sociobiologists would probably be inclined to say that far more human behaviors have a genetic basis than scholars have ever discovered or even imagined.

The "nature versus nurture" argument is hardly a neutral topic of purely academic interest. The argument is closely tied to another issue that has been a topic of great dispute in biology: biological determinism. Throughout history, some people have argued that humans behave the way they do because of the "nature" component of the "nature versus nurture" interaction. That is, many of the features of human society exist because of a fundamental biological bias in that direction.

Perhaps the most obvious illustration of this view is the position prominent in human society for so long that men are somehow genetically superior to women. That view has long been used to justify the subjugation of women to men in most societies. Men rule because they are biologically predetermined to dominant women, according to this theory.

Many of Wilson's critics feared that his theory could be used to justify society as it existed. That is, what if it could be discovered that men *did* carry a gene or set of genes that predetermined them to greater aggressiveness, dominance, or some other trait not possessed by women? How would a discovery of that kind affect social policy?

Another aspect of that argument is that Wilson's theory of sociobiology is an evolutionary theory. That is, the genes that exist within the human gene pool today are there because they are better adapted to the environment than are genes that no longer exist. That would suggest that the human society that exists is also better adapted for survival than other conceivable societies that might have existed in the past or that might be competing with existing societies. In this regard, sociobiology might be perceived as a very conservative theory, one that tends to justify the status quo in human society.

Finally, critics also pointed out the failure to find any evidence for any gene or any set of genes that could be correlated with the existence of any human behavior. In fact, one might argue that there might not even be a scientific method by which the existence of such genes could ever be found.

In spite of these criticisms, Wilson's ideas have found a very substantial place in evolutionary research today, particularly in research on lower organisms. The furor over its applications to human societies has, to a large extent, calmed down, although the philosophical issues it aroused are not likely to be disappear from the debate as to what it is that makes people "human."

See also: Biological Determinism; Nature versus Nurture

References

Ann Arbor Science for the People Editorial Collective. *Biology as a Social Weapon.* Minneapolis: Burgess Publishing, 1977, 131–149.

Fisher, Arthur. " Sociobiology: Science or Ideology?" *Society,* July/August 1992, 67–79.

Holden, Constance. "The Genetics of Personality." *Science,* 7 August 1987, 598–601.

Kitcher, P. "Precis of *Vaulting Ambition: Sociobiology and the Quest for Human Nature. Behavioral and Brain Sciences*, 1987, 61–71. This article is followed by critics supporting and opposing Kitcher's views.

Lewontin, Richard C. "Sociobiology as an Adaptationist Program." *Behavioral Science*, 1979, 5–14.

Maxwell, Mary, ed. *The Sociobiological Imagination*. Albany: State University of New York Press, 1991.

Wilson, Edward O. *Sociobiology: The New Synthesis*. Cambridge, MA: Belknap Press, 1975.

———. *On Human Nature*. Cambridge: Harvard University Press, 1978.

———. "Science and Ideology." *Academic Questions*, Summer 1995. Also at http://lrainc.com/swtaboo/taboos/wilson01.html. 30 June 1998.

Wilson, Edward O., and Charles J. Lumsden. *Promethean Fire: Reflections on the Origin of the Mind*. Cambridge: Harvard University Press, 1983.

Space Station

The space station is a structure designed to orbit the Earth and provide a setting in which near-zero-gravity experiments can be performed. The station is generally regarded as the next step forward in human exploration of space. It is being financed and built by a sixteen-member consortium led by the United States and Russia. It is designed to be fully operational by 2002.

The space station will consist of a number of components, including (1) a living space that will house a crew of six, (2) seven laboratories for research experiments, (3) a power plant, (4) an emergency escape system, (5) multiple solar wings, and (6) a robot-operated crane for external use. The station was scheduled to be assembled in a series of seventy-three rocket launches carried out over a period of fifty-five months. The first launch was scheduled to have occurred in 1997 but was delayed when Russia was unable to complete the space tug that makes up the first component of the station.

Purposes of the Space Station

The U.S. National Aeronautics and Space Administration (NASA) has listed nine major objectives of the space station:

1. To create a permanent orbiting science institute in space capable of performing long-duration research in the materials and life sciences areas in a nearly gravity-free environment.
2. To conduct medical research in space.
3. To develop new materials and processes in collaboration with industry.
4. To accelerate breakthroughs in technology and engineering that will have immediate, practical applications for life on Earth and will create jobs and economic opportunities today and in the decades to come.
5. To maintain U.S. leadership in space and in global competitiveness and to serve as a driving force for emerging technologies.
6. To forge new partnerships with the nations of the world.
7. To inspire our children, foster the next generation of scientists, engineers, and entrepreneurs, and satisfy humanity's ancient need to explore and achieve.
8. To invest for today and tomorrow. Every dollar spent on space programs returns at least $2 in direct and indirect benefits.
9. To sustain and strengthen the United States' strongest export sector—aerospace technology—which in 1995 exceeded $33 billion. (NASA 1998)

This list of goals was written by NASA at least partly in an effort to obtain funding and support for the space station. A reality of the program is that it is, to a large extent, an exercise in basic research. As objective 7 notes, humans have long been driven by

Critics ask whether the benefits to be gained from the International Space Station are worth the enormous financial investment needed to complete the project. (Courtesy of NASA)

the desire to know more about the natural world. As scientists discover more and more about planet Earth, outer space looms more and more as the next great unknown calling for discovery and exploration. The next round of research in space, however, is likely to be enormously expensive, with projects like the space station costing tens of billions of dollars to build and operate. It is at least partly for this reason that the space station is a multinational effort, involving the cooperative financial and technical support of sixteen nations.

Politics of the Space Station

The history of the U.S. space program has been permeated with political issues since its creation in 1961 by President John F. Kennedy. In announcing the Apollo Project, Kennedy was driven as much by domestic and international political considerations as he was by the desire to promote ongoing space research. Throughout its history, NASA has had to argue with the U.S. Congress, a variety of presidential administrations, and its critics in the general public about the expenditure of billions of tax dollars on basic research, research that is designed primarily to find out more about space and only secondarily to provide improvements in human life. NASA has been more or less successful in arguing this point, partly to the extent that its programs have been able to capture the enthusiasm of the U.S. public. At all points, however, it has also made promises about improvements the space program might bring to the everyday lives of ordinary people, even when those benefits occurred as a side effect of the main program itself.

In planning for the space station, the agency has had to continue this line of argument. While the space station may very well "forge new partnerships," "inspire our children," and "sustain and strengthen the United States' strongest export sector," these goals may not be the primary reason for building the space station, nor is the space station necessarily the only or best way to achieve these objectives.

The debate over the space station has also been complicated by President Bill Clinton's 1993 decision to invite Russia to participate in the construction and operation of the station. For most adult Americans, Russia (in the form of the Soviet Union) had traditionally been the United States' competitor in space, not its partner. Clinton's invitation created a new research and partnership scenario with which many Americans were uncomfortable.

Some of the strongest warnings about U.S.-Russia cooperation on the space station came from Representative F. James Sensenbrenner Jr. (R-Wisc.), chairman of the House Science Committee. Sensenbrenner repeatedly warned that the United States could not trust Russia to follow through on the commitments it had made to the space station. In fact, Sensenbrenner's view very nearly won the day. In 1993, Representatives Tim Roemer (D-Ind.) and Dick Zimmer (R-N.J.) joined together to offer an amendment to the NASA authorization bill calling for a cancellation of the space station. That amendment failed by a single vote, 216 to 215. Since that time, other amendments have been proposed to various bills calling for an end to the space station program. All have failed, and usually by much larger margins.

Technical Problems

The space station has been beleaguered by technical as well as political problems. At the end of 1996, for example, engineers at the Boeing Company discovered that connecting nodes joining sections of the station had been improperly designed. The nodes had to be redesigned and rebuilt at an additional cost of $100 million and a delay of six months in the overall project.

Questions continued to arise about the ability of other partners in the space station consortium to complete their portion of the necessary work. For example, the Russian Space Agency (RSA) announced in 1997 that it would not be able to complete on time the FGB (Functional Cargo Block) tug that constituted one of the first elements of the station. The Russian government, suffering from severe financial problems, had provided only one-tenth of the money promised and needed for completion of the module. As a result, launch of the station's first unit, the Russian-built tug, was delayed a year.

Finally, in November 1998, the FGB tug was launched on a Russian Proton rocket. A month later, the United States Space Shuttle delivered its first part of the station, a unit called Node 1. The station finally began to take shape, then, when the FGB tug and Node 1 were joined to each other.

Latest plans called for delivery of the first living quarters, constructed entirely by the RSA, in mid-1999. After two more assembly flights by crews from the Space Shuttle, the first crew to occupy the station was scheduled to be launched in a Russian Soyuz capsule sometime in 2000.

See also: Research, Basic and Applied

References

Cooper, Mary H.. "Expanding NASA's Horizons." *Oregonian*, 14 May 1997, A14.

Davis, Brett. "Russia's Delay in Space Station Project Worries NASA." *Oregonian*, 8 March 1997, A15.

Dunn, Marcia. "Space Station Promise Is Hard to Keep." *San Francisco Chronicle*, 14 September 1996, A3.

"International Space Station: It's about Life on Earth . . . and Beyond." http://spaceflight.nasa.gov/station/reference/factbook/beyond.html. 22 May 1999.

"Russia Space Station Proposal Encounters Heavy Opposition." http://www.aip.org/enews/fyi/1994/fyi94.060.txt. 27 June 1998.

"Space Legislation: House Votes." http://www.ari.net/nss/voters_guide/page8.html. 26 June 1998.

"Why a Space Station?" http://spaceflight.nasa.gov/station/reference/factbook/why.html. 22 May 1999.

Sterilization, Human

Human sterilization refers to any procedure by which a man or woman is made incapable of having children. A common method of sterilization in women is tubal ligation. In this procedure, the tubes that carry eggs to the uterus are sealed off. In men, a common method of sterilization is a vasectomy. In a vasectomy, the vas deferens is cut. The vas deferens is the tube through which sperm is conducted into the semen. Both methods of sterilization are highly effective, but not perfect, at preventing fertilization during sexual intercourse.

Sterilization as a Birth Control Method

Sterilization has become a popular method of contraception largely because of its effectiveness. Men and women generally do not have to worry about other methods of birth control if one partner has had a sterilization procedure. Sterilization does have its disadvantages as a method of birth control, however. For one thing, the procedure is unsuccessful in a very small percentage of cases. Also, some individuals later change their minds about having been sterilized and wish to have the procedure reversed. This operation can be attempted in men and is sometimes successful.

Sterilization is now used by about one-third of all women in the United States and about 10 percent of all men in the United States as their means of birth control. The procedure is used widely in other parts of the world also. According to one estimate, 70 million men and women in China have been sterilized as a method of birth control.

Social and Ethical Issues

Sterilization has been used widely in modern times in connection with programs of eugenics. Eugenics is a science developed by Sir Francis Galton in the 1880s. Galton was strongly influenced by Darwin's theory of natural selection and believed that the human race had the ability to control its future destiny. Specifically, he thought that society should exert control over the reproductive practices of its members, encouraging sexual intercourse between "better" individuals and discouraging (or preventing) intercourse between "less desirable" individuals.

One of the most popular methods of discouraging reproduction among the "less desirable" was sterilization. Men and women who were judged to be socially undesirable or burdens on society were forcefully sterilized. Those most likely to be included in this category of "undesirables" were the mentally retarded, alcoholics, criminals, sexual deviates, the rebellious, and otherwise unsatisfactory citizens of the community.

In the late 1990s, for example, Sweden admitted that it had sterilized about 62,000 men and women between 1937 and 1974. The goal of this program was to improve the overall gene pool of the Swedes by weeding out those with low intelligence, those who were promiscuous or rebellious, or those of mixed blood. Other nations known to have pursued similar policies are Australia, Belgium, Germany, Norway, and Switzerland.

One of the largest and most aggressive programs of sterilization, however, was carried out in the United States. At one time, twenty-five states had laws allowing the mandatory sterilization of "socially inadequate" individuals. The most common people affected by this law were poor white men and women who were accused of immorality, low intelligence, or alcoholism. The laws in general were modeled after a statute originally proposed by Harry Laughlin, a U.S. citizen and adherent of the German Nazi Party. Laughlin's model statute called for the mandatory sterilization of epileptics, syphilitics, the deformed, retarded, blind, deaf, handicapped, alcoholics, unwed mothers, runaways, illegitimate children, dwarfs, habitual criminals, and "flagrantly immoral" individuals.

The use of sterilization to improve racial qualities has long had a certain appeal to the general public. In 1937, for example, a poll conducted by *Fortune* magazine found that more than 50 percent of those asked favored Laughlin-like laws. The U.S. Supreme Court had also given its approval for such legislation in two separate cases, one in 1916 and one in 1927.

Sterilization Today

The use of sterilization by the Nazis in Germany in the 1930s illuminated a whole new aspect of the practice. People around the world had a glimpse of the very worst consequences that could result from the adoption of a national sterilization policy. Since that time, sterilization has become a much less widely accepted method for dealing with racial improvements, although it has not totally disappeared from some countries.

The question that still arises today is whether certain individuals who may present difficulties for society should ever be sterilized without their consent. For example, should a thirty-year-old woman with the mental capacity of a four-year-old be sterilized without her consent? On the one hand, the procedure might prevent the woman from having children who might (or might *not*) themselves be mentally retarded. It might save both individual families and society the cost and heartache of caring for another generation of mentally retarded people.

On the other hand, many people with diminished mental capacities get along very well in society. They often have children who have at least average mental capacities and, often, much higher. The issue may be that society has to learn to incorporate a greater variety of people with different traits rather than wiping out those with whom it finds most difficult to deal.

Finally, the United States in particular has long set great store on individual freedom and the right of self-determination. How does one even provide that kind of freedom to an individual who may be unable to understand intellectually what the meaning of sterilization is and what choices he or she may have in the matter?

References

"Blessed Are the Barren." *Brown Daily Herald,* 8 September 1997; also at http://www.theherald.org/issues/090897/kaza.f.html. 6 July 1998.

Burritt, Chris. "Criticism for Pusher of Sterilization Pills." *San Francisco Chronicle,* 26 June 1998, A8.

"Eugenics and Sterilization Laws in the U.S." http://www.oakland.edu/~crouch/102/eugenics.txt. 22 May 1999.

Lush, Tamara. "Decision for Sterilization Becomes 12-year Fight." *USA Today,* 24 April 1998, 10A.

Steroids

As commonly used, the term *steroid* refers to a class of chemical compounds related to the sex hormone testosterone. Over the past half century, steroids have become popular among athletes in a number of sports at levels ranging from high school to the professional because steroids have been perceived as being capable of improving body traits that contribute to success in those sports. Many critics have pointed out possible health effects from the use of steroids, however, and an ongoing debate has developed over the fairness of allowing their use by competitors. Currently, most amateur and professional sports organizations ban the use of steroids in their competitions.

Physiological Effects of Steroids

Steroid is commonly taken to refer to a class of synthetic compounds known collectively as the anabolic androgenic steroids. The term *anabolic* refers to those biochemical processes that occur in cells that result in the building of new body parts, especially protein molecules. The term *androgenic* refers to properties of maleness, such as increased body mass, the appearance of facial and body hair, a deeper voice, and an increase in aggressive behavior. The term *steroid* refers to a particular chemical structure found in molecules of many important chemicals that occur naturally in the human body, such as the male sex hormone testosterone.

Steroids have long been popular among weight lifters and bodybuilders because of the compounds' ability to increase body mass, especially muscle mass. It is not unusual for a person to be able to add a dozen pounds or more in a matter of weeks and to increase his or her weight-lifting ability by twenty kilograms (fifty pounds) simply by taking steroids and maintaining a normal exercise routine.

The success of steroid use among weight lifters and bodybuilders has prompted athletes in other sports to experiment with the drugs. The expectation has been that steroids will help one to run or swim faster, jump higher or farther, or develop greater overall strength and endurance. Some studies have shown that a relatively constant 2–3 percent of high school male athletes use steroids, while a much higher fraction of college and professional athletes use the drugs. A 1997 study showed that steroid use had also spread to females, with an estimated 175,000 high school girls having admitted to steroid use at least once.

Arguments against the Use of Steroids

Most health scientists have long argued against the use of steroids by athletes. One reason for this concern has been based on physiology and psychology. The connection between steroid use and athletic performance is not as clear as outlined in the preceding paragraph, they point out. Steroids may increase muscle mass, but they do not necessarily improve speed, reaction time, or other qualities needed in most sports. Furthermore, extended use of steroids may cause a number of health problems, including stunted growth in adolescents, headaches, bone pain, nausea, changes in bowel and urinary functions,

and, on a long-term basis, damage to the cardiovascular and reproductive systems and to the liver and kidneys.

In addition, use of steroids has long been thought to cause severe mood swings. At one extreme, these mood swings may result in violent and uncontrollable behavior called "'roid rage." Some attorneys have presented a "'roid rage" defense for clients accused of aggravated assault, rape, or other violent crimes, claiming that their clients' use of steroids made them incapable of controlling their behavior.

Athletic officials have generally argued, in addition, that the use of steroids gives an athlete an unfair advantage in competition. The practice is similar, they say, to giving a runner or a swimmer a two-meter (five yards) head start or a weight lifter a ten-kilogram (twenty-five pound) reduction in a lifting competition. This view is somewhat at odds with the medical argument presented above that steroids often do *not* provide such advantages. It has been influential, nonetheless, among sports governing bodies, and most sports organizations from high school athletic leagues, through the National Collegiate Athletic Association, to professional associations have now banned the use of steroids in competition.

Arguments for the Use of Steroids

The debate over physiological and psychological effects of steroid use has at times been lacking in sound scientific data. In July 1996, however, researchers at Charles R. Drew University in Los Angeles reported the first comprehensive research showing that steroid use does increase muscle mass and strength. Men in the Drew study gained an average of 6 kilograms (thirteen pounds) of weight, virtually all of it muscle, and increased their ability to bench press by twenty-two kilograms (forty-eight pounds) after a ten-week program of steroid injections. The same study also found no evidence of a "'roid rage" among those who participated in the study. The Drew study did provide, therefore, some scientific support for some arguments that steroids can enhance some kinds of athletic performance.

A second factor involved in the debate over steroid use is a possible disparity that might exist between public remarks about steroid use made by coaches, managers, athletic trainers, and other team officials and the actual behavior of such individuals. After all, winning is an important goal for many competitors and their trainers, and the advantage provided by steroid use—even if minimal—could provide the margin of success and victory over failure and defeat. It might not be impossible to suppose that some athletic officials would actually provide steroids for their players or, at the very least, look the other way when those players choose to use steroids.

Some athletic officials go even further. They point to the widespread use of drugs among athletes for other purposes, an injection to relieve the pain of a damaged knee in a valuable quarterback during a crucial game, for example. They ask how the carefully controlled use of steroids in a program of performance enhancement is any different from this practice.

See also: Creatine

References
Catlin, Don, et al. "Assessing the Threat of Anabolic Ssteroids." *The Physician and Sportsmedicine,* August 1993, 37.

Duda, Marty. "Do Anabolic Steroids Pose an Ethical Dilemma for US Physicians?" *The Physician and Sportsmedicine,* November 1986, 173–175.

Fost, Norman. "Drugs Should Not Be Banned from Sports." In Julie S. Bach, *Drug Abuse: Opposing Viewpoints.* St. Paul, MN: Greenhaven Press, 1988, 137–144.

Haupt, H. A., and G. D Rovere. "Anabolic Steroids: A Review of the Literature." *American Journal of Sports Medicine,* November-December 1984, 469–484.

Kinney, David. "Steroids Are No Longer Just for Boys." *San Francisco Chronicle,* 15 December 1997, A9.

Millar, Anthony P. "Anabolic Steroids: Should They Be Used to Improve Athletic Performance?" *Current Therapeutics,* October 1985, 17–21. This article was reprinted in Julie S. Bach. *Drug Abuse: Opposing Viewpoints.* St. Paul, MN: Greenhaven Press, 1988, 157–161.

Steroids: Just the Facts. http://www.tcada.state.tx.us/research/facts/steroids.html. 9 January 1998.

"Steroids Really Do Build up Muscles, Scientists Say." *San Francisco Chronicle,* 4 July 1996, A3.

Yesalis, Charles E., ed. *Anabolic Steroids in Sport and Exercise.* Windsor, Ont.: Human Kinetics Publishers, 1993.

Stream Channelization

Stream channelization is the process by which the course of a river or stream is straightened, widened, and/or deepened in order to reduce flooding or soil erosion or to drain wetlands in upper regions of the river or stream. Stream channelization was practiced for many years by the U.S. Army Corps of Engineers as a highly effective way of dealing with these problems.

Stream channelization has its undesirable effects also. Downstream areas may be made more subject to flooding, sedimentation, and bank erosion when stream paths are changed. Habitats for fish, birds, and other animals may be severely disrupted by channelization. Finally, stream channelization can have dramatic aesthetic effects, changing an area from a beautiful natural landscape to a bare, lifeless, human-made canal.

References
Wagner, Richard H. *Environment and Man.* 3d ed. New York: W. W. Norton, 1978, 80–81.

White, Kim. "Restoration of Channelized Streams to Enhance Fish Habitat." http://www.ies.wisc.edu/research/ies900/kimchannelization.htm. 20 June 1998.

Superconducting Super Collider

See Particle Accelerators

Superfund

Superfund is the nickname for the Comprehensive Environmental Response, Com-

Superfund legislation was an exciting effort on the part of farsighted legislators to solve some of the nation's most serious hazardous waste problems, but it has become bogged down in an almost endless series of political disputes. (Corbis/Galen Rowell)

pensation, and Liability Act (CERCLA). CERCLA was enacted by the U.S. Congress in 1980 as part of a long-term, comprehensive effort to deal with the nation's hazardous waste problems. An important factor in making this effort possible—and probably inevitable—was the Love Canal incident of 1976. In that case, the residents of a suburb of Niagara Falls, New York, discovered that their community had been built on top of a hazardous waste site and that an epidemic of health problems had begun to occur within the community, presumably because of that fact.

The general principle behind Superfund was that polluters should pay for the pollution they cause. That is, those who have dumped hazardous wastes into landfills should be responsible for cleaning up those landfills. The original CERCLA bill also created a trust fund—the Superfund—to pay for the cleanup of "orphan sites," that is, sites where the original users could no longer be identified, had gone out of business, or were financially incapable of paying for cleanup activities. Money for Superfund was to come from taxes on certain chemical and petrochemical industries and from general tax revenues. Individual states were also required to contribute to the cost of cleanups.

In 1986, CERCLA was amended and modified by the Congress. The new act, the Superfund Amendments and Reauthorization Act (SARA), extended most provisions of CERCLA to 1995.

Reauthorization of Superfund

As the prescribed completion date for CERCLA and SARA approached (1995), the massive failure of those acts became obvious. Less than 20 percent of the nation's most seriously polluted waste dumps had been cleaned up at a cost of more than $25 billion. Cleanup efforts had cost an average of $20 to $30 million per site, with anywhere from 40 to 80 percent of that cost going to consultants, attorneys, and others not directly involved in the removal of hazardous wastes. In frustration over this state of affairs, Representative Michael G. Oxley (R-Ohio) had called Superfund "a scholarship program for lawyers' kids" (Smargon 1997).

Attempts to adopt Superfund reauthorization legislation in the 104th and 105th Congresses were thwarted by the complexity of the issues involved as well as by the relatively wide divisions between Democrats and Republicans and between environmentalists and business on a number of crucial issues. Three issues appeared to be the most controversial: liability provisions, state roles, and treatment criteria.

Liability provisions: Probably the most contentious point in the Superfund reauthorization debate was the question as to who should pay for hazardous waste site cleanups. According to the original CERCLA and SARA bills, the cost of cleanup was to be divided among all "potentially responsible parties" (PRPs). That is, anyone who had any part in the development of a hazardous waste site had to pay for its cleanup.

The list of PRPs might include the company that built and/or operated the landfill or dump site, the industry that generated the wastes, and the firm that transported wastes to the dump site. The law ensnared everyone involved, even if they were engaged in activities that were legal at the time they occurred. CERCLA and SARA involved the principle of joint-and-several liability, in which the cost of cleanup was to be divided among all PRPs in a ratio equal to that of their original contribution to the development of the hazardous waste site.

It had been the concept of joint-and-several liability that had led to much of the legal and administrative costs of Superfund. At almost every site, the U.S. Environmental Protection Agency (EPA) had to decide who the PRPs were and what their relative contributions had been to development of the site. That information was needed to determine the relative share of cleanup costs among all PRPs. Naturally, such decisions involved enormous studies of historical records and prolonged disputes over EPA's decisions.

The major bills on Superfund reform in the 104th and 105th Congresses (S1285, S8, and HR2500) all called originally for dramatic changes in the joint-and-several liability aspect of cleanups, excusing PRPs whose participation preceded the enactment of CERCLA or SARA or whose activities had been legal at the time. In an attempt to get approval for these bills, those provisions were rewritten and modified a number of times, reducing their changes in the original act.

State roles: In keeping with the general political philosophy of the time, Republicans in the 104th and 105th Congresses also attempted to shift more of the decision-making procedures for site cleanups to the states. In fact, the EPA was expressly forbidden in some bills from overturning state decisions about site cleanups. The three original bills also reflected this trend by providing for significant decreases in Superfund funding in the years up to and following 2000. Congressional Democrats, however, saw this change as an unprecedented prohibition on the prerogatives of the federal government.

Treatment criteria: Finally, Republicans also sought to reduce somewhat the standards for cleanup at a hazardous waste site. Instead of demanding removal of *all* hazardous wastes, their reauthorization legislation called for a more limited and highly specific standard: that any wastes remaining cause no more than one additional case of cancer in a population of 10,000 to 1,000,000 people exposed to that waste in a lifetime. Democrats responded that such a highly specific standard did not take into account the greater sensitivity to hazardous wastes of certain groups of people, such as children, the elderly, and those with certain types of health problems.

As the debate over Superfund reauthorization continued, Congress provided annual appropriations for continuation of cleanup programs. By early 1998, however, Republicans announced that those appropriations would not be renewed unless a new bill were adopted by 15 May 1998.

As the debate over Superfund reauthorization continued, Congress continued to provide annual appropriations for continuation of cleanup programs. By early 1999, legislative action on this issue had still not been taken. Threats to withhold funding for the program in 1998 were put on hold as the U.S. Congress became embroiled in the Clinton impeachment controversy. Legislators had still not found common ground on which a reauthorization bill could be passed.

References

Henderson, Rick. "Spinal Tap: Checking the Backbone of the New Congress." *Reason,* April 1997, 5–7.

Miller, G. Tyler, Jr. *Living in the Environment.* 4th ed. Belmont, CA: Wadsworth Publishing, 1985, E77–78.

Pickering, Kevin T., and Lewis A. Owen. *An Introduction to Global Environmental Issues.* London: Routledge, 1994, 269–270.

Raber, Linda. "Democrats Blast Revised Superfund." *Chemical and Engineering News,* 15 September 1997, 22–23.

Smargon, Adam J. "Superfund and Retroactive Liability: Is It Really Fair?" http://www.afin.org/~recycler/politics. html. 12 September 1997.

U.S. Senate Committee on Environment and Public Works. *Superfund Cleanup Acceleration Act: Hearing before the Subcommittee on Superfund, Waste Control, and Risk Assessment.* 105th Cong., 1st sess., 5 March 1997.

Additional information on Superfund is available at http://www.epa.gov/superfund/.

Three Mile Island

Three Mile Island is the name of a nuclear power plant operated in Middletown, Pennsylvania, near Harrisburg, by General Utilities Corporation. On 28 March 1979, the most serious accident to have taken place in a U.S. nuclear power plant occurred. Some experts believe that the plant was within an hour of a complete meltdown, perhaps of the kind experienced at Chernobyl nearly a decade later.

Sequence of Events

The accident at the Three Mile Island plant was very complex. Analysts eventually identified about forty separate events leading to the final disaster. The accident was triggered during a routine procedure, a change in a water purifier used in the cooling system.

Although this procedure is a relatively simple and standard process, an error was made that allowed air to escape into the cooling system. Pumps designed to correct for such an error did not operate properly, however, because they were being serviced. Repair tags on the pumps hid warning lights that would have notified operators that the pumps were not responding properly. As a result, operators misinterpreted events that occurred next, and the actions they took made the problem worse rather than better.

Eventually, a relief valve opened to allow superheated steam in the cooling system to escape. As it did so, part of the reactor core was exposed. It rapidly became overheated and began to melt. Nearly three-quarters of the core was damaged and about half melted and collected at the bottom of the reactor. At the same time, radioactive steam escaped from the top of the reactor. The steam spread out over the surrounding area and was dissipated into the atmosphere.

Response to the Disaster

Initial response to the disaster by plant operators and government agencies was designed to assure the general public. Announcements told residents that an accident had occurred, but that no one was at risk. At virtually the same time, however, about 50,000 people were evacuated from their homes within a ten-mile radius of the plant. Another 130,000 people were told to stay inside their homes.

Within weeks of the Three Mile Island accident, attorneys had filed a class action suit on behalf of all businesses and residents within a twenty-five-mile radius of the plant. They named as defendants in the suit the Metropolitan Edison Company, a subsidiary of General Public Utilities. In addition, more than 2,000 individual suits were filed by persons claiming that their health had been damaged by exposure to the Three Mile Island radiation. To simplify matters, the Pennsylvania District Court consolidated these suits into ten discrete cases. All of these cases worked their way through state and federal courts before finally being decided in June 1996 by Sylvia Rambo, chief judge for the Middle District of Pennsylvania. Judge Rambo ruled that the defendants had not provided sufficient

While acknowledging the seriousness of disasters like the one that struck the Three Mile Island nuclear plant, proponents of nuclear power ask how many human lives nuclear power has *saved* by reducing the amount of air pollution caused by fossil fueled plants. (Corbis/George Lepp)

evidence that their health had been damaged by the Three Mile Island accident. She denied all claims against Metropolitan Edison.

On a larger scale, the Three Mile accident had a profound effect on the development of nuclear power production in the United States. The general public appears to have developed both a fear of the nuclear power production process and a distrust of the agencies responsible for production and control of nuclear power. The collapse of interest in nuclear power production within the electrical industry can be traced to a large extent to this specific event.

See also: Chernobyl

References

Freiman, Fran Locher and Neil Schlager. *Failed Technology*. Detroit: Gale Research, 1994.

"Inside Three Mile Island." http://www.wowpage.com/tmi/main.html. 26 June 1998.

Moss, Thomas H., and David L. Sills, eds. *The Three Mile Island Nuclear Accident: Lessons and Implications*. New York: New York Academy of Sciences, 1981.

President's Commission on the Accident at Three Mile Island. *Report of the President's Commission on the Accident at Three Mile Island*. Washington, DC: Government Printing Office, 1979.

"Three Mile Island: The Judge's Ruling." http://www.pbs.org/wgbh/pages/frontline/shows/reaction/readings/tmi.html. 22 June 1998.

"The Three Mile Island (THI-2) Recovery and Decontamination Collection." http://www.libraries.psu.edu/crsweb/tmi/help.htm. 26 June 1998.

Wood, Sandra, and S. Schultz. *Three Mile Island: A Selectively Annotated Bibliography*. New York: Greenwood Press, 1988.

Tris

Tris is the abbreviated name of the chemical tris(2,3-dibromopropyl) phosphate. The chemical was first prepared in the early 1950s and was found to be a highly effective flame retardant. Fabrics soaked in tris do not catch fire when exposed to a flame, but smolder or char. The product is sold commercially under a number of brand names, including Apex, Flammex, Firemaster, and Fyrol.

One of the products in which tris was once widely used was children's sleepwear. As of the late 1970s, some authorities had estimated that more than 45 million children in the United States slept in clothing that had been treated with tris. The use of a flame retardant was and is, of course, highly desirable in all kinds of clothing, especially children's sleepwear. If the child should be exposed to a fire, his or her clothing will not catch fire, providing a level of safety from harm for the child.

During the early 1970s, however, a number of research studies showed that tris was a mutagen, a substance that causes changes in chromosomal material. Some scientists suspected that the compound might also be a carcinogen.

In 1976, the Environmental Defense Fund (EDF) asked the Consumer Product Safety Commission (CPSC) to require warning labels on all children's sleepwear treated with tris. The EDF asked for a statement that the clothing should not be worn until it had been washed at least three times. The reason for this warning is that washing tends to remove tris from the product. The CPSC declined to issue the warning.

A year later, the EDF appeared before the CPSC again, this time asking for a ban in the United States on the sale of sleepwear treated with tris. This time, the commission agreed with EDF and issued the requested prohibition. It has not been legal to sell tris-treated sleepwear in the United States since that date.

Clothing manufacturers found themselves with a large stock of tris-treated sleepwear that they could no longer sell in this country. Instead, they decided to ship the sleepwear to other nations, where a similar ban had not yet been adopted. For about a year after the U.S. ban, tris-treated sleepwear continued to be sold in countries around the world.

Some critics have questioned the practice of selling in other countries products that have been banned in the United States. If those products are not safe for Americans to use, they say, how can we justify their sale to people in other nations? Is the need to make a profit so powerful that companies can put the lives of people around the world at risk?

Companies accused of this practice argue that U.S. health standards are often unreasonably high. The fact that a product is banned by the CPSC, the Food and Drug Administration, the Environmental Protection Agency, or some other governmental body does not necessarily mean it is a serious threat to human health, they say. Regulatory agencies are too likely to go "overboard," they argue, in imposing unreasonably severe standards on products.

See also: Hazardous Wastes, International Dumping

References

"EDF Petitions CPSC to Ban Carcinogen in Children's Sleepwear." http://www.edf.org/pubs/EDF-Letter/1977/Jan/a_cpsc.html. 22 July 1998.

"Suit Seeks Protection for Babies." http://www.edf.org/pubs/EDF%2Dletter/1976/May/I%5Fbabysuit.html. 22 May 1999.

Tuskegee Syphilis Study

See Human Experimentation

Violence

See Criminality and Heredity

Vitamins and Minerals

Vitamins are organic compounds that are essential in very small amounts for the normal growth, development, and good health of plants and animals. Minerals are inorganic substances that have the same function in plants and animals.

The discovery of the biochemical role of vitamins and minerals over the past century has led to a situation in which many diseases can now be prevented or cured by dietary means. For example, scurvy was once a common disorder characterized by anemia, muscle weakness, bleeding gums, and loosening of teeth. We now know that scurvy is caused by a lack of vitamin C in a person's diet. Vitamin C occurs naturally in citrus fruits, such as oranges, lemons, and limes, and in green leafy vegetables. It is also possible to purchase tablets that contain vitamin C obtained from natural or synthetic sources. If the vitamin is pure, it is chemically the same no matter the source from which it comes.

Most Americans have grown up hearing, and probably believing, that a person needs to have certain amounts of about two dozen different vitamins and minerals in order to stay healthy. Many people take vitamin supplements to ensure that they get the proper amount of vitamins and minerals. Vitamin supplements are pills engineered by a manufacturer to contain one or more of the essential vitamins. They may also contain other products that the manufacturer believes—or wants consumers to believe—will contribute to a person's good health.

Issues with Vitamin and Mineral Supplements

One might imagine that there is almost nothing bad that one can say about vitamins and minerals. Yet, a number of nutrition experts have expressed concerns about the way that consumers think about and use vitamins and minerals. One problem may be that people are likely to think that "more is better." That is, if one vitamin C tablet is said by the federal government to be good for you, then two tablets might be twice as good, three tablets three times as good, and so on.

In fact, taking too much of a vitamin or mineral is relatively difficult to do. But "overdosing" on supplements is possible, and it can cause serious health problems. A health condition that develops as the result of taking too much of a vitamin is known as *hypervitaminosis*. Harmful effects from hypervitaminosis with different vitamins have now been identified. For example, very high intake of vitamin D may lead to kidney failure. Excess intake of vitamin C may cause diarrhea that, in turn, can lead to abnormal dehydration. An excess of iodine in the diet can lead to an enlargement of the thyroid gland that resembles a goiter.

Some vitamins and minerals can interact with other drugs to produce dangerous conditions. For example, vitamin D amplifies the effect of digoxin, a heart medication. When taken together, the two substances can produce an irregular heart pattern. The minerals calcium, magnesium, iron, and zinc interact with tetracycline, a popular antibiotic, in such as way as to decrease the ability of the body to absorb the drug. Vitamin B_6 interferes with the action of L-dopa, a drug used to treat Parkinson's disease.

Regulation and Economics

At one time, vitamin and mineral supplements were regulated by the U.S. Food and Drug Administration (FDA). The Dietary Supplement Health and Education Act of 1994 changed that situation. The act removed vitamin and mineral supplements from the list of products over which the FDA has any control. Today, the only time FDA can act on a vitamin or mineral supplement is in those cases in which it has received confirmed reports of some harmful effects from the use of the supplement. In other words, the vitamin and mineral supplements available for sale to consumers are essentially unregulated in the United States.

Perhaps the most fundamental issue to some nutrition authorities is that vitamin and mineral supplements are seldom necessary for any person who eats even a minimally nutritious diet. That diet will provide the person with all of the vitamins and minerals needed for good health. The enormous sale of vitamin and mineral supplements—a business estimated to be worth hundreds of millions of dollars—tends to reflect the sales and advertising expertise of the companies who produce the supplements more than the needs of consumers themselves.

The bottom line appears to be that vitamin and mineral supplements taken in moderate amounts are not likely to be harmful in most instances. But neither is their expense likely to be justified by health reasons alone.

References

Editors of Consumer Reports Books. *Health Quackery.* Mount Vernon, NY: Consumers Union, 1980.

Farley, Dixie. "Look for 'Legit' Health Claims on Foods." http://www.fda.gov/fdac/special/foodlabel/health.html. 15 July 1998.

Health Media of America and Elizabeth Somer. *The Essential Guide to Vitamins and Minerals.* New York: Harper Perennial, 1992.

Leaf-Brock, Suzanne. "Use Caution and Knowledge When Using Vitamins and Supplements to Help Improve Your Health." *Mayo Clinic Health Information News,* 1 June 1997; also at http://www.mayo.edu/comm/health_info_products/news/news_133.html. 14 July 1998.

"Nutritional Supplements Misuse among Older Adults." http://www.ag.ohio-state.edu/~ohioline/ss-fact/ss–126.html. 15 July 1998.

O'Neill, Patrick. "Vitamins Spell Controversy for Doctors." *Oregonian,* 13 May 1998, A1+.

"Vitamin and Mineral Supplements in the Headlines." *Mayo Clinic Health Letter,* June 1997; also at http://www.mayohealth.org/mayo/9707/htm/me_3sb.htm. 15 July 1998.

"The Vitamin Pushers." *Consumer Reports,* March 1986, 170–175.

Water Rights

There may be no natural resource that humans tend to take more for granted (with the possible exception of air) than water. Water is essential to nearly every aspect of human life: drinking, cooking, cleaning, removing sewage and other wastes, growing crops, and recreation, for example. Nearly every human society has had to develop laws determining how the water resources available to that society were to be apportioned among its members and its various activities.

Riparian and Appropriated Systems

In the United States, water rights are controlled by a staggering variety of federal, state, tribal, and local laws and regulations. It is very difficult to make many generalizations about water rights laws for any part of the country.

In general, however, most water laws east of the Mississippi River tend to follow *riparian* customs. Under riparian customs, water rights belong to those who live adjacent to a river, stream, lake, or other body of water. Those who own water under riparian rights are allowed to remove as much water as they like from the body of water provided that they do not disturb the natural flow or quality of the water. In many cases, state and local governments have applied additional conditions to the use of water under riparian rights. Water use may be restricted, for example, in times of drought.

West of the Mississippi, water rights are controlled by a different doctrine, known as *prior appropriation*. Under the principle of prior appropriation, the first person to use water from a stream, river, or lake for some beneficial purpose (such as watering crops) receives priority for use of the water from that resource.

The prior appropriation doctrine allows for the possibility that two or more individuals may own rights to water use at the same time. However, a priority exists among these water owners. That is, ten landowners adjacent to a stream may all have earned the right to take and use water from the stream. But some landowners have a prior right to that of others. In time of drought, the landowner at the top of the list gets the first opportunity to take as much water as has been allotted to him or her. Any water that is left then passes to the second landowner on the list, then to the third landowner, and so on down the line. In contrast to the riparian practice, where every owner usually gets to draw some water from a source, the prior appropriation may result in many people *not* having access to their water.

Water Rights and the Development of the West

Water rights tend to be a more difficult issue in the western states than in the states east of the Mississippi. The simple reason for this fact is rainfall. The eastern states receive a considerably larger annual rainfall than do the western states. Serious droughts are rarities in the East, while they are very common in the West.

Development of the western states in the late nineteenth century could never have

In many parts of the western United States and other parts of the world, the question as to who owns the available water is the most serious of all socioscientific questions. (U.S. Department of the Interior, Bureau of Land Management)

occurred without the development of systems for finding and using water. For agriculture, that meant building dams to collect and store water, from which it was then transferred through huge and complex irrigation systems to farms. The development of such systems turned areas that are naturally arid into blooming gardens. The Central Valley of California is perhaps the best single example of the way an abundant supply of water, taken from hundreds of miles away, has turned a desert into one of the richest agricultural regions in the world.

Battles over water use have been a part of western history for more than a century. At first, it was a battle between farmers and cattle ranchers, each laying claim to the water needed to sustain their operations.

Then, a new factor was introduced into the western water battles: urbanization. As large cities began to grow up in the West, they too laid claim to the water supplies needed to operate any urban area. Cities such as Los Angeles, San Francisco, San Jose, and Phoenix all had to find ways to buy and ship water from sources hundreds of miles away to meet urban needs in regions that seldom saw more than a dozen inches of rain a year.

Competing Interests

Today, a number of different users are clamoring for the right to use a portion of the West's limited water resources. These users include cities of all sizes, farmers, ranchers, Native American tribes, and ordinary people interested in fishing, swimming, boating, and other activities. In addition, concerns are being raised about providing adequate water supplies to maintain aquatic life traditionally found in western lakes and rivers. By some estimates, more than 200 species of fish are now at risk because of reduced water flow in rivers. Some of these species have been listed as endangered, others will soon be so listed, while others will probably become extinct before listing can occur.

In 1992, the U.S. Congress commissioned a large-scale study of the West's water resources that was to include recommendations as to how water should be allocated in the future. In 1998, the Western Water Policy Review Commission issued its report. It said that significant changes in water use policy would be needed in the future. Farms and ranches, which currently use about three-quarters of all water available, would have to give up more of that resource to growing cities and urban areas in the seventeen western states. Nine of the ten fastest growing states in the United States are located in this region, and enhanced water supplies would be needed to keep up with the growth expected in the region.

The report did not meet with unanimous approval, either within the commission itself or among the general public. Farmers and ranchers objected, as might be expected, to having their "lifeblood" diverted to growing metropolitan areas. Native American tribes objected that water supplies in the West were one of their original natural resources, and they deserved more input as to what was to happen to that water in the future. And environmentalists were concerned that diverting even more water to the big cities would result in even greater threats to the natural habitats of fish in the Pacific Northwest.

Possibly the most difficult problem of all in deciding about the use of water resources is the complex, decentralized nature of water ownership law and practice in the West. Each town, county, or water district has its own rules and regulations, its own traditions, as to how water is to be distributed within its areas. Government policy announcements from Washington may sound all right in principle but applying those principles to the solution of specific disputes in towns throughout the West is likely to be difficult and fraught with controversy.

References

Barnum, Alex. "Study Says Western Farms Must Yield Water to Cities." *San Francisco Chronicle,* 4 July 1998, A1+.

Coats, Robert. "The Colorado River: River of Controversy." *Environment,* March 1984, 6–13+.

Lacey, W. C. "Water: A Treasure in Trouble." *National Wildlife,* February/March 1984, 7–21.

Monroe, Bill. "A Question of Access." *Oregonian,* 11 April 1996, F1.

Petulla, Joseph M. *American Environmental History.* San Francisco: Boyd and Fraser Publishing, 1977, 227–228.

"The Prior Appropriation System." http://water.state.co.us/prior.htm. 9 August 1998.

Thomas, Pete. "Water-flow Battle Could Leave Fishery up a Creek." *Los Angeles Times,* 28 November 1997, 10.

Wagner, Richard H. *Environment and Man.* 3d ed. New York: W. W. Norton & Company, 1978, Chapter 4.

"Water Rights." http://phylogeny.arizona.edu/AZWATER/rights.html. 9 August 1998.

"Water Rights and Allocations for Sound Resource Management." http://www.awwa.org/govtaff/watripap.htm. 11 August 1998.

Wetlands

The term *wetlands* is generally used to describe ecosystems that are covered with water for all or most of the year. Wetlands are often subdivided into two categories depending on their location. Coastal wetlands are wetlands connected with or close to the oceans. They include saltwater marshes, mangrove swamps, and tidal basins. Coastal wetlands tend to be covered with salty, briny, or brackish water. Inland

Efforts to protect the nation's wetlands have often been hampered because people seldom see how wetlands have any value to humans. (USDA—Soil Conservation Service)

wetlands generally contain fresh water. Examples of inland wetlands are bogs, swamps, sloughs, potholes, and freshwater marshes. As with all ecosystems, wetlands are characterized by distinctive plant and animal populations that have become adapted to surviving in very moist conditions that may include water containing high concentrations of salts.

Values of Wetlands

Wetlands serve a number of important functions. For one thing, they serve as the breeding ground for a great variety of animals, including many types of fish and shellfish, ducks and geese, and alligators. Some experts estimate that nearly three-quarters of the fish caught by commercial fishermen are spawned and hatched in coastal wetlands.

Wetlands are also an important source of flood control. The soil beneath a swamp, bog, saltwater marsh, or other wetland acts like a sponge. It is able to soak up large amounts of water that pour into it. In areas where wetlands have been destroyed, rainwater is less likely to be absorbed by soil and, instead, flows across the surface of the ground, producing flooding conditions.

Wetlands also act as natural water purifiers. As contaminated water moves through a wetland, pollutants are removed from the water by soil particles and plant roots. In regions where wetlands have been drained, it has been necessary to install water purification plants to replace the work formerly done naturally by a wetland.

Coastal wetlands also tend to protect the shoreline from erosion. The force of wind and waves is absorbed by a saltwater marsh or its equivalent, reducing the impact of these forces on dry land farther inland.

Finally, wetlands are an important source of recreation for humans. Many people enjoy bird-watching, boating, fishing, and hunting in or around wetland areas.

Trends in Wetland Loss

Humans have traditionally had little interest in maintaining wetlands. Swamps, bogs, and marshes have often been seen as useless wastes of land that could be put to better use. For example, a farmer in Kansas is likely to think of a slough on his or her land as an inconvenience. That land could be put to better use, may be his or her thinking, if the water were drained and the dry land planted to crops. Indeed, many farmers have and continue to hold this view of wetlands.

The result of such thinking has been an enormous loss of natural wetlands throughout the world. In the United States, for example, experts believe that nearly 60 percent of the wetlands that existed in the 1600s have now been converted to dry land. They estimate that more than 100,000 acres of wetlands are being lost every year.

This trend differs dramatically for various parts of the country. Midwestern states such as Indiana, Illinois, Missouri, Kentucky, and Ohio have lost more than 80 percent of their original wetlands. In California and Iowa, that fraction is more than 99 percent.

Threats to Wetlands

Wetlands have been and are being destroyed for a number of reasons. One is to increase the amount of agricultural land available. As in the Kansas example, most farmers can see little or no value in having a bog or swamp or pothole on their land. With a little effort, the water in one of these wetland areas can be drained, and they will have more room on which to grow crops or to graze cattle.

As with all wetlands destruction, a number of hidden losses are associated with the conversion of wetlands to agricultural land. For example, many migratory birds use freshwater wetlands for feeding and nesting. The U.S. Environmental Protection Agency has estimated that the draining of potholes in the United States and Canada is a major factor in the 50–80 percent decline in certain waterfowl populations between 1955 and 1995.

Urbanization has also been a major threat to wetlands. Imagine a small town built alongside a swamp. Townspeople may not realize the ecological and economic value of the swamp. When pressures to expand the urban area appear, developers may be tempted just to pave over the swamp to make room for new streets, homes, schools, office buildings, parking lots, and other structures. In such cases, an urban area is likely to experience the consequences of its actions. For example, the town may experience more frequent and more serious flooding. It may find that its public water source is less pure than it had been previously.

Wetlands are also often drained or disrupted to change water patterns in an area. For example, there may be a demand to straighten out and widen a natural, meandering river. This process of *stream channelization* is very popular when a river is needed to transport goods or people, to reduce the dangers of flooding (for at least one area on the river), or to make more land available for development. When stream and river patterns are changed, however, so is the structure of the underlying water table in the area. Modifying a water table, in turn, may result in the loss of water from an adjacent wetland, causing it to dry up.

In many cases, the water content of a wetlands area may not be altered but its chemical composition may be. For example, wetlands in agricultural areas are likely to become contaminated with fertilizer and pesticide runoffs from farms. Swamps in the vicinity of a mine may accumulate heavy metals, compounds of sulfur, and other toxic chemicals from mine wastewater.

Growing Concerns about Wetland Protection

The debate over wetland use has been a classic study in environmental studies over the past four decades. Scientists can point to a number of very important ecological functions performed by wetlands. They can easily make a case for the importance of wetlands in the maintenance of a healthy environment. Yet, the forces acting to destroy wetlands are very strong economic forces: developers, farmers, mining companies, and large industrial corporations, for example. It is almost always easier to put a dollar value on the advantage of converting a wetland to productive use than it is to specify the exact economic value of keeping it in its native condition. One can hardly be surprised, therefore, at the rate of wetlands destruction.

Yet, there has been a growing feeling among legislators and the general public that a more conservative approach to wetlands use is necessary. The 1985 Farm Act, for example, contained a provision known as the Swampbuster Law that prohibited the payment of agricultural subsidies to farmers who converted wetlands to croplands. A few years later, then-president George Bush declared it the government's policy that there should be "no net loss" of wetlands in the nation. For each acre of wetlands converted to commercial use, Bush said, a new acre of wetlands elsewhere would be created.

For some time, however, it was difficult to see any effects from this new outlook on wetlands. In 1987, for example, two years after the Farm Act had discouraged the destruction of wetland, more wetland draining occurred than in the preceding two decades.

Gradually, however, it appears that the forces operating to preserve wetlands are making progress. A 1997 report by the U.S. Fish and Wildlife Service announced that the net loss of wetlands between 1985 and 1995 averaged 117,000 acres per year. By comparison, the net loss of wetlands in the preceding decade had been 290,000 acres per year and between 1965 and 1975, 458,000 acres per year. A discouraging aspect of the report was that, despite earlier efforts, 80 percent of all wetland loss was still taking place on agricultural land.

References

Finlayson, M., and M. Moser. *Wetlands*. New York: Facts on File, 1991.

Freedman, Bill. "Wetlands." In *Gale Encyclopedia of Science*, edited by Bridget Travers, 3931–3936. Detroit: Gale Research, 1996.

"If You Protect America's Wetlands: They Will Protect You." http://www.sierraclub.org/wetlands/fact2.html. 1 July 1998.

Mears, Teresa. "Hard Slogging for a Swamp." *San Francisco Chronicle*, 20 April 1998, A3.

Mitsch, William J., and James G. Gosselink. *Wetlands*. New York: Van Nostrand Reinhold, 1986.

"Rate of Wetlands Loss Is Slowing." *San Francisco Chronicle*, 18 September 1997, A13.

U.S. Congress. Office of Technology Assessment. *Wetlands: Their Use and Regulation*. Washington, DC: U.S. Government Printing Office, 1984.

"Wetland Protection." http://h2osparc.wq.ncsu.edu/info/wetlands/protect.html. 1 July 1998.

"Wetlands Loss and Degradation." http://h2osparc.wq.ncsu.edu/info/wetlands/wetloss.html. 1 July 1998.

Two excellent general sources of wetlands information can be found on the Internet at http://www.wetlands.ca/ and http://www.anca.gov.au/environm/wetlands/nwpindex.htm.

White Abalone

The white abalone is a member of the genus Haliotis (*Haliotis sorenseni*), of which there are seven other species. The white abalone has traditionally been consumed in smaller amounts than its relative, the red abalone, but is regarded as a more desirable form of the genus for cooking and eating.

Rise and Fall of the White Abalone

The first specimens of the white abalone were discovered in the 1940s off the coast of

California. One reason that they remained hidden so long was their habitat. They tend to live at the greatest depths of any abalone, ranging from about 25 to 100 meters below the surface of the ocean. For another two decades, they were harvested only infrequently, largely because the equipment needed to dive to their depths was not generally available.

By the 1970s, however, that situation had changed. Commercial diving companies had developed equipment and methods that allowed deep-sea diving, and the harvest of white abalone increased dramatically. Annual yields jumped from about 20 metric tons in 1971 to a peak of 65 metric tons in 1972. They then began to fall slowly until 1980, when harvests dropped to almost zero.

Marine wildlife managers recognized the problem of declining white abalone populations by the late 1970s and implemented a program to deal with it. They released young abalone animals to the mollusks' natural habitat, assuming that the presence of more individuals would increase reproduction rates. They chose to take this action rather than place restrictions on commercial fishing for abalone.

By the early 1990s, marine biologists realized that the conservation plan implemented in the 1970s had failed. Surveys of white abalone habitats found an astonishingly small number of animals. In a three-hectare (6.5-acre) section of the ocean floor, researchers found only three live animals. They had expected to find about 12,000.

Fate of the White Abalone

The white abalone is now listed as an endangered species. Many marine biologists fear that this listing has come too late to prevent extinction of the species, however. They point to a combination of factors that has led to the probable demise of the species. First, biologists knew too little about the animal, particularly its reproductive patterns, to take necessary action early. As biologists were discovering that white abalones mate only relatively late in life,

commercial fishing companies were harvesting all of those older animals.

Second, consumer demand for abalones of all species had become so great that they were an irresistible target for fishing companies. Even today, the Mexican government regards white abalones in its territorial waters as potential cash crops, not endangered species. They have expressed no interest in acting to ensure the survival of the species.

Third, legislative and regulatory agencies sometimes worked at cross purposes to save the abalone. While the California Department of Fish and Game (DFG) was developing plans to protect the abalone, the state legislature was devising its own ideas on the same topic, often with less expertise and information than the DFG.

By 1998, a few white abalone had been captured for possible use in a captive breeding program. One such program had three males, but no females. Scientists operating this program were posting notices on the Internet, asking divers to look for and capture any white abalones they could find to promote this program. Other biologists had given up hope of saving the white abalone from extinction. They argued that the most that could be done was to examine carefully how this animal was discovered, harvested, and then driven into extinction so rapidly.

References

Martin, Glen. "White Abalone Poised on Brink of Extinction." *San Francisco Chronicle,* 7 July 1998, A1+.

"Research Update, July-August 1997." http://danr.ucop.edu/calag/JA97-w/update1.html. 7 August 1998.

"Single White Abalone Seeks Male." http://www.science.org/abs/. 8 August 1998.

Wild Horses and Burros

The history of horses and burros on the North American continent is not well understood. Fossil remains suggest that ancestors of these animals may have thrived on the continent more than 60 million years

Some observers question whether programs to save wild horses and burros have any real value beyond making humans feel better that they are protecting a small number of endangered animals. (Corbis)

ago. But there is no evidence of their populating the continent when Europeans arrived here in the eighteenth century. At that time, both species were introduced (or reintroduced) by Italian and Spanish explorers. Both animals soon became an integral part of the European conquest of the continent, the horses as a means of carrying men and their goods and the burros as beasts of burdens for miners.

Decline of Horses and Burros

As the period of the "Wild West" came to an end, there was less need for both horse and burro. Thousands of animals were abandoned by miners, ranchers, and the military. Scientists believe that these horses and burros returned to the wild, filling a niche in the ecosystem that their ancestors may have filled millions of years earlier. They constitute the herds of wild horses and burros that still remain throughout the American West.

For many decades, about the only people to care about wild horses and burros were mustangers, men who hunted and killed the animals. Sometimes they were destroyed simply because they were regarded as competitors for domesticated cattle and sometimes to be sold for dog and cat food.

During the 1960s, the general public became increasingly better informed about wild horses and burros, and a movement to protect these animals began to grow. Perhaps the best known leader of that movement was Velma Johnston, popularly known as "Wild Horse Annie." Johnston had been largely responsible for the adoption by Congress in 1959 of Public Law 86-234. This law prohibited the use of motorized vehicles in capturing wild horses and the poisoning of water holes used by the animals.

In response to growing public demands for protection of wild horses and burros,

Congress passed the Wild Horse and Burro Act (WHBA) of 1971. That act required that the Bureau of Land Management (BLM) take an active role in protecting the two species and ensuring that populations were kept at a level that would guarantee their survival into the future. Perhaps the best-known offshoot of the WHBA was the Adopt-a-Horse or Burro Program instituted in 1976. This program allows for private citizens to adopt wild horses or burros that are regarded as excess.

Conflicts over Wild Horses and Burros

Legislative history might suggest that wild horses and burros have been gaining more and more protection in the U.S. West. That conclusion is not necessarily accurate. In the first place, the BLM has been accused of not acting aggressively enough to carry out the mandate of the WHBA. Critics say that the BLM is more inclined to listen to farmers and ranchers than to those interested in protecting these two species.

Probably more important is the opposition to wild horses and burros mounted by ranchers, who say the animals compete with their cows for forage. They say that their own business is severely impacted by herds of wild horses and burros that roam the open range and deprive domestic cows of their food. They also blame wild horses and burros for ruining public lands by overgrazing.

Supporters of the WHBA disagree with this position. They point out that horses, burros, and cows have different feeding patterns. Whereas cows tend to graze within a mile of a lake, river, or stream, horses and burros tend to graze at higher elevation in more rugged terrain. It is rare that the regions in which these three animals feed overlap, experts say. Besides, scientific studies have shown that the majority of damage caused by overgrazing results from feeding by cattle rather than by horses or burros.

The issue of what to do with wild horses and burros would appear to be one of many scientific and technological issues in which science itself may be of less importance than political factors. In the state legislatures of western states, ranchers and farmers have a powerful influence. Legislators may be less inclined to listen to scientific arguments as to what is best for wild horses and burros than to listen to lobbyists for ranchers, on whom they may depend for financial support.

Legislation introduced in Nevada in 1997, for example, appeared to call for the gradual removal of all wild horses and burros in the state. The threat these animals posed to cattle herds, farms, and private homes was too severe, according to some legislators, to justify continuing the protection programs mandated by WHBA. Environmentalists argued that the state's 23,000 wild horses could hardly be thought to pose a serious obstacle to the more than 2 million cattle grazing in the state.

References

"Achievements." http://ourworld.compuserve.com/homepages/ispmb/webdoc1.htm#Achievements. 22 May 1999.

Bellisle, Martha. "Nevada Lawmakers Ready to Mount Wild Horse Battle." *Sunday Oregonian,* 9 February 1997, A21.

BLM Wild Horse and Burro Internet Adoption. http://www.adoptahorse.blm.gov/. 8 August 1998.

Frazier, Deborah. "BLM to Cut Wild Horse Herds in State." *Rocky Mountain News,* 14 August 1997, A23.

Gilbert, Bil. "Our Dealings with the Species Have Been Complicated." *Sports Illustrated,* 17 June 1991, 96+.

"Wild Horses: Exposing the Myths." http://www.api4animals.org/WildHorseFactSheet.htm. 8 August 1998.

Wildlife Management

Wildlife management is the term used to describe all the methods used by humans to control the kind and number of animals in an area. For example, a particular community may decide that it has too many or too few deer, raccoons, or other animals in its

Though hunting has long been an essential tool of wildlife management, some critics wonder whether the practice is essentially just a system for saving wild animals so that humans can hunt them. (U.S. Fish and Wildlife Service)

area. It then has a number of options as to what can be done to increase or decrease the number of those animals in the area.

History

Wildlife management is sometimes seen as a relatively modern practice. However, its roots can be traced to very early stages in the development of human civilization. For example, Marco Polo told about such practices he observed in the Mongol empire of the Kublai Khan in the thirteenth century. Hunting was restricted in specific areas during breeding season to ensure that adequate numbers of animals would survive for food and sport in succeeding years.

Wildlife management often becomes an important practice when the numbers of certain specific species begin to decline.

For example, the concept of wildlife management would probably have seemed absurd to Europeans arriving in North America in the seventeenth century. To those pioneers, it probably seemed that the continent was filled to overflowing with more game animals than could ever be harvested by humans. Certainly, there are few or no records indicating that early pioneers worried about conserving wildlife for future generations.

And the results of that philosophy have been clear. Species after species was hunted to extinction, or almost so. For example, early settlers reported that the wild turkey was among the most common of all birds. Some observers reported seeing thousands of such birds a day, and they estimated that more than 10 million turkeys lived in the United States. Yet, by the end of the nineteenth century, the wild turkey had been completely eliminated from the New England states, Michigan, Wisconsin, and Minnesota by hunting. The bird was also close to extinction in most other parts of the nation.

Methods of Wildlife Management

Prior to the 1930s, wildlife management made use of a relatively small number of unsophisticated techniques. If there were too many animals in an area, wildlife managers killed off the excess population or allowed hunters to do so. If too few animals were present, laws were passed to prevent hunting or removing the threatened species.

A leader of the modern wildlife management movement was Aldo Leopold, a naturalist, ecologist, and forester. Leopold argued that the environment should be regarded as an interrelated whole. No one species of plant or animal should be regarded as existing in isolation from other plants and animals around it. He suggested that controlling the numbers of animals in an area, for example, could be accomplished by a host of methods, including management of the animal's food supply, the diseases with which it is afflicted, its natural predators, and other factors that affect the size of its population.

Today, wildlife managers use a number of techniques to control the size of animal populations. These include fire, vegetation management, water developments, predator control, and hunting. For example, there is an inverse relationship between the number of wolves and coyotes and the number of elk that live in an area. Wolves and coyotes prey on elk, and as their numbers increase, the number of elk decreases. Wildlife managers can control the number of elk in an area, therefore, by killing off or otherwise removing the wolves and coyotes in an area or by restoring them to an area from which they have been removed.

Goals of Wildlife Management

Wildlife managers have three primary goals. First, they attempt to protect endangered species from extinction. Some of the most exciting stories in the annals of wildlife management are those that tell of the saving of a species, such as the American bison, that seemed on its road to extinction.

A second goal of wildlife management is the protection of human resources—their land, domestic animals, and their own lives—from predators. As humans have moved into more and more remote areas, the battle between themselves and native animals has become more severe. In many cases, wildlife managers have developed schemes for eliminating potential predators from an area that humans now wish to dominate. For example, stories are often told about invasions of bears into human communities at the very outposts of civilization. Wildlife managers may be called in to kill those "intruders" or to remove them to even more remote regions.

A third goal of wildlife management is to guarantee a supply of game animals to meet the needs of hunters. Numerous examples exist of the effects of unbridled hunting on animal species. Perhaps the classical example is that of the passenger pigeon. In the mid-1800s, the passenger pigeon was among the most common of all birds in North America. Observers reported that the sun was sometimes blocked out by the flocks of passenger pigeons passing overhead. There is one report of a nesting site in Wisconsin in 1871 at which an estimated 136 million birds were present.

Hunters found the passenger pigeon an irresistible target. At one time, special trains were created to carry hunters into areas where the pigeons could be found. Thousands were shot by hunters as they passed through on the train. Many hunters never left the train or bothered to pick up any of the birds they killed. By the early 1900s, the passenger pigeon had become an endangered species and, in 1914, the last remaining member of the species died in the Cincinnati Zoo.

A very substantial effort is maintained today, therefore, to make sure that hunters do not kill off all the members of a game species and that enough will remain to provide hunting forever into the future. Organizations such as Ducks, Unlimited, have been created to provide some control over the hunting of ducks and guarantee a supply for future generations.

Disputes over Land Management

Some of the goals and methods of wildlife management evoke relatively little controversy. For example, there are few individuals who would object to efforts to save species that are threatened or endangered. However, other wildlife management activities have been criticized for a variety of reasons.

For example, some critics reject the notion that humans are always to be accorded a standing higher than that of other species. If humans choose to move into a remote area that is already home to bears, wolves, coyotes, or other animals, those humans have an obligation to accommodate themselves to the native species, these critics say. Humans do not have any inherent right to drive out all species with whom they come into conflict.

Some individuals object also to the concept that some species are "better" than others. For example, some predatory species, such as coyotes, fox, and wolves, are sometimes regarded as "bad" compared to other species, such as deer and elk. Wildlife management is reasonable, these critics say, if efforts are made to maintain some kind of natural balance within an area but not to promote one species over another.

Finally, the role of wildlife management in meeting the needs of hunters often receives serious criticism. There is something weirdly illogical, critics say, about developing a sophisticated system of preserving a species (such as ducks) so that they can be killed by human hunters. This issue has become a highly controversial one that often leads to strong rhetoric and, occasionally, personal and violent confrontations between those on either side of the issue.

The issue is also very complex. For example, questions have been raised about the development of personal wildlife refuges where somewhat exotic game animals are bred solely for the purpose of being hunted by individuals who pay very large amounts of money for the privilege. Such refuges are not under the control of any governmental body and have sometimes been criticized as the worst possible

examples of wildlife management practices. Other wildlife managers are less concerned about "fee-for-hunting" refuges. After all, they say, we live in an environment of which humans have largely taken control. If some kinds of hunting are to survive at all, they may have to take place on such private refuges.

References

Kupchella, Charles E., and Margaret C. Hyland. *Environmental Science: Living within the System of Nature.* Boston: Allyn and Bacon, 1986, 539–541.

Miller, J. A. "Wildlife Refuge System: Refuge for Whom?" *Science News,* 22 March 1986, 183.

Newton, David E. *Hunting.* New York: Franklin Watts, 1992, Chapter 3.

Purdom, P. Walton, and Stanley H. Anderson. *Environmental Science: Managing the Environment.* Columbus, OH: Charles E. Merrill, 1980, 223–228.

Robinson, William L., and Eric G. Bolden. *Wildlife Ecology and Management.* New York: Macmillan, 1984.

"Wildlife Habitat Management." http://www.nhq.nrcs.usda.gov/BCS/wild/wildlife.html. 22 July 1998.

"Wildlife Management." http://bluegoose.arw.r9.fws.gov/NWRSFiles/WildlifeMgmt/WildlifeMgmtIndex.html. 22 July 1998.

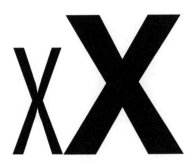

XYY Controversy

See Criminality and Heredity

Selected Bibliography

"A Citizen's Guide to Radon." 2d ed. (EPA Document #402-K92-001) Washington, DC: U.S. Environmental Protection Agency, 1992.

Abernethy, Virginia D., and Garrett Hardin. *Population Politics: The Choices that Shape Our Future.* New York: Plenum Publishing, 1993.

Abrams, Richard. *Electroconvulsive Therapy.* New York: Oxford University Press, 1997.

Abrams, Stan. *The Complete Polygraph Handbook.* Lexington, MA: Lexington Books, 1989.

Alpern, K. D., ed. *The Ethics of Reproductive Technology.* New York: Oxford University Press, 1992.

Alston, Dana, ed. *We Speak for Ourselves: Social Justice, Race and the Environment.* Washington, DC: Panos Institute, 1990.

American Lung Association, Environmental Protection Agency, Consumer Product Safety Commission, American Medical Association. *Indoor Air Pollution: An Introduction for Health Professionals.* Washington, DC: U.S. Government Printing Office, Publication No. 1994-523-217/81322, 1994.

Andryszewski, Tricia, and Victoria Sherrow. *Abortion: Rights, Options, and Choices.* New York: Millbrook Press, 1996.

The Ann Arbor Science for the People Editorial Collective. *Biology as a Social Weapon.* Minneapolis: Burgess Publishing Company, 1977.

Annas, George J. *Informed Consent to Human Experimentation: The Subject's Dilemma.* Cambridge, MA: Ballinger Publishing, 1977.

The Assisted Reproductive Technology Workbooks. Somerville, MA: Resolve, 1993.

Bach, Julie S. *Drug Abuse: Opposing Viewpoints.* St. Paul, MN: Greenhaven Press, 1988.

Bahnsen, Greg L. *Homosexuality: A Biblical View.* Grand Rapids, MI: Baker Book House, 1978.

Bailey, Derrick Sherwin. *Homosexuality and the Western Christian Tradition.* London: Longman, Green, 1955.

Baird, Robert M., and Stuart E. Rosenbaum, eds. *Animal Experimentation: The Moral Issues.* Amherst, NY: Prometheus Books, 1991.

Barbour, Ian G. *Religion and Science: Historical and Contemporary Issues.* New York: HarperCollins, 1997.

Barlett, Donald L., and James B. Steele. *Forevermore: Nuclear Waste in America.* New York: W. W. Norton, 1985.

Benedick, Richard E. *Ozone Diplomacy: New Directions in Safeguarding the Planet.* Cambridge: Harvard University Press, 1991.

Bishop, J., and M. Waldholz. *Genome: The Story of the Most Astonishing Scientific Adventure of Our Time—The Attempt to Map All the Genes in the Human Body.* New York: Simon and Schuster, 1990.

Black, David L., ed. *Drug Testing in Sports.* Las Vegas, NV: Preston Publishing, 1996.

Block, Eugene B. *Lie Detectors: Their History and Use.* New York: D. McKay Company, 1977.

Boswell, John. *Christianity, Social Tolerance, and Homosexuality.* Chicago: University of Chicago Press, 1980.

Bracewell, Ronald N. *Intelligent Life in Outer Space.* San Francisco: W. H. Freeman and Company, 1975.

Breggin, Peter. *Talking Back to Ritalin: What Doctors Aren't Telling You about Stimulants for Children.* Monroe, ME: Common Courage Press, 1998.

Brenner, David J. *Radon: Risk and Remedy.* Salt Lake City, UT: W. H. Freeman, 1989.

Brookins, Douglas G. *The Indoor Radon Problem.* Irvington, NY: Columbia University Press, 1990.

Bryant, Bunyan, ed. *Environmental Justice: Issues, Policies, and Solutions*, Washington, DC: Island Press, 1995.

Bryant, Bunyan, and Paul Mohai. *Race and the Incidence of Environmental Hazards: A Time for Discourse*. Boulder, CO: Westview Press, 1992.

Bullard, Robert D. *Dumping in Dixie: Race, Class, and Environmental Quality*. Boulder, CO: Westview Press, 1994.

Bullard, Robert D., ed. *Confronting Environmental Racism: Voices from the Grassroots*. Boston: South End Press, 1993.

Butler, J. Douglas. "Abortion and Reproductive Rights: A Comprehensive Guide to Medicine, Ethics, and the Law" (CD-ROM). Phoenix: Oryx Press, 1997.

Center for Investigative Reporting and Bill Moyers. *Global Dumping Ground*. Cambridge: The Lutterworth Press, 1991.

Cheney, Glen Alan. *Chernobyl: The Ongoing Story of the World's Deadliest Nuclear Disaster*. New York: New Discovery Books, 1993.

Chernousenko, V. M. *Chernobyl: Insight from the Inside*. New York: Springer Verlag, 1991.

Clawson, Marion. *Forests for Whom and for What?* Baltimore: Johns Hopkins University Press, 1975.

Conner, Gary L., ed. *Symposium: Sexual Preference and Gender Identity*. A special issue of the *Hasting Law Journal,* March 1979.

Conrad, Chris. *Hemp for Health: The Medicinal and Nutritional Uses of Cannabis Sativa*. Los Angeles: Inner Traditions International, 1997.

"Consumer's Guide to Radon Reduction." (EPA Document #402-K92-003) Washington, DC: U.S. Environmental Protection Agency, 1992.

Coombs, Robert H., and Louis Jolyon West, eds. *Drug Testing: Issues and Options*. New York: Oxford University Press, 1991.

Cranor, C., ed. *Are Genes Us? The Social Consequences of the New Genetics*. Brunswick, NJ: Rutgers University Press, 1994.

Dana, Samuel T., and Sally K. Fairfax. *Forest and Range Policy: Its Development in the United States*. 2d edition. New York: McGraw-Hill, 1980.

Daugherty, Helen Ginn, and Kenneth C. W. Kammeyer. *An Introduction to Population*. 2d ed. New York: The Guilford Press, 1995.

Day, Nancy. *Abortion: Debating the Issue.* Hillside, NJ: Enslow Publishers, 1995.

———. *Animal Experimentation: Cruelty or Science?* Hillside, NJ: Enslow Publishers, 1994.

Delgado, Jose M. R. *Physical Control of the Mind, Toward a Psychocivilized Society*. New York: Harper and Row, 1969.

Devall, Bill. *ClearCut: The Tragedy of Industrial Forestry*. San Francisco: Earth Island Press, 1993.

Dossey, Larry. *Healing Words: The Power of Prayer and the Practice of Medicine*. San Francisco: Harper San Francisco, 1995.

Dunlap, Thomas R. *DDT Scientists, Citizens, and Public Policy*. Princeton, NJ: Princeton University Press, 1981.

Durning, Alan Thein, and Christopher D. Crowther. *Misplaced Blame: The Real Roots of Population Growth*. Seattle, WA: Northwest Environment Watch, 1997.

Duster, T. *Backdoor to Eugenics*. New York: Routledge, Chapman, and Hall, 1990.

Editors of Consumer Reports Books. *Health Quackery*. Mount Vernon, NY: Consumers Union, 1980.

Editors of the *Harvard Law Review. Sexual Orientation and the Law*. Cambridge: Harvard University Press, 1990.

Editors of Time-Life Books. *Lords of the Plains*. Alexandria, VA: Time-Life Books, 1993.

Efron, Edith. *The Apocalyptics: How Environmental Politics Controls What We Know about Cancer*. New York: Simon and Schuster, 1984.

Ehrlich, Paul, and Anne Ehrlich. *The Population Explosion*. New York: Doubleday, 1990.

Fincher, Jack. *The Brain: Mystery of Matter and Mind*. Washington, DC: U.S. News Books, 1981.

Finlayson, M., and M. Moser. *Wetlands*. New York: Facts on File, 1991.

Freiman, Fran Locher, and Neil Schlager. *Failed Technology*. Detroit: Gale Research, 1994.

Freund, Paul Abraham, ed. *Experimentation with Human Subjects*. New York: G. Braziller, 1970.

Fyson, Nance Lui. *World Population*. New York: Franklin Watts, 1998.

Gale, Anthony, ed. *The Polygraph Test: Lies, Truth, and Science*. London: Sage Publications, 1988.

Gallant, Roy A. *Beyond Earth: The Search for Extraterrestrial Life*. New York: Four Winds Press, 1977.

Galton, Francis. *Essays in Eugenics*. London: Scott Townsend Publishers, 1996.

Gibson, James N. *Nuclear Weapons in the United States: An Illustrated History*. Atglen, PA: Schiffer Publications, 1996.

Gilliom, John. *Surveillance, Privacy, and the Law: Employee Drug Testing and the Politics of Social Control.* Ann Arbor: The University of Michigan Press, 1994.

Gorney, Cynthia. *Articles of Faith: A Frontline History of the Abortion Wars.* New York: Simon and Schuster Books, 1998.

Gramick, Jeannine, and Pat Furey. *The Vatican and Homosexuality.* New York: Crossroad, 1988.

Greenberg, Daniel S. *The Politics of Pure Science.* New York: The New American Library, 1967.

Gribbin, John. *The Hole in the Sky: Man's Threat to the Ozone Layer.* New York: Bantam Books, 1988.

Grinspoon, Lester, and James B. Bakalar. *Marijuana, the Forbidden Medicine.* New Haven: Yale University Press, 1997.

Guither, Harold D. *Animal Rights: History and Scope of a Radical Social Movement.* Carbondale: Southern Illinois University Press, 1998.

Hart, Harold. *Organic Chemistry: A Short Course,* 8th ed. Boston: Houghton Mifflin, 1991.

Health Media of America and Elizabeth Somer. *The Essential Guide to Vitamins and Minerals.* New York: Harper Perennial, 1992.

Herrnstein, Richard, and Charles Murray. *The Bell Curve.* New York: The Free Press, 1994.

Hewett, Charles E., and Thomas E. Hamilton, eds. *Forests in Demand: Conflicts and Solutions.* Boston: Auburn Publishing House, 1982.

Hoffman, Roald, and Shira Leibowitz Schmidt. *Old Wine, New Flasks: Reflections on Science and Jewish Tradition.* New York: W. H. Freeman, 1997.

Holtzman, Neil A., and Michael S. Watson, eds. *Promoting Safe and Effective Genetic Testing in the United States: Final Report of the Task Force on Genetic Testing.* Baltimore: Johns Hopkins University Press, 1998.

Horowitz, E. C. J. *Clearcutting.* Washington, DC: Acropolis Books, 1974.

Hughes, Liz Rank. *Reviews of Creationist Books.* 2d ed. Berkeley, CA: The National Center for Science Education, 1992.

Humber, James M., Robert F. Almeder, and Gregg A. Kasting. *Physician-Assisted Death.* Totowa, NJ: Humana Press, 1994.

Hundert, Edward M. *Lessons from an Optical Illusion.* Cambridge: Harvard University Press, 1995.

Indoor Air Pollution/Formaldehyde. Washington, DC: American Lung Association, 1995.

"Indoor Air Quality: Basics for Schools" (EPA Document #402-F-96-004). Washington, DC: U.S. Environmental Protection Agency, 1996.

"The Inside Story—A Guide to Indoor Air Quality" (EPA Document #402-K-93-007). Washington, DC: U.S. Environmental Protection Agency, 1995.

Joesten, Melvin D., et al. *World of Chemistry.* Philadelphia: Saunders College Publishing, 1991.

Jones, James H. *Bad Blood: The Tuskegee Syphilis Experiment.* New York: Collier Macmillan, 1981.

Kamin, Leon. *The Science of Politics and I.Q.* New York: Halsted Press, 1974.

Kevles, Daniel J. *In the Name of Eugenics: Genetics and the Uses of Human Heredity.* Cambridge: Harvard University Press, 1995.

Kevles, Daniel J., and Leroy Hood, eds. *Code of Codes: Scientific and Social Issues in the Human Genome Project.* Cambridge: Harvard University Press, 1993.

Kohl, Marvin, ed. *Beneficent Euthanasia.* Buffalo: Prometheus Books, 1970.

Kuhl, Stefan. *The Nazi Connection: Eugenics, American Racism, and German National Socialism.* New York: Oxford University Press, 1994.

Kupchella, Charles E., and Margaret C. Hyland. *Environmental Science: Living within the System of Nature.* Boston: Allyn and Bacon, 1986.

Larson, Edward J. *Sex, Race, and Science: Eugenics in the Deep South.* Baltimore: Johns Hopkins University Press, 1995.

Leahy, Michael P. T. *Against Liberation: Putting Animals in Perspective.* New York: Routledge Press, 1994.

Leopold, Aldo. *A Sand County Almanac.* New York: Oxford University Press, 1949.

Leslie, G. B., and F. W. Lunau, eds. *Indoor Air Pollution.* New York: Cambridge University Press, 1994.

Ligocki, Kenneth. *Drug Testing: What We All Need to Know.* Bellingham, WA: Scarborough Publishing, 1996.

Lindberg, David C., and Ronald L. Numbers, eds. *God and Nature: Historical Essays on the Encounter Between Christianity and Science.* Berkeley: University of California Press, 1986.

Lombroso, Cesare. *Criminal Man.* New York: G. P. Putnam's, 1911.

Lorenz, Konrad. *On Aggression.* New York: Harcourt Brace Jovanovich, 1966.

Lutz, Wolfgang. *The Future of World Population.* Washington, DC: Population Reference Bureau, 1994.

Macdonald, Scott, and Peter Roman. *Drug-testing in the Workplace.* New York: Plenum Press, 1994.

Manahan, Stanley E. *Environmental Chemistry.* 6th ed. Boca Raton, FL: Lewis Publishers, 1994.

Markley, O. W., ed. *Twenty-First Century Earth.* San Diego: Greenhaven Press, 1996.

Matté, James Allen. *Forensic Psychophysiology Using the Polygraph.* Williamsville, NY: J.A.M. Publications, 1996.

Maxwell, Mary, ed. *The Sociobiological Imagination.* Albany: State University of New York Press, 1991.

Mazur, Laurie Ann, and Timothy Wirth. *Beyond the Numbers: A Reader on Population, Consumption, and the Environment.* Washington, DC: Island Press, 1994.

McClure, Frank J. *Water Fluoridation: The Search and the Victory.* Washington, DC: National Institutes of Health, U.S. Department of Health, Education, and Welfare, 1970.

McCullagh, Peter. *Brain Dead, Brain Absent, Brain Donors: Human Subjects or Human Objects?* New York: John Wiley and Sons, 1993.

McFalls, Joseph A., Jr. *Population: A Lively Introduction.* Washington, DC: Population Reference Bureau, 1995.

Meadows, Donella H., et al. *The Limits to Growth.* New York: Universe Books, 1972.

————. *Groping in the Dark: The First Decade of Global Modeling.* New York: John Wiley, 1982.

"Measuring Air Quality: The Pollutant Standards Index" (EPA Publication 451/K-94-001).

Medvedev, Zhores A. *The Legacy of Chernobyl.* New York: W. W. Norton, 1992.

Mesarovic, Mihajlo, and Eduard Pestel. *Mankind at the Turning Point,* New York: E. P. Dutton, 1974.

Miller, G. Tyler, Jr. *Living in the Environment.* 4th ed. Belmont, CA: Wadsworth Publishing, 1985.

————. *Environmental Science: Sustaining the Earth.* 3d ed. Belmont, CA: Wadsworth Publishing, 1991.

Mitsch, William J., and James G. Gosselink. *Wetlands.* New York: Van Nostrand Reinhold, 1986.

Moffat, Donald W. *Handbook of Indoor Air Quality Management.* Englewood Cliffs, NJ: Prentice Hall, 1997.

Moran, Joseph M., Michael D. Moran, and James H. Wiersma. *Introduction to Environmental Science.* 2d ed. New York: W. H. Freeman and Company, 1986.

Moss, Thomas H., and David L. Sills, eds. *The Three Mile Island Nuclear Accident: Lessons and Implications.* New York: New York Academy of Sciences, 1981.

Nash, Roderick. *The American Environment: Readings in the History of Conservation.* Reading, MA: Addison-Wesley, 1968.

Newton, David E. *Particle Accelerators: From the Cyclotron to the Superconducting Super Collider.* New York: Franklin Watts, 1989.

————. *Taking a Stand against Environmental Pollution.* New York: Franklin Watts, 1990.

————. *AIDS Issues: A Handbook.* Hillside, NJ: Enslow Publishers, 1992.

————. *Hunting.* New York: Franklin Watts, 1992.

————. *Population: Too Many People?* Hillside, NJ: Enslow Publishers, 1992.

————. *Global Warming: A Reference Handbook.* Santa Barbara, CA: ABC-CLIO, 1993.

————. *The Ozone Dilemma: A Reference Handbook.* Santa Barbara, CA: ABC-CLIO, 1995.

————. *Environmental Justice: A Reference Handbook.* Santa Barbara: ABC-CLIO, 1996.

————. *Drug Testing in the Workplace.* Springfield, NJ: Enslow Publishers, 1997.

Norris, Ruth, ed. *Pills, Pesticide and Profits: The International Trade in Toxic Substances.* Croton-on-Hudson, NY: North River Press, 1982.

Oltmans, Willem L. *On Growth.* New York: G. P. Putnam, 1974.

Pearson, Mark A. *Christian Healing: A Practical and Comprehensive Guide.* Chicago: Fleming H. Revell, 1995.

Petrikin, Jonathan. *Environmental Justice.* San Diego: Greenhaven Press, 1995.

Petulla, Joseph M. *American Environmental History.* San Francisco: Boyd and Fraser Publishing, 1977.

Pickering, Kevin T., and Lewis A. Owen. *An Introduction to Global Environmental Issues.* London: Routledge, 1994.

Pittock, A., et al. *Environmental Consequences of Nuclear War. Vol. 1: Physical and Atmospheric Effects.* New York: Wiley, 1985.

Polkinghorne, John C. *Belief in God in an Age of Science*. New Haven, CT: Yale University Press, 1998.

President's Commission on the Accident at Three Mile Island. *Report of the President's Commission on the Accident at Three Mile Island*. Washington, DC: Government Printing Office, 1979.

"Protect Your Family and Yourself from Carbon Monoxide Poisoning" (EPA Document #402-F-96-005). Washington, DC: U.S. Environmental Protection Agency, 1996.

Purdom, P. Walton, and Stanley H. Anderson. *Environmental Science: Managing the Environment*. Columbus, OH: Charles E. Merrill, 1980.

Reagan, Leslie J. *When Abortion Was a Crime: Women, Medicine, and Law in the United States, 1867–1973*. Berkeley: University of California Press, 1997.

Regan, T. *The Case for Animal Rights*. Berkeley: University of California Press, 1983.

Revkin, A. *The Burning Season: The Murder of Chico Mendes and the Fight for the Amazon Rain Forest*. London: Collins, 1990.

Robinson, William L., and Eric G. Bolden. *Wildlife Ecology and Management*. New York: Macmillan, 1984.

Rosenbert, H. S., and Y. M. Epstein. *Getting Pregnant When You Thought You Couldn't*. New York: Warner Books, 1993.

Rothenberg, Karen H., and Elizabeth J. Thomson, eds. *Women and Prenatal Testing: Facing the Challenges of Genetic Technology*. Columbus: Ohio State University Press, 1994.

Rowan, A., and F. M. Loew. *The Animal Research Controversy*. Boston: Tufts Center for Animals and Public Policy, 1995.

Rubin, Eva R., ed. *The Abortion Controversy: A Documentary History*. Westwood, CT: Greenwood Publishing Group, 1994.

Russell, Bertrand. *Religion and Science*. London: Oxford University Press, 1960.

Sagan, Carl, and Richard P. Turco. *A Path Where No Man Thought: Nuclear Winter and the End of the Arms Race*. London: Century Press, 1990.

Sagan, Carl. *The Cosmic Connection: An Extraterrestrial Perspective*. New York: Dell Publishing, 1973.

Sagan, Scott Douglas, and Kenneth Neal Waltz. *The Spread of Nuclear Weapons: A Debate*. New York: W. W. Norton, 1995.

Schneider, Stephen. *Global Warming: Are We Entering the Greenhouse Century?* New York: Random House, 1989.

Schroeer, Dietrich. *Science, Technology and the Nuclear Arms Race*. New York: John Wiley and Sons, 1984.

"Secondhand Smoke" (EPA Document #402-F-93-004). Washington, DC: U.S. Environmental Protection Agency, 1993.

Sherry, Clifford J. *Animal Rights: A Reference Handbook*. Santa Barbara, CA: ABC-CLIO, 1994.

Shklovskii, I. S., and Carl Sagan. *Intelligent Life in the Universe*. San Francisco: Holden-Day, 1966.

"Sick Building Syndrome (SBS) (EPA Document #6607J; Indoor Air Facts No. 4 [revised]). Washington, DC: U.S. Environmental Protection Agency, 1991.

Siegel, Ronald K. *Intoxication: Life in the Pursuit of Artificial Paradise*. New York: E. P. Dutton, 1989.

Singer, Peter, and Susan Reich. *Animal Liberation*. New York: New York Review of Books, 1990.

Skinner, B. F. *Walden Two*. New York: Macmillan, 1948.

Smith, J. David. *The Eugenic Assault on America: Scenes in Red, White, and Black*. Washington, DC: Georgetown University Press, 1993.

Smith, Kenneth. *Alar: Five Years Later*. New York: American Council on Science and Health, 1994.

Solinger, Rickie, Faye Ginsburg, and Patricia Anderson. *Abortion Wars: A Half Century of Struggle, 1950–2000*. Berkeley: University of California Press, 1998.

Southwick, Charles H. *Global Ecology in Human Perspective*. New York: Oxford University Press, 1996.

Steffen, Lloyd, ed. *Abortion: A Reader*. Cleveland, OH: Pilgrim Press, 1996.

Steinbock, Bonnie, and Alastair Norcross. *Killing and Letting Die*. 2d ed. New York: Fordham University Press, 1994.

Swisher, Karin L. *Drug Abuse: Opposing Viewpoints*. San Diego: Greenhaven Press, 1994.

"Targeting Indoor Air Pollution: EPA's Approach and Progress" (EPA Document #400-R-92-012). Washington, DC: U.S. Environmental Protection Agency, 1993.

Thomasma, David C., and Thomasine Kushner, eds. *Birth to Death: Science and Bioethics.* Cambridge: Cambridge University Press, 1996.

Thompson, Larry. *Correcting the Code: Inventing the Genetic Cure for the Human Body.* New York: Simon and Schuster, 1994.

Travers, Bridget, ed. *The Gale Encyclopedia of Science.* Detroit: Gale Research, 1996.

Turco, Richard P. *Earth Under Siege.* Oxford: Oxford University Press, 1997.

Turk, Jonathan, and Amos Turk. *Environmental Science.* 4th ed. Philadelphia: Saunders College Publishing, 1988.

Turner, Stanfield. *Caging the Nuclear Genie: An American Challenge for Global Security.* Boulder, CO: Westview Press, 1997.

U.S. Congress. *Environmental Justice.* Hearings before the Subcommittee on Civil and Constitutional Rights of the House Committee on the Judiciary, 103d Congress, 1st Session, 3–4 March 1993.

U.S. Congress, Office of Technology Assessment. *Wetlands: Their Use and Regulation.* Washington, DC: U.S. Government Printing Office, 1984.

U.S. Congress, Office of Technology Assessment. *Biological Effects of Power Frequency Electric and Magnetic Fields: Background Paper.* 1989.

U.S. Department of Agriculture. National Resources Conservation Service. *Agricultural Waste Management Field Handbook.* Springfield, VA: National Technical Information Services Division, U.S. Department of Commerce.

U.S. Environmental Protection Agency. Environmental Equity Workgroup. *Environmental Equity: Reducing Risks for All Communities.* Washington, DC: Environmental Protection Agency, 1992.

U.S. Environmental Protection Agency. *Water Pollution from Feedlot Waste: An Analysis of Its Magnitude and Geographic Distribution.* Washington, DC: Environmental Protection Agency, 1993.

U.S. House Committee on Energy and Commerce. *Basel Convention on the Export of Waste: Hearing before the Subcommittee on Transportation and Hazardous Materials.* 102d Cong., 1st sess., 10 October 1991.

U.S. House Committee on Science. *Scientific Integrity and Public Trust: The Science behind Federal Policies and Mandates: Case Study 1— Stratospheric Ozone: Myths and Realities: Hearing before the Subcommittee on Energy and Environment.* 104th Cong., 1st sess., 20 September 1995.

U.S. Senate Committee on Environment and Public Works. *Superfund Cleanup Acceleration Act: Hearing before the Subcommittee on Superfund, Waste Control, and Risk Assessment.* 105th Cong., 1st sess., 5 March 1997.

Wagner, Richard H. *Environment and Man.* 3d ed. New York: W. W. Norton & Company, 1978.

Wallechinsky, David, and Irving Wallace. *The People's Almanac #2.* New York: Morrow, 1978.

Weinberg, Martin S., and Alan P. Bell. *Homosexuality: An Annotated Bibliography.* New York: Harper and Row, 1972.

Wekesser, Carol, ed. *Euthanasia: Opposing Viewpoints.* San Diego: Greenhaven Press, 1995.

White, Andrew Dickson. *A History of the Warfare of Science with Theology in Christendom.* New York: Dover Publications, 1896, 1960.

Wilkie, T. *Perilous Knowledge: The Human Genome Project and Its Implications.* Berkeley: University of California Press, 1993.

Wilson, Edward O. *Sociobiology: The New Synthesis.* Cambridge, MA: Belknap Press, 1975.

———. *On Human Nature.* Cambridge: Harvard University Press, 1978.

Wilson, Edward O., and Charles J. Lumsden. *Promethean Fire: Reflections on the Origin of the Mind.* Cambridge: Harvard University Press, 1983.

Yaroshinska, Alla, et al. *Chernobyl: The Forbidden Truth.* Lincoln: University of Nebraska Press, 1995.

Yesalis, Charles E., ed. *Anabolic Steroids in Sport and Exercise.* Windsor, Ont.: Human Kinetics Publishers, 1993.

Zimmer, Lynn, and John P. Morgan. *Marijuana Myths Marijuana Facts: A Review of the Scientific Evidence.* New York: Lindesmith Center, 1997.

Index

Abbott Laboratories, 54–55
Abortion, 217
   early-term surgical abortion,
     71–72
   and fetal tissue research, 94
   intact dilation evacuation, 145
   "morning-after" pill, 166–167
   RU486, 240–242
Acquired immunodeficiency
   syndrome. *See* AIDS
Acupuncture, 8
Adaptive Management Areas, 181
Adenosine triphosphate (ATP),
   47
Adopt-A-Horse or Burro
   Program, 285
Agricultural Council of the
   European Union, 30
AIDS, 124–129, 175–177
   and AZT, 66
Alar, 5–7
Allen, Paul, 255
Allied-Signal Chemical
   Company, 115
Alternative therapies, 7–9
Altruism, 259
America Online (AOL), 223
American Academy of Pediatrics,
   127
American Association for the
   Advancement of Science
   (AAAS), 8
American Bar Association, 176
American Booksellers
   Association, 223
American Civil Liberties Union
   (ACLU), 224
American College of
   Obstetricians and
   Gynecologists, 127
American Eugenics Society, 84
American Lung Association, 65,
   168
American Medical Association
   (AMA), 127, 176

American peregrine falcon, 78
American Psychological
   Association (APA), 130
American Society for
   Reproductive Medicine,
   Ethics Committee, 41
American Speech, Language and
   Hearing Association, 178
Amniocentesis, 102
Anderson, French, 136
Anencephalic babies. *See*
   Anencephaly
Anencephaly, 9
Animal rights, 10–13, 65
Animal welfare, 11–13, 65
Animal Welfare Act, 12
Antiballistic Missile Treaty of
   1972, 190
Apex. *See* Tris
Apollo space program, 235
Apple Corporation, 223
Applied research, 234–236
Appropriate technology, 13
Arabian oryx, 31
Archer Daniels Midland, 82
Aromatherapy, 8
Artificial Insemination by Donor
   (AID) program, 83
Aryan race, 84
As Man Becomes Machine, 74
Assisted reproductive tech-
   nologies (ART), 14–17, 41
Assisted suicide. *See* Right to Die
   Movement
Astrobiology, 253
Atomic bombs. *See* Nuclear
   weapons
Attention Deficit Disorder
   (ADD), 239–240 *See also*
   ADHD
Attention Deficit Hyperactivity
   Disorder (ADHD), 239–240
   See also ADD
Automatic Teller Machines
   (ATMs), 26–27

Avicel, 198
AZT, 66

Babbitt, Bruce, 113
Bald eagle, 78
Barnard, Christiaan, 199
Barrier islands, 19–20
Basel Convention on the Control
   of Transboundary
   Movements of Hazardous
   Waste and Their Disposal,
   117–118
Basic research, 140, 209–210,
   234–236, 254–255, 261–264
Beckwith, Jon, 131
Behavior modification, 21–24
The Bell Curve, 147
Berger, Hans, 73
BGH. *See* Bovine somatotropin
Billings, Paul, 131
Binet, Alfred, 146
Biofeedback training, 8
Biological determinism, 24–26,
   259–260
Biological education, 49–51
Biometric identification, 26–27
Bison, 27–29
Blaese, Michael, 136
Boeing Company, 263
Bovine somatotropin (BSH),
   29–30, 105
Brain surgery. *See* Psychosurgery
Brave New World, 41, 137
Breggin, Peter, 227
Breyer, Stephen G., 238
Bridlewood Residents Hydro
   Line Committee, 122
Briggs, Robert, 39–40
British Eugenics Society, 84
Brown pelican, 78
Brucellosis, 28–29
Bryan, William Jennings, 50
BSH. *See* Bovine somatotropin
BST. *See* Bovine somatotropin
Buck, Carrie, 84

Buckley, William F., Jr., 153
Bulkhandling, Inc., 116
Bureau of Land Management
	(BLM), 180, 186, 244, 285
Bureau of Reclamation, 249
Burkhardt, Gottlieb, 226
Burros, 283–85
Burt, Sir Cyril, 146–147
Bush, George, 94, 282

Calcium magnesium acetate. See
	CMA
Calgene Flavr Savr tomato, 104
California condor, 31
California Department of
	Environmental Quality, 211
California Psychiatric
	Association, 34
"Cancer alley (LA)," 79
Cape Hatteras, North Carolina,
	19
Captiva Island, Florida, 19
Captive breeding programs, 31
Carbon monoxide, 143
Carson, Rachel, 58
Cassini spacecraft, 31–33
Cato Institute, 82–83
CBS Television, 6
Center for Food Safety and
	Applied Nutrition (FDA),
	55
Center for Reproductive Law
	and Policy, 4
Center for Science in the Public
	Interest, 198
Centre Européen pour la
	Recherche Nucléaire
	(CERN), 210
CFCs. See Chlorofluorocarbons
Chavis, Benjamin, 79–80
Chelation therapy, 8
Chemical air pollutants, 143–144
Chemical castration, 33–35
Chemische Fabrik Kalb, 212
Chernobyl nuclear power plant
	accident, 35–36, 182–183
Child molestation, 33–35
Child Online Protection Act of
	1998 (COPA), 224
Chlorofluorocarbons (CFCs),
	36–37, 141–142, 166,
	204–206
Chorionic villus sampling (CVS),
	102
Christian Coalition, 3
Christian Science, 89, 91
Church of the First Born, 91
Circle 4 Farms (Utah), 91
Citizens Internet Empowerment
	Coalition, 223–224

Clearcutting, 37–39
Clinton, Bill
	and the Child Online
		Protection Act (COPA) of
		1998, 224
	on cloning, 41
	and the Communications
		Decency Act (CDA) of 1996,
		223
	and the Comprehensive Test
		Ban Treaty (CTBT), 191
	and environmental justice, 80
	and fetal tissue research, 94
	and food safety laws, 61
	and intact dilation evacuation,
		145
	and medical use of marijuana,
		162
	and needle exchange
		programs, 177
	and the Northwest Forest
		Plan, 179–181
	and the space station, 263
Cloning, 39–42
Cloning Prohibition Act of 1997,
	41
Club of Rome, 155–156
CMA, 60
Coats, Dan, 95
Coho salmon, 120, 248–249
Cold fusion, 42–45
Cold Fusion Times, 44
Columbia River Forest Reserve
	(Belize), 232
Commission on Racial Justice
	(United Church of Christ),
	80
Communications Decency Act of
	1996 (CDA), 223–224
Communities of color, 78–81
Community-acquired immune
	deficiency syndrome
	(CAIDS), 124
Comprehensive Environmental
	Response, Compensation,
	and Liability Act
	(CERCLA). See Superfund
Comprehensive Test Ban Treaty
	(CTBT) of 1996, 191
Conditioning, aversive, 23
Conditioning, classical, 21–22
Conditioning, operant, 22
Conditioning, Pavlovian. See
	Conditioning, classical
Conditioning, psychological,
	21–24
Conservation of natural
	resources, 45–47
Conservation versus
	preservation, 179–182

Consumer Product Safety
	Commission (CPSC), 273
Consumers Union, 6
Controlled burn. See Prescribed
	burn
Controlled substances, 160
Copenhagen Amendments to the
	Montreal Protocol on
	Substances that Deplete the
	Ozone Layer, 142, 165, 166,
	205
Correlation versus cause-and-
	effect, 122–123
Council of Europe, 41
Council on Scientific Affairs
	(AMA), 127
Creat-TEEN, 47
Creatine, 47–48
Creationism, 48–51
Crick, Francis, 101, 138
"Crime Times," 53
Criminality and heredity, 51–54
Crockett, George W., 153
Cyclamates, 54–56
Cystic fibrosis, 41

Daminozide. See Alar
Darrow, Clarence, 50
Darwin, Charles, 49, 83, 252
DDT, 57–59
Death, criteria for, 200
"Death with Dignity." See Right
	to Die Movement
Deicing roads, 59–60
Delaney, James J., 60
Delaney Clause, 60–61
DeLay, Tom, 177
Delgado, José M. R., 73–74
Demographic principles, 216–217
Demographic transition, 218–220
Deoxyribonucleic acid. See DNA
Department of Fish and Game
	(California), 283
Depo Provera
	(medroxyprogesterone
	acetate), 33–35
The Descent of Man, 49
Dichlorodiphenyltrichloroethane
	. See DDT
Dietary Supplement Health and
	Education Act of 1994, 276
DNA, 29–30
	in genetic testing, 101–104
	in human gene therapy,
		135–137
	in the Human Genome
		Project, 138–139
	See also DNA fingerprinting
DNA fingerprinting, 61–63
Dobson, George M., 64

Dobson unit, 64
Dolly (cloned sheep), 39–41
Domenici, Pete, 113
Donor egg in vitro fertilization (DEIVF), 15–17
Doomsday Clock (*Bulletin of the Atomic Scientists*), 191
Dorfman, D. D., 147
Dow Chemical Company, 212
Drake, Frank, 255
Drug abuse, 67–69, 153–155, 160–162, 175–177
Drug Enforcement Administration (DEA), 162
Drug legalization. *See* Legalization of drugs
Drug testing. *See* Testing of drugs
Drug testing in the workplace, 67–69
Ducks Unlimited, 287

E. I. du Pont de Nemours and Company, 205
Early-term surgical abortions, 71–72
Earth Summit. *See* United Nations Conference on Environment and Development
Echinacea, 72–73
ECT. *See* Electroconvulsive shock therapy
Edwards, Jerry, 71
Edwards v. Aguillard, 50
Ehrlich, Paul, 216
Electrical stimulation of the brain (ESB), 73–74
Electroconvulsive shock therapy (ECT), 74–76
Electromagnetic fields, health hazards, 121–124
Electromagnetic radiation, 121
Electronic surveillance, 222–225
Elephant seal, 31
Emergency contraception. *See* "Morning-after pill"
Emissions trading, 109–110
Endangered species, 76–78, 249
bison, 27–29
captive breeding programs, 31
and logging, 119–120, 179–182
white abalone, 282–283
Endangered Species Act of 1973, 76–77
Endangered Species Conservation Act of 1969, 76
Endangered Species Preservation Act of 1966, 76
Environmental Defense Fund (EDF), 273

Environmental inequity, 78–81
Environmental justice, 78–81
Environmental racism, 79–81
Epperson v. Arkansas, 50
ESB. *See* Electrical stimulation of the brain
Estrada-Oyuela, Raùl, 110
Ethanol, 81–83
Eugenics, 83–85, 146, 173, 264–265
Eugenics Review, 84
European Union policy on genetically manipulated foods, 105
Euthanasia, 85–87
Evolution, 48–51
Executive Order 12898, 80

Facial recognition, 26
Fahlberg, Constantine, 243
Faith healing, 89–91
"False positive," 68, 215
Farm Act of 1985, 282
Farmer, James, 204
Fauntroy, Walter E., 80
Federal Food, Drug, and Cosmetic Act of 1958, 60
Federal Land Policy and Management Act of 1976 (FLPMA), 245
Feedlot pollution, 91–93
Fertility drugs, 16–17
Fetal tissue research, 93–95
Fingerprinting, 26–27
Firemaster. *See* Tris
Firemaster BP-6, 212
First Amendment to the U.S. Constitution, 50, 89, 224
First National People of Color Environmental Leadership Summit, 80
Fission weapons. *See* nuclear weapons
Flammex. *See* Tris
Fleischmann, Martin, 42–44
Fluoridation, 95–97
Food irradiation. *See* Irradiation of food
Food Quality Protection Act of 1996, 61
Forest Plan for the Pacific Northwest. *See* Northwest Forest Plan
Forest use policy, U.S., 45–47, 230–233
multiple use/sustained yield, 169–171
Northwest Forest Plan, 179–182
prescribed burn, 220–222

Formaldehyde, 98–99, 144
Forrester, Jay, 155–156
Fossil fuel combustion, 107
Fourth Amendment to the U.S. Constitution, 27
Frankenstein monster, 41
"Freedom of Choice" movement, 2–5
Freedom to Read Foundation, 223
Freeman, Walter J., 226
Fridgedaire Corporation, 36, 204
Friedman, Milton, 153, 154
Fulton, John F., 226
Fusion, cold. *See* Cold fusion
Fusion, nuclear. *See* Nuclear fusion
Fusion weapons. *See* nuclear weapons
Fyrol. *See* Tris

Galilei, Galileo, 250–52
Galton, Sir Francis, 83–85, 173, 264
Gamete intrafallopian transfer (GIFT), 15–17
Gas chromatography/mass spectrometry (GC/MS), 68
Gasohol, 82–83
Gattaca, 137
Gay-related immune deficiency syndrome (GRID), 124
Gel electrophoresis, 62–63
General Utilities Corporation, 271
Genes, 62, 259
Genesis, 46, 48, 250
Genetic testing, 101–104
Genetically manipulated foods, 104–106
Genetics and Education, 147
Gerber Baby Food Company, 6
Global climate change. *See* Global warming
Global Climate Coalition, 111
Global warming, 106–111
Goldin, Dan, 255
Gorton, Slade, 249
Grand Canyon National Park, 178
Grant, Ulysses S., 45
Grants Pass (OR) Irrigation District, 248–249
Gray wolf, 78
Grazing legislation, 111–113
Great Lakes Chemical Company, 115
Greenhouse effect, 106–111
Greenhouse gases, 107
Grest, Jeffrey, 53

Grossman, Barry, 168
Gulf War of 1990–91, 66
Gurdon, John, 40

Halons, 115
Hamer, Dean, 131
Hamidi, Ken, 225
Hathaway, Janet, 169
Hawaiian goose. *See* Nene
Hazardous waste dumping, 115–118
Headwaters Forest, 118–121
Heller, Jean, 134
Helms, Jesse, 191
Herrington, John, 210
Herrnstein, Richard, 147
Herschel, Sir William J., 26
Hewlett, William, 255
Hitler, Adolf, 84
HIV (human immunodeficiency virus), 124–129, 175–177
Hoge, Bill, 34
Holmes, Oliver Wendell, 84
Homeopathy, 8
Homosexual behavior, 76, 129–132, 217–218
and HIV and AIDS, 124–129
HotWired, 223
Human Clone Clinic, 42
human experimentation, 65, 128, 132–135
Human Fertilization and Embryology Authority (England), 17
Human gene therapy (HGT), 135–137
Human Genome Project, 85, 138–141
Human Genome Project, Ethical, Legal, and Social Issues (ELSI) of the, 140–141
Human immunodeficiency virus. *See* HIV
Human sterilization. *See* Sterilization, human
Hurwitz, Charles, 119–120
Huxley, Aldous, 41, 137
Huygens space probe, 32
Hydrochlorofluorocarbons (HCFCs), 37, 141–142, 166, 206
Hydrogen bombs. *See* nuclear weapons
Hydrosphygmograph, 214

In vitro fertilization (IVF), 14–17
Indoor air pollution, 143–145, 229–230
Infinite Energy, 44

Informed consent, 65, 76, 133–134, 227
Institute of Medicine, 176
Insulin-like growth factor–1 (IGF-1), 29–30
Intact dilation evacuation, 145
Intel Corporation, 225
Intelligence and race. *See* IQ
Internal Revenue Service (IRS), 162
International Agency for Research on Cancer, 6
Internet, 222–225
Intoxication: Life in Pursuit of Artificial Paradise, 154
Intracytoplasmic sperm injection (ICSI), 15–17
IQ, 146–149, 174
Irradiation of food, 149–150

Jacobs, Patricia, 52
Jacobson, Caryle G., 226
Jensen, Arthur, 147
John Paul II (Pope), 131
Johnson, "Magic," 126
Johnston, Velma ("Wild Horse Annie"), 284
"Junk DNA," 62

Kaposi's sarcoma, 124
Kassa Island, Guinea, 116
Kassirer, Jerome P., 162
Keeler, Leonarde, 214
Kennedy, John F., 235
Kennewick Man, 151–152
Kepler, Johannes, 48
Kerekou, Mathieu, General, 117
Kervorkian, Jack, 238
King, T. J., 39–40
KS. *See* Kaposi's sarcoma
Kyoto Conference of 1997, 109

Lancaster County (PA) Prison, 27
Large hadron collider (LHC), 210
Larson, John A., 214
Laughlin, Harry, 264–265
Law of the Sea Treaty of 1982, 195
Lawrence, E. O., 208
"Leaded" gasoline, 168
legalization of drugs, 153–155
Leopold, Aldo, 46, 286
LeVay, Simon, 131
Lie detectors. *See* Polygraph
Lilja, Larry, 48
Limited Test Ban Treaty of 1963, 190
The Limits to Growth, 155–156
Linet Study of 1997, 123
List, Bob, 245

Lobotomy. *See* Prefrontal lobotomy
Logging roads, 156–157
Lombroso, Cesare, 51–52, 214
London Amendments to the Montreal Protocol on Substances that Deplete the Ozone Layer, 166, 205
Love Canal, New York, 268
Lumbering practices in the United States, 45–47
Lyerly, J. G., 226

Macular degeneration of the retina, 94
Maginnis, Robert, 177
Magnetic-field therapy, 8
Malathion, 163–164
Male factor procedures, 15
Man and Nature, 45
Manhattan Project, 188
Marbled murrelet, 119–120
Marijuana, medical use, 160–163
Marine Mammal Protection Act of 1972, 76
Marsh, George Perkins, 45
Masked bobwhite quail (*Colinus virginianus ridgwayi*), 159–160
Maxxam Corporation, 119–120
McCaffrey, General Barry, 162
McDonald, Gail, 111
McFadden, Dennis, 131
Meadows, Dennis, 155–156
Meadows, Donella, 155–156
Medfly. *See* Mediterranean fruit fly
Medical Center for Federal Prisoners (Springfield, MO), 24
Medical use of marijuana. *See* Marijuana, medical use
Mediterranean fruit fly (*Ceratitis capitata*), 162–163
Meninick, Jerry, 152
Methyl bromide, 164–166
Methyl-tertiary-butyl ether. *See* MTBE
Metropolitan Edison Company, 271
Miami Beach, Florida, 19
Microsoft Corporation, 223
Midgley, Thomas, Jr., 36, 204
Mifepristone. *See* RU486
Molina, Mario, 204
Moniz, C. de A. F. Egas, 226
Monsanto Company, 29–30, 105
Montagnier, Luc, 125
Montreal Protocol on Substances that Deplete the Ozone

Layer of 1987, 115, 165, 166, 204–205
Moore, Gordon, 255
"Morning-after pill," 166–167
MTBE, 167–169
Muir, John, 46
Müller, Paul, 57
Multiple use, sustained yield forest policy, 46, 169–171, 180
Murphy, Patrick V., 153
Murray, Charles, 147
Murray, Joseph, 199

National Aeronautics and Space Administration (NASA), 31–33, 235, 254–255, 261–264
National Cancer Institute, 123
National Cancer Society, 65
National Cattlemen's Beef Association, 113
National Collegiate Athletic Association (NCAA), 266
National Commission for the Protection of Human Subjects of Biomedical and Behavioral Research, 227
National Environmental Justice Advisory Council, 80
National Food Processors Association, 6
National Football League, 47
National Forest Management Act of 1976, 180
National Institute for Occupational Safety and Health (NIOSH), 99
National Institutes of Health (NIH), 75, 139
National Marine Fisheries Service, 249
National Park Service, 28, 152
National parks, 45–46
National Research Council (National Academy of Sciences), 123
National Resources Defense Council (NRDC), 6, 169
National Right to Life Committee, 4, 71
National Toxicology Program (National Institute of Environmental Health Sciences), 244
Native American Graves and Repatriation Act, 151
Natural philosophy, 250
Natural resources, policies regarding, 45–47

"Nature versus nurture," 24–26, 173–175, 259–260
Naturopathic medicine, 8
Nazi Party (Germany), 84–85, 86, 265
Needle exchange programs, 175–177
Nellis Air Force Base, 186
Nene, 31
Neuroaugmentation. See Electrical stimulation of the brain
New building syndrome. See "Sick building syndrome"
Nixon, Richard, 76
Noise pollution, 178–179
Nonpoint sources of pollution, 196
Northern spotted owl, 77
Northwest Economic Adjustment Initiative, 181
Northwest Forest Plan (NWFP), 179–182
Northwest Power Planning Council, 249
Nuclear fission, 182–185
Nuclear fusion, 42–45
Nuclear Non-Proliferation Treaty of 1968, 190
Nuclear power plants, 271–272, 182–185
Nuclear Waste Policy Act of 1982, 186–87
Nuclear wastes, 184, 185–188
Nuclear weapons, 188–192, 192–193
Nuclear winter, 192–193

Ocean City, Maryland, 19
Oceans, 195–197
Old growth forests, 118–121
Olds, James, 73
Olean®. See Olestra
Olestra, 197–199
On Human Nature, 259
On the Origin of Species, 49
Oregon Water Resources Commission, 249
organ transplantation, 9, 199–202
Organization of Petroleum Exporting Countries (OPEC), 82
"Orphan sites" (hazardous waste disposal), 268–269
Osteen, William L., 256–257
Oxley, Michael, G., 223, 269
Oxy gas. See Ethanol; MTBE
Oxy-Busters, 168
Ozone, stratospheric, 37
Ozone depletion, 165, 202–206

Ozone depletion potential (ODP), 206
Ozone "hole," See Ozone depletion

Pacific Lumber Company. See Palco
Packard, David, 255
Padre Island, Texas, 19
Palco, 118–121
Parkinson's disease, 94
Partial-birth abortion. See Intact dilation evacuation
Particle accelerators, 207–210
Passenger pigeon, 77, 287
Passive smoking. See Secondhand smoke
Pavlov, Ivan, 21–22
PCP, 124
Pelosi, Nancy, 177
Persistent vegetative state (PVS), 9
Personal identification number (PIN), 27
Pesticides. See DDT; Mediterranean fruit fly
Pike, Joseph, 241
Pinchot, Gifford, 46, 170
Planned Parenthood Foundation, 71, 85
Plato, 83
Plutonium, 31–33
Pneumocystis carinii pneumonia. See PCP
Pollutant standard index (PSI), 211
Pollution alerts, 211
Polybrominated biphenyls (PBBs), 211–212
Polychlorinated biphenyls (PCBs), 79, 212–214
Polygraph, 214–216
Pons, Stanley, 42–44
The Population Bomb, 216
Population Council, 241–242
Population issues, 155–156, 216–220
Posilac®. See Bovine somatotropin
Prefrontal lobotomy, 226
Prescribed burn, 220–222
Preservation of natural resources, 45–47
Principles of Environmental Justice, 80
Prior appropriation water rights, 277–278
Privacy issues, 222–225
Proctor & Gamble Company (P&G), 197–198

Project Beta, 255
Project Big Ear, 255
Project Meta, 255
Project Phoenix, 255
Project Serendip, 255
Psychosurgery, 225–228
Public Law 86–234 (wild horse and burro legislation), 284
Public Rangelands Improvement Act of 1978, 113
Pure research. See Basic research
Purkinje, J. E., 26
Purple coneflower. See Echinacea
Pyridostigmine bromine (PB), 66

Qigong, 8

Radon, 143, 229–230
Rain forests, 230–233
Rainforest Action Network (RAN), 232
Rambo, Sylvia, 271–272
Reagan, Nancy, 67
Reagan, Ronald, 179, 210, 235, 245
   and basic research, 235
   and commercial development of forests, 171
   and noise pollution, 179
   and the Sagebrush Rebellion, 245
   and the superconducting super collider, 210, 235
   and the war on drugs, 67
Recombinant DNA (rDNA) technology, 29–30
Reed, Lowell, 224
Religion and science. See Science and religion
Reno, Janet, 162
Replacement rate (population), 219
The Republic, 83
Republican Party, 3
Reserve Areas, 180–181
Retinography, 26
Richardson, Bill, 186–187
Right to Die movement, 86, 236–239
"Right to Life" movement, 2–5
Right to privacy. See Privacy and the Internet
Riparian water rights, 277–278
Ritalin, 239–240
Robertson, John, 41
Robertson, Pat, 252
Roe v. Wade, 1–5
Roemer, Tim, 263
"'Roid rage," 266
Roman Catholic Church, 250–253

Rorvik, David, 74
Roslin Institute, 40–41
Roussel Uclaf, 240–241
Rowland, F. Sherwood, 204
Royal Adelaide Hospital ELF Mice Study of 1998, 123
RU486, 240–242
Russian Space Agency, 263

Saccharin, 54–55, 243–244
"Safe level" policies, 61
Sagan, Carl, 153
Sagebrush Rebellion, 244–246
Salmon. See Coho salmon
Salton Sea, 246–248
Salvage logging, 120
Salvage Rider Bill of 1995, 181
Sanger, Margaret, 85
Sanibel Island, Florida, 19
Sarawak (Malaysia) rain forests, 232
Savage Rapids Dam, 248–250
Schedules of Reinforcement, 22
Schmidt, H., 213
Schmoke, Kurt, 153
Schulz, George, 153, 213
Schumacher, E. F., 13
Science and religion, 250–253
Scientific creationism, 50–51
Scopes, John, 50
Scopes trial of 1924, 50
Search for extraterrestrial intelligence, 253–255
Secondhand smoke, 255–257
Seed, Richard, 42
Seed-tree cutting, 38–39
Selenium, 247–248
Sensenbrenner, F. James, 263
Sesco-Gibraltar, 117
SETI. See Search for extraterrestrial intelligence
SETI Institute, 255
Sex education, 216–217
Sex offenders, 33–35
Shalala, Donna E., 162
Shelterwood cutting, 38–39
Shifflett, Wayne, 160
Shock therapy. See Electroconvulsive shock therapy
"Sick building syndrome," 144
Siegel, Ronald K., 154
Sierra Club, 46, 169
Silent Spring, 58
Simplesse, 198
Singer, Fred, 205
Sixty Minutes (television program), 6
Skinner, B.F., 22–23
Skinner box, 23

Slash-and-burn agriculture, 232–233, 257–258
Small Is Beautiful: Economics as if People Mattered, 13
Smith, Bob, 113
Smoking. See Tobacco smoke
Snowmobiling, 28–29
Society of Professional Journalists, 223
Sociobiology, 258–261
Sociobiology: The New Synthesis, 259
Soman (nerve gas), 66
Somers (CT) State Prison, 24
Space station, 261–264
"Spamming," 225
Special Treatment and Rehabilitative Training (START) program, 24
Spemann, Hans, 39
Stanford Linear Accelerator Center (SLAC), 208
Starzl, Thomas, 199
Stereotactic neurosurgery, 226–227
Sterilization, human, 264–265
Sterilization (of women), 83–84
Steroids, 265–267
Strategic Arms Limitation Treaty (SALT I) of 1972, 190–191
Strategic Arms Limitation Treaty (SALT II) of 1979, 191
Strategic Arms Reduction Treaty (START I) of 1991, 191
Strategic Arms Reduction Treaty (START II) of 1993, 191
Stream channelization, 267, 281
Sucaryl®, 54–55
Summers, Lawrence, 116
Superconducting super collider (SSC), 209–210, 235
Superfund, 267–270
Superfund Amendments and Reauthorization Act of 1986 (SARA), 268–269
Supersonic transport (airplane; SST), 203
Surrogacy, 15–17
Sveda, Michael, 54
Swampbuster Law, 282
Sweet, Robert, 153
Sweet 'N Low®, 243

Taylor, Edward T., 112
Taylor Grazing Act of 1934, 112
Tenth Amendment to the U.S. Constitution, 245
Terman, Lewis, 146
testing of drugs, 64–67

Tetraethyl lead. *See* "Leaded" gasoline
Tetrahydrocannabinol (THC). *See* Marijuana, medical use
Thin-layer chromatography (TLC), 68
Thomas, Clarence, 215
Thorndike, Edward, 22
Three Mile Island Nuclear Power Plant, 183, 271–272
Timber industry. *See* Lumbering practice in the United States
Titan (rocket), 32–33
Titan (Saturn moon), 31–33
Tobacco smoke, 143
Total fertility rate, 219
Toxic Substances Control Act of 1976, 213
Transgenic animals, 40–41
Trephination, 225
Tribe, Laurence, 225
Tris, 272–273
Trisomic XXY condition, 52
Trout Unlimited, 249
Tuskegee Syphilis Study, 133–134

U.S. Army Corps of Engineers, 151, 267
U.S. Bureau of Land Management, 113
U.S. Centers for Disease Control and Prevention (CDCP), 134, 149, 176
U.S. Clean Air Act, 165, 168
U.S. Conference of Mayors, 176
U.S. Department of Defense (DOD), 66, 244
U.S. Department of Energy (DOE), 139, 186–187, 210
U.S. Department of Health and Human Services (HHS), 65, 201
U.S. Environmental Protection Agency (EPA), 169
and Alar, 6
and DDT, 58
The Delaney Clause, 61
environmental justice, 80
hazardous wastes, examples of, 115
and noise pollution, 179
and nuclear wastes, 187
and the Pollutant Standard Index, 211
and radon exposure, 229
and the Savage Rapids Dam, 249
on secondhand smoke, 256
superfund sites, 269

wetlands, 281
U.S. Fish and Wildlife Service (FWS), 78, 244–245
restoration of the masked bobwhite quail, 159
and the Savage Rapids Dam, 249
wetlands, 282
U.S. Food and Drug Administration (FDA)
and bovine somatotropin, 29
and cyclamates, 55
and DDT, 58
Drug Testing, 64–65, 132, 134
and genetically manipulated products, 105, 106
irradiation of food, 149
medical uses of marijuana, 161
"morning-after" pill, 167
and Olestra, 197
and Ritalin, 239
and RU486, 241
and saccharin, 243
vitamins and minerals, 276
U.S. Forest Service, (USFS), 46, 156, 180, 245
and clearcutting, 38–39
forest use policy, 170
Multiple Use/Sustained Yield Act of 1960, 170–171
prescribed burns, 221
U.S. General Accounting Office (GAO), 80, 82, 176
U.S. National Oceanic and Atmospheric Administration (NOAA), 205
U.S. National Toxicology Program, 6
U.S. Public Health Service (PHS), 133–134
U.S. Rubber Company. *See* Uniroyal
U.S. Social Security Administration, 8
UDMH, 5–6
ultraviolet (UV) radiation, 203
Umatilla tribe, 151–152
Uniform Anatomical Gift Act of 1968, 95, 200
Uniform Brain Death Act of 1978, 200
Uniroyal, 5
United Nations Conference on Environment and Development, 109
United Network for Organ Sharing, 201

University of Michigan School of Natural Resources, 80
Unsymmetrical 1,1-dimethyl-hydrazine. *See* UDMH
Uranium Mill Tailings Radiation Control Act of 1978, 229

V. I. Lenin Nuclear Power Facility, 35, 182–83
Vajpayee, Atal Bihari, 191
Vitamins and minerals, 275–276
Voice authentication, 26
Volatile organic compounds (VOCs), 144

Walden Two, 23
Ward Transformer Company, 79
Warren County (NC) protest, 79
Waste Isolation Pilot Plant (WIPP), 184, 186–187
Water rights, 277–279
Water Watch of Oregon, 249
Watson, James, 101, 138
Watt, James, 46, 171, 245
Watts, James W., 226
Watts riots of 1967, 52
Western Water Policy Review Commission, 279
Wetlands, 279–282
White abalone, 77, 282–283
Whooping crane, 31
Wild Horse and Burro Act of 1971, 285
Wild horses, 283–285
Wilderness areas, 45–46
Wildlife management, 27–29, 285–288
Wilmut, Ian, 40–41
Wilson, Edward O., 258–260
Wilson, James Q., 154
Wilson, Pete, 34
Wired, 223
"Wise use" ordinances, 246
Woodley, Richard W., 122
World Resources Institute, 111
World Wide Web. *See* Internet

Xenotransplantation, 199–201
XYY males, 52–53

Yellowstone National Park, 28, 45
Yoga, 8
Yucca Mountain, Nevada, 186

"Zero tolerance" policies, 61
Zidovudine. *See* AZT
Zimmer, Dick, 263
Zygote intrafallopian transfer (ZIFT), 15–17